TABLE OF CONTENTS

W9-CHP-784

Chapter 1   Functions, Graphs, and Models; Linear Functions

Algebra Toolbox 1

1.1 Functions and Models 3

1.2 Graphs of Functions 7

1.3 Linear Functions 13

1.4 Equations of Lines 17

Chapter 1 Skills Check 23

Chapter 1 Review 25

Chapter 1 Group Activities/Extended Applications 28

Chapter 2   Linear Models, Equations, and Inequalities

Algebra Toolbox 31

2.1 Algebraic and Graphical Solution of Linear Equations 34

2.2 Fitting Lines to Data Points: Modeling Linear Functions ... 42

2.3 Systems of Linear Equations in Two Variables 46

2.4 Solutions of Linear Inequalities 55

Chapter 2 Skills Check 61

Chapter 2 Review 64

Chapter 2 Group Activities/Extended Applications 70

Chapter 3   Quadratic, Piecewise-Defined, and Power Functions

Algebra Toolbox 72

3.1 Quadratic Functions; Parabolas 76

3.2 Solving Quadratic Equations 84

3.3 Piecewise-Defined Functions and Power Functions 94

3.4 Quadratic and Power Models 101

Chapter 3 Skills Check 107

Chapter 3 Review 111

Chapter 3 Group Activities/Extended Applications 117

Chapter 4   Additional Topics with Functions

    Algebra Toolbox   119

    4.1 Transformations of Graphs and Symmetry   121

    4.2 Combining Functions; Composite Functions   127

    4.3 One-to-One and Inverse Functions   134

    4.4 Additional Equations and Inequalities   138

    Chapter 4 Skills Check   149

    Chapter 4 Review   150

    Chapter 4 Group Activities/Extended Applications   154

Chapter 5   Exponential and Logarithmic Functions

    Algebra Toolbox   158

    5.1 Exponential Functions   160

    5.2 Logarithmic Functions; Properties of Logarithms   165

    5.3 Exponential and Logarithmic Equations   170

    5.4 Exponential and Logarithmic Models   177

    5.5 Exponential Functions and Investing   181

    5.6 Annuities; Loan Repayment   186

    5.7 Logistic and Gompertz Functions   190

    Chapter 5 Skills Check   194

    Chapter 5 Review   196

    Chapter 5 Group Activities/Extended Applications   202

Chapter 6   Higher-Degree Polynomial and Rational Functions

Algebra Toolbox 204

6.1 Higher-Degree Polynomial Functions 210

6.2 Modeling with Cubic and Quartic Functions 216

6.3 Solution of Polynomial Equations 222

6.4 Polynomial Equations Continued; Fundamental Theorem
     of Algebra 227

6.5 Rational Functions and Rational Equations 234

6.6 Polynomial and Rational Inequalities 242

Chapter 6 Skills Check 247

Chapter 6 Review 252

Chapter 6 Group Activities/Extended Applications 259

Chapter 7   Systems of Equations and Matrices

Algebra Toolbox 261

7.1 Systems of Linear Equations in Three Variables 264

7.2 Matrix Solution of Systems of Linear Equations 274

7.3 Matrix Operations 284

7.4 Inverse Matrices; Matrix Equations 290

7.5 Determinants and Cramer's Rule 300

7.6 Systems of Nonlinear Equations 307

Chapter 7 Skills Check 313

Chapter 7 Review 320

Chapter 7 Group Activities/Extended Applications 331

Chapter 8   Special Topics in Algebra

8.1   Systems of Inequalities   **334**

8.2   Linear Programming: Graphical Methods   **348**

8.3   Sequences and Discrete Functions   **357**

8.4   Series   **360**

8.5   Preparing for Calculus   **363**

8.6   The Binomial Theorem   **366**

8.7   Conic Sections: Circles and Parabolas   **368**

8.8   Conic Sections: Ellipses and Hyperbolas   **374**

Chapter 8 Skills Check   **381**

Chapter 8 Review   **388**

Chapter 8 Group Activity/Extended Application   **396**

CHAPTER 1
Functions, Graphs, and Models; Linear Functions

Toolbox Exercises

1. $\{1,2,3,4,5,6,7,8\}$ and
   $\{x\,|\,x<9, x\in N\}$
   Remember that $x\in N$ means that x is a natural number.

2. Yes.

3. No. A is not a subset of B. A contains elements 2, 7, and 10 which are not in B.

4. No. $N=\{1,2,3,4,...\}$. Therefore, $\frac{1}{2}\notin N$.

5. Yes. Every integer can be written as a fraction with the denominator equal to 1.

6. Yes. Irrational numbers are by definition numbers that are not rational.

7. Integers. However, note that this set of integers could also be considered as a set of rational numbers. See question 5.

8. Rational numbers

9. Irrational numbers

10. $x>-3$

11. $-3\le x\le 3$

12. $x\le 3$

13. $(-\infty,7]$

14. $(3,7]$

15. $(-\infty,4)$

16.

17. Note that $5>x\ge 2$ implies $2\le x<5$, therefore:

18.

19.

20.

21.

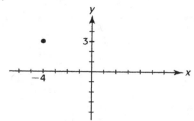

22.

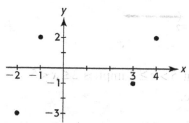

23.

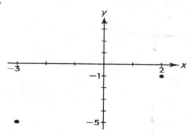

24. Yes, it is a polynomial with degree 4.

25. No, it is not a polynomial since there is a variable in a denominator.

26. No, it is not a polynomial since there is a variable inside a radical.

27. Yes, it is a polynomial with degree 6.

28. The x^2 term has a coefficient of -3. The x term has a coefficient of -4. The constant term is 8.

29. The x^4 term has a coefficient of 5. The x^3 term has a coefficient of 7. The constant term is -3.

30. $\left(z^4 - 15z^2 + 20z - 6\right) + \left(2z^4 + 4z^3 - 12z^2 - 5\right)$
$= \left(z^4 + 2z^4\right) + \left(4z^3\right) + \left(-15z^2 - 12z^2\right) +$
$\quad (20z) + (-6 - 5)$
$= 3z^4 + 4z^3 - 27z^2 + 20z - 11$

31. $3x + 2y^4 - 2x^3y^4 - 119 - 5x - 3y^2 + 5y^4 + 110$
$= \left(2y^4 + 5y^4\right) - 2x^3y^4 - 3y^2 +$
$\quad (3x - 5x) + (-119 + 110)$
$= 7y^4 - 2x^3y^4 - 3y^2 - 2x - 9$

32. $4\left(p + d\right) = 4p + 4d$

33. $-2\left(3x - 7y\right) = -6x + 14y$

34. $-a\left(b + 8c\right)$
$= -ab - 8ac$

35. $4\left(x - y\right) - \left(3x + 2y\right)$
$= 4x - 4y - 3x - 2y$
$= 1x - 6y$
$= x - 6y$

36. $4\left(2x - y\right) + 4xy - 5\left(y - xy\right) - \left(2x - 4y\right)$
$= 8x - 4y + 4xy - 5y + 5xy - 2x + 4y$
$= \left(8x - 2x\right) + \left(4xy + 5xy\right) + \left(-4y - 5y + 4y\right)$
$= 6x + 9xy - 5y$

37. $2x\left(4yz - 4\right) - \left(5xyz - 3x\right)$
$= 8xyz - 8x - 5xyz + 3x$
$= \left(8xyz - 5xyz\right) + \left(-8x + 3x\right)$
$= 3xyz - 5x$

38. $(4x^2y^3)(-3a^2x^3) = -12a^2x^5y^3$

39. $2xy^3(2x^2y + 4xz - 3z^2)$
$= 4x^3y^4 + 8x^2y^3z - 6xy^3z^2$

40. $(x-7)(2x+3) = 2x^2 + 3x - 14x - 21$
$= 2x^2 - 11x - 21$

41. $(k-3)^2 = (k-3)(k-3)$
$= k^2 - 3k - 3k + 9$
$= k^2 - 6k + 9$

42. $(4x - 7y)(4x + 7y) = 16x^2 + 28xy - 28xy - 49y^2$
$= 16x^2 - 49y^2$

43. $\dfrac{12x - 5x^2}{x} = 12 - 5x$

44. $\dfrac{8x^2 + 2x}{2x} = 4x + 1$

Section 1.1 Skills Check

1. Using Table A

   a. −5 is an x-value and therefore is an input into the function $f(x)$.

   b. $f(-5)$ represents an output from the function.

   c. The domain is the set of all inputs. $D:\{-9,-7,-5,6,12,17,20\}$. The range is the set of all outputs. $R:\{4,5,6,7,9,10\}$

   d. Each input x into the function f yields exactly one output $y = f(x)$.

3. $f(-9) = 5$
$f(17) = 9$

5. No. In the given table, x is not a function of y. If y is considered the input variable, one input will correspond with more than one output. Specifically, if $y = 9$, then $x = 12$ or $x = 17$.

7. **a.** $f(2) = -1$, since $x = 2$ in the table corresponds with $f(x) = -1$.

   b. $f(2) = 10 - 3(2)^2$
$= 10 - 3(4)$
$= 10 - 12$
$= -2$

   c. $f(2) = -3$, since $(2, -3)$ is a point on the graph.

9.

x	y
0	2
−2	−4

11. Recall that $R(x) = 5x + 8$.

   a. $R(-3) = 5(-3) + 8 = -15 + 8 = -7$

   b. $R(-1) = 5(-1) + 8 = -5 + 8 = 3$

   c. $R(2) = 5(2) + 8 = 10 + 8 = 18$

13. Yes. Each input corresponds with exactly one output. The domain is $\{-1, 0, 1, 2, 3\}$. The range is $\{-8, -1, 2, 5, 7\}$.

15. No. The graph fails the vertical line test. Each input does not match with exactly one output.

17. Yes. The graph passes the vertical line test. Each input matches with exactly one output.

19. No. If $x = 3$, then $y = 5$ or $y = 7$. One input yields two outputs. The relation is not a function.

21. **a.** Not a function. If $x = 4$, then $y = 12$ or $y = 8$.

    b. Yes. Each input yields exactly one output.

23. **a.** Not a function. If $x = 2$, then $y = 3$ or $y = 4$.

    b. Function. Each input yields exactly one output.

25. The domain is the set of all inputs. $D: \{-3, -2, -1, 1, 3, 4\}$. The range is the set of all outputs. $R: \{-8, -4, 2, 4, 6\}$

27. Considering y as a function of x, the domain is the set of all inputs, x. Therefore the domain is $D: [-10, 8]$. The range is the set of all outputs, y. Therefore, the range is $R: [-12, 2]$.

29. Considering y as a function of x, the domain is the set of all inputs, x. Therefore the domain is $D: (-\infty, \infty)$. The range is the set of all outputs, y. Therefore, the range is $R: [-4, \infty)$.

31. The input is the number of years after 2000, therefore 2015 is 15 years after 2000 and the input would be $x = 15$. Similarly, 2022 is 22 years after 2000, and the input would be $x = 22$.

33. No. If $x = 0$, then $(0)^2 + y^2 = 4 \Rightarrow y^2 = 4 \Rightarrow y = \pm 2$. So, one input of 0 corresponds with 2 outputs of -2 and 2. Therefore the equation is not a function.

35. $F = \dfrac{9C}{5} + 32$, where F is the Fahrenheit temperature and C is the Celsius temperature.

37. C is found by using the steps; a. subtract 32; b. multiply by 5; c. divide by 9.

Section 1.1 Exercises

39. **a.** Yes. Each input (a, age in years) corresponds with exactly one output (p, life insurance premium). The independent variable is a, age in years, and the dependent variable is p, life insurance premium.

    b. No. One input of $11.81 corresponds with six outputs (a, age in years).

41. Yes. Each input (y, year) corresponds with exactly one output (p, percent). The independent variable is y, the year, and the dependent variable is p, the percent of Americans who are obese.

43. **a.** Yes. Each input (the barcode) corresponds with exactly one output (an item's price).

    b. No. Every input (an item's price) could correspond with more than one output (the barcode). Numerous items can have the same price but different barcodes.

45. Each input (*x*, years) corresponds with exactly one output (*V*, value of the property). The graph of the equation passes the vertical line test.

47. **a.** Yes. Each input (day of the month) corresponds with exactly one output (weight).
b. The domain is the first 14 days of May or
D:$\{1,2,3,4,5,6,7,8,9,10,11,12,13,14\}$.
c. The range is
$\{171,172,173,174,175,176,177,178\}$.

d. The highest weights were on May 1 and May 3.

e. The lowest weight was on May 14.

f. Three days from May 8 until May 11.

49. **a.** $P(3) = \$1096.78$. If the car is financed over three years, the payment is \$1096.78.

b. $C(5) = \$42,580.80$.

c. $t = 4$. If the total cost is \$41,014.08, then the car has been financed over four years,

d. Since $C(5) = \$42,580.80$, and $C(3) = \$39,484.08$, the savings would be $C(5) - C(3) = \$3096.72$.

51. **a.** Approximately 22 million

b. $f(1930) = 11$. Approximately 11 million women were in the work force in 1930.

c. D: $\{1930,1940,1950,1960,1970,$
$1980,1990,2000,2005,2010,2015\}$

d. Increasing. As the year increases, the number of women in the work force also increases.

53. **a.** $f(1890) = 26.1$
$f(2015) = 29.2$

b. $g(1940) = 21.5$
$g(2013) = 26.6$

c. If $x = 1980$, then $f(x) = 24.7$. The median age at first marriage for men in 1980 was 24.7 years.

d. $f(2015) = 29.2 > 22.8 = f(1960)$. Therefore, the median age at first marriage for men increased between 1960 and 2015.

55. **a.** In 2020, 5.7 million U.S. citizens age 65 and older are expected to have Alzheimer's disease.

b. $f(2030) = 7.7$. In 2030, 7.7 million U.S. citizens age 65 and older are expected to have Alzheimer's disease.

c. 2040. $f(2040) = 11$.

d. The function is increasing. Since 2000, the number of U.S citizen's 65 and older that are expected to have Alzheimer's disease in going up.

57. **a.** $f(1995) = 56.0$. In 1995, the birth rate for U.S. girls ages 15 to 19 was 56.0 per 1000 girls.

b. 2005. $f(2005) = 40.5$.

c. The birth rate appears to be at its maximum (59.9) in the year 1990.

d. In the years 2005 to 2009, the birth rate appears to have increased until 2007, then decreased.

59. a. $R(200)=32(200)=6400$. The revenue generated from selling 200 golf hats is $6400.

b. $R(2500)=32(2500)=\$80,000$

61. a. $f(1000) = 0.857(1000)+19.35$
$= 857+19.35$
$= 876.35$
The monthly charge for using 1000 kilowatt hours is $876.35.

b. $f(1500) = 0.857(1500)+19.35$
$= 1285.50+19.35$
$= 1304.85$
The monthly charge for using 1500 kilowatt hours is $1304.85.

63. a. $P(100)=32(100)-0.1(100)^2-1000$
$= 3200-1000-1000$
$= 1200$
The daily profit from the production and sale of 100 Blue Chief bicycles is $1200.

b. $P(160)=32(160)-0.1(160)^2-1000$
$= 5120-2560-1000 = 1560$
The daily profit from the production and sale of 160 Blue Chief bicycles is $1560.

65. a. $0.3+0.7n=0$
$0.7n=-0.3$
$\dfrac{0.7n}{0.7}=\dfrac{-0.3}{0.7}$
$n=-\dfrac{3}{7}$
Therefore the domain of $R(n)$ is all real numbers except $-\dfrac{3}{7}$ or
$\left(-\infty,-\dfrac{3}{7}\right)\cup\left(-\dfrac{3}{7},\infty\right).$

b.   In the context of the problem, n represents the factor for increasing the number of questions on a test. Therefore it makes sense that n is positive ($n>0$).

67. a.   Since p is a percentage, $0\le p\le 100$. However in the given function, the denominator, $100-p$, cannot equal zero. Therefore, $p\ne 100$. The domain is $0\le p<100$ or, in interval notation, $[0,100)$.

b. $C(60)=\dfrac{237,000(60)}{100-60}=355,500$
$C(90)=\dfrac{237,000(90)}{100-90}=2,133,000$

69. a. $V(12) = (12)^2(108-4(12))$
$= 144(108-48)$
$= 144(60)$
$= 8640$
$V(18) = (18)^2(108-4(18))$
$= 324(108-72)$
$= 324(36)$
$= 11,664$

b.   First, since x represents a side length in the diagram, x must be greater than zero. Second, to satisfy postal restrictions, the length (longest side) plus the girth must be less than or equal to 108 inches. Therefore,
Length + Girth ≤ 108
Length $+4x\le 108$
$4x\le 108-\text{Length}$
$x\le\dfrac{108-\text{Length}}{4}$
$x\le 27-\dfrac{\text{Length}}{4}$

Since x is greatest if the longest side is smallest, let the length equal zero to find the largest value for x.

$$x \leq 27 - \frac{0}{4}$$

$$x \leq 27$$

Therefore the conditions on x are $0 < x \leq 27$. If $x = 27$, the length would be zero and the package would not exist. Therefore, in the context of the question, $0 < x < 27$ and the corresponding domain for the function $V(x)$ is $(0, 27)$.

c.

x	Volume
10	6800
12	8640
14	10192
16	11264
18	**11664**
20	11200
22	9680

The maximum volume occurs when $x = 18$. Therefore the dimensions that maximize the volume of the box are 18 inches by 18 inches by 72 inches, a total of 108 inches.

Section 1.2 Skills Check

1. a.

x	$y = x^3$	(x, y)
−3	−27	(−3, −27)
−2	−8	(−2, −8)
−1	−1	(−1, −1)
0	0	(0, 0)
1	1	(1, 1)
2	8	(2, 8)
3	27	(3, 27)

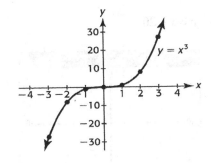

b.

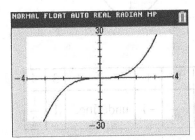

c. Your hand-drawn graph in part a) by plotting points from the table should match the calculator-drawn graph in part b).

3.

x	-2	-1	0	1	2
y	-7	-4	-1	2	5

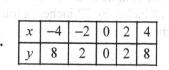

x-intercept
y-intercept

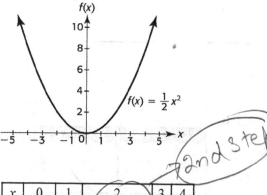

$f(x) = 3x - 1$

5.

x	-4	-2	0	2	4
y	8	2	0	2	8

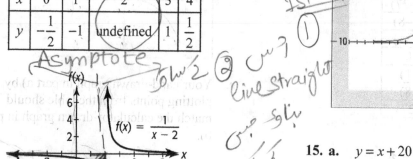

$f(x) = \frac{1}{2}x^2$

7.

x	0	1	2	3	4
y	$-\frac{1}{2}$	-1	undefined	1	$\frac{1}{2}$

2nd step

1st step

solv2 @ one ? ①

line straight

one solv

one در 3

Asymptote

$f(x) = \frac{1}{x-2}$

Note:-
$\frac{1}{x-2}$
$x-2$
$x+2$

3rd step   check on paper

Asymptote

4th   Plug in the the x-value in
the question find the y-value

e.g $x = 3$ $y = \frac{1}{x-2}$ $\Rightarrow$ $\frac{1}{3-2}$ $\Rightarrow$ $\frac{1}{1}$ $y = 1$
$x = 3, y = 1$

9. $y = x^2 - 5$, yes there is a turning point in this window

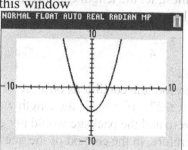

11. $y = x^3 - 3x^2$, yes there are 2 turning points in this window

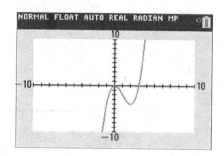

13. $y = 9/(x^2 + 1)$, yes there is a turning point in this window

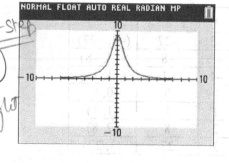

15. a. $y = x + 20$

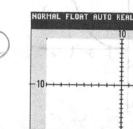

b. $y = x + 20$

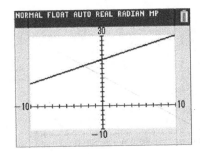

View b) is better.

17. a. $y = \dfrac{0.04(x - 0.1)}{x^2 + 300}$

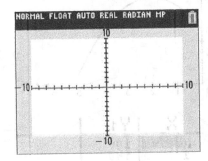

b. $y = \dfrac{0.04(x - 0.1)}{x^2 + 300}$

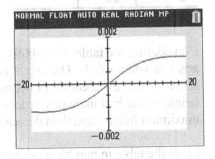

View b) is better.

19. When $x = -8$ or $x = 8$, $y = 114$. When $x = 0$, $y = 50$. Letting y vary from -5 to 100 gives one view.

$$y = x^2 + 50$$

The turning point is at $(0, 50)$ and is a minimum point.

20. $y = x^2 + 60x + 30$

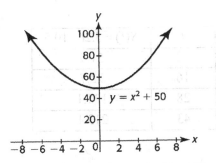

21. When $x = -10, y = -250$. When $x = 10, y = 850$. When $x = 0, y = 0$. Letting y vary from -200 to 300 gives one view.

$$y = x^3 + 3x^2 - 45x$$

The turning points are at $(-5, 175)$, a maximum, and at $(3, -81)$, a minimum.

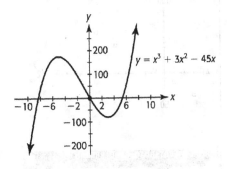

23.

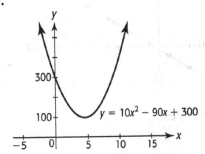

Note that answers for the window may vary.

25.

t	$S(t) = 5.2t - 10.5$
12	51.9
16	72.7
28	135.1
43	213.1

27.

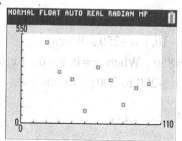

29. a.

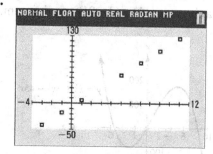

b.

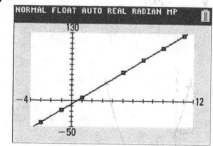

c. Yes. Yes.

31. a. $f(20) = (20)^2 - 5(20)$
$= 400 - 100$
$= 300$

b. $x = 20$ implies 20 years after 2000. Therefore the answer to part a) yields the millions of dollars earned in 2020.

Section 1.2 Exercises

33.

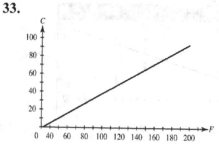

35. a.

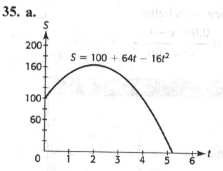

b.

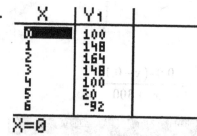

Considering the table, $S = 148$ feet when x is 1 or when x is 3. The height is the same for two different times because the height of the ball increases, reaches a maximum height, and then decreases.

c. From the table in part b), it appears the maximum height is 164 feet, occurring 2 seconds into the flight of the ball.

37.

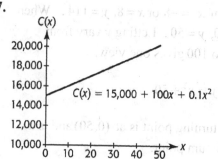

39. a. $P = 0.79x + 20.86$

$\qquad = 0.79(70) + 20.86$

$\qquad = 55.3 + 20.86$

$\qquad = 76.16$

The model predicts that there will be 76.16 million women in the workforce in 2020.

b. $80; 2030 - 1950 = 80$

$P = 0.79x + 20.86$

$\qquad = 0.79(80) + 20.86$

$\qquad = 63.2 + 20.86$

$\qquad = 84.06$

The model predicts that there will be 84.06 million (or 84,060,000) women in the workforce in 2030.

41. a. $t = \text{Year} - 2005$

For 2010, $t = 2010 - 2005 = 5$

For 2016, $t = 2016 - 2005 = 11$

b. $P = f(10)$ represents the value of P in 2015 $(2005 + 10 = 2015)$.

$f(10) = 1.84(10) + 60.89 = 79.29$

79.29 represents the percent of households with Internet access in 2015.

c. $x_{\min} = 2005 - 2005 = 0$

$x_{\max} = 2020 - 2005 = 15$

43. a.

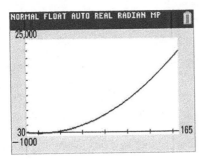

b.   Use the Trace feature, and when $x = 130$, $P = 10,984,200$.

45. a.

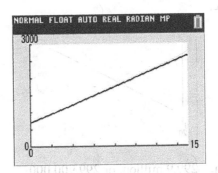

b. $y = 130.7(13) + 699.7 = 2398.8$

In 2023, the balance of federal direct student loans will be 2398.8 billion dollars.

c. $y = 130.7(19) + 699.7 = 3183$

In 2029, the balance of federal direct student loans will be 3183 billion dollars.

47.

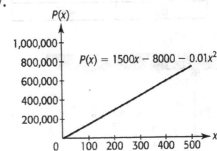

49. a. $x = \text{Year} - 2000$

For 2000, $x = 2000 - 2000 = 0$

For 2030, $x = 2030 - 2000 = 30$

b.   For $x = 0$.

$y = 0.665(0) + 23.4$

$\qquad = 23.4$

For $x = 30$.

$y = 0.665(30) + 23.4$

$\qquad = 43.35$

c.

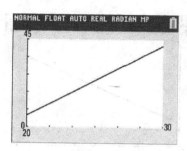

51. a. 299.9 million, or 299,900,000

b.

Years after 2000	Population (millions)
0	275.3
10	299.9
20	324.9
30	351.1
40	377.4
50	403.7
60	432

c.

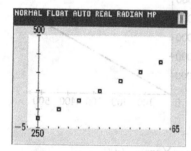

53. a.

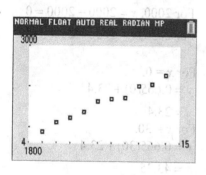

b.

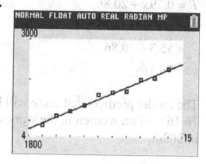

55. a. In 2026 the crude oil production will be 2.26 billion barrels.

b.

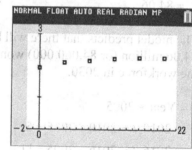

c.

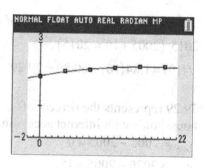

Section 1.3 Skills Check

1. Recall that linear functions must be in the form $f(x) = ax + b$.

    a. Not linear. The equation has a 2nd degree (squared) term.

    b. Linear.

    c. Not linear. The x-term is in the denominator of a fraction.

3. $m = \dfrac{y_2 - y_1}{x_2 - x_1} = \dfrac{-6 - 6}{28 - 4} = \dfrac{-12}{24} = -\dfrac{1}{2}$

5. The given line passes through $(-2, 0)$ and $(0, 4)$. Therefore the slope is

$$m = \frac{y_2 - y_1}{x_2 - x_1} = \frac{4 - 0}{0 - (-2)} = \frac{4}{2} = 2$$

7.   **a.** x-intercept: Let $y = 0$ and solve for x.

$$5x - 3(0) = 15$$
$$5x = 15$$
$$x = 3$$

    y-intercept: Let $x = 0$ and solve for y.

$$5(0) - 3y = 15$$
$$-3y = 15$$
$$y = -5$$

    x-intercept: (3, 0), y-intercept: (0, −5)

  b.

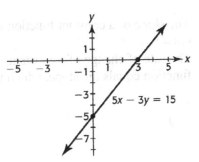

9.   **a.** x-intercept: Let $y = 0$ and solve for x.

$$3(0) = 9 - 6x$$
$$0 = 9 - 6x$$
$$0 + 6x = 9 - 6x + 6x$$
$$6x = 9$$
$$x = \frac{9}{6}$$
$$x = \frac{3}{2} = 1.5$$

    y-intercept: Let $x = 0$ and solve for y.

$$3y = 9 - 6(0)$$
$$3y = 9$$
$$y = 3$$

    x-intercept: (1.5, 0), y-intercept: (0, 3)

  b.

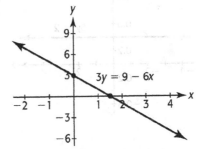

11. Horizontal lines have a slope of **zero**. Vertical lines have an **undefined** slope.

13.   **a.** Positive. The graph is rising.

    b. Undefined. The line is vertical.

14. ?

15. a. $m = 4, b = 8$

b.

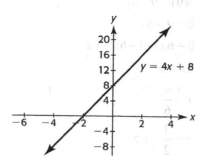

17. a. $5y = 2$

$y = \dfrac{2}{5}$, horizontal line

$m = 0, b = \dfrac{2}{5}$

b.

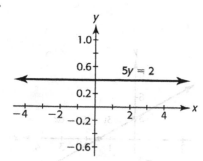

19. a. $m = 4, b = 5$

b. Rising. The slope is positive

c.

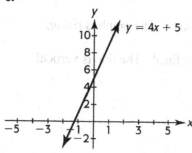

21. a. $m = -100, b = 50{,}000$

b. Falling. The slope is negative.

c.

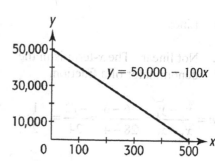

23. For a linear function, the rate of change is equal to the slope. $m = 4$.

25. For a linear function, the rate of change is equal to the slope. $m = -15$.

27. For a linear function, the rate of change is equal to the slope.

$m = \dfrac{y_2 - y_1}{x_2 - x_1} = \dfrac{-7 - 3}{4 - (-1)} = \dfrac{-10}{5} = -2$.

29. a. The identity function is $y = x$. Graph *ii* represents the identity function.

b. The constant function is $y = k$, where k is a real number. In this case, $k = 3$. Graph *i* represents a constant function.

31. a. The slope of a constant function is zero ($m = 0$).

b. The rate of change of a constant function equals the slope, which is zero.

Section 1.3 Exercises

33. Yes, the function is linear since it is written in the form $y = mx + b$. The independent variable is x, the number of years after 1990.

35. a. The function is linear since it is written in the form $y = ax + b$.

b. The slope is 0.077. The life expectancy is projected to increase by approximately 0.1 year per year.

37. a. x-intercept: Let $y = 0$ and solve for x.

$$132x + 1000y = 9570$$
$$132x + 1000(0) = 9570$$
$$132x = 9570$$
$$x = 72.5$$

The x-intercept is $(72.5, 0)$.

b. y-intercept: Let $x = 0$ and solve for y.

$$132x + 1000y = 9570$$
$$132(0) + 1000y = 9570$$
$$1000y = 9570$$
$$y = 9.57$$

The y-intercept is $(0, 9.57)$.
In 1980, the marriage rate for unmarried women was about 10 per 1000.

c. Integer values of $x \ge 0$ on the graph represent years 1980 and after.

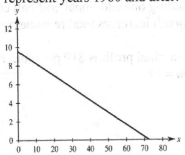

39. x-intercept: Let $R = 0$ and solve for x.

$$R = 3500 - 70x$$
$$0 = 3500 - 70x$$
$$70x = 3500$$
$$x = \frac{3500}{70} = 50$$

The x-intercept is $(50, 0)$.

R-intercept: Let $x = 0$ and solve for R.

$$R = 3500 - 70x$$
$$R = 3500 - 70(0)$$
$$R = 3500$$

The R-intercept is $(0, 3500)$.

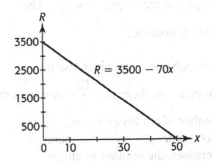

41. a. The data can be modeled by a constant function. Every input x yields the same output y.

b. $y = 11.81$

c. A constant function has a slope equal to zero.

d. For a linear function the rate of change is equal to the slope. $m = 0$.

43. a. The slope is 0.328.

b. From 2010 through 2040, the disposable income is projected to increase by $328 million per year.

45. a. The P – intercept is 20.86. Approximately 21 million women were in the workforce in 1950.

b. The slope is 0.79.

c. Rate of change of the number of women in the workforce is 790 thousand per year.

47. a. $m = 0.057$

b. From 1990 to 2050, the percent of the U.S. population that is black increased by 0.057 percentage points per year.

49. a. For a linear function, the rate of change is equal to the slope. $m = \dfrac{12}{7}$. The slope is positive.

b. For each one degree increase in temperature, there is a $\dfrac{12}{7}$ increase in the number of cricket chirps per minute. More generally, as the temperature increases, the number of chirps increases.

51. a. The rate of change of revenue for call centers in the Philippines from 2006 to 2010, was 0.975 billion dollars per year.

b. 2010 corresponds to
$x = 2010 - 2000 = 10$. When $x = 10$,
$R = 0.975(10) - 3.45$
$R = 9.75 - 3.45 = 6.30$

Thus in 2010, the revenue for call centers in the Philippines was 6.3 billion dollars.

c. No, it would not be valid since the result would be a negative number of dollars.

53. a. $m = \dfrac{y_2 - y_1}{x_2 - x_1}$

$= \dfrac{700,000 - 1,310,000}{20 - 10}$

$= \dfrac{-610,000}{10}$

$= -61,000$

b. Based on the calculation in part a), the property value decreases by \$61,000 each year. The annual rate of change is –\$61,000 per year.

55. Marginal profit is the rate of change of the profit function.

$m = \dfrac{y_2 - y_1}{x_2 - x_1}$

$= \dfrac{9000 - 4650}{375 - 300}$

$= \dfrac{4350}{75}$

$= 58$

The marginal profit is \$58 per unit.

57. a. $m = 0.56$

b. The marginal cost is \$0.56 per unit.

c. Manufacturing one additional golf ball each month increases the cost by \$0.56 or 56 cents.

59. a. $m = 1.60$

b. The marginal revenue is \$1.60 per unit.

c. Selling one additional golf ball each month increases total revenue by \$1.60.

61. The marginal profit is \$19 per unit. Note that $m = 19$.

Section 1.4 Skills Check

1. $m = 4$, $b = \dfrac{1}{2}$. The equation is $y = 4x + \dfrac{1}{2}$.

3. $m = \dfrac{1}{3}$, $b = 3$. The equation is $y = \dfrac{1}{3}x + 3$.

5. $y - y_1 = m(x - x_1)$

$y - (-6) = -\dfrac{3}{4}(x - 4)$

$y + 6 = -\dfrac{3}{4}x + 3$

$y = -\dfrac{3}{4}x - 3$

7. $x = 9$

9. Slope: $m = \dfrac{y_2 - y_1}{x_2 - x_1} = \dfrac{7 - 1}{4 - (-2)} = \dfrac{6}{6} = 1$

Equation: $y - y_1 = m(x - x_1)$

$y - 7 = 1(x - 4)$

$y - 7 = x - 4$

$y = x + 3$

11. Slope: $m = \dfrac{y_2 - y_1}{x_2 - x_1} = \dfrac{2 - 2}{5 - (-3)} = \dfrac{0}{8} = 0$

The line is horizontal. The equation of the line is $y = 2$.

13. With the given intercepts, the line passes through the points $(-5, 0)$ and $(0, 4)$. The slope of the line is

$m = \dfrac{y_2 - y_1}{x_2 - x_1} = \dfrac{4 - 0}{0 - (-5)} = \dfrac{4}{5}$.

Equation: $y - y_1 = m(x - x_1)$

$y - 0 = \dfrac{4}{5}(x - (-5))$

$y = \dfrac{4}{5}(x + 5)$

$y = \dfrac{4}{5}x + 4$

15. $3x + y = 4$

$y = -3x + 4$

$m = -3$

Since the new line is parallel with the given line, the slopes of both lines are the same.

Equation: $y - y_1 = m(x - x_1)$

$y - (-6) = -3(x - 4)$

$y + 6 = -3x + 12$

$y = -3x + 6$

17. $2x + 3y = 7$

$3y = -2x + 7$

$\dfrac{3y}{3} = \dfrac{-2x + 7}{3}$

$y = -\dfrac{2}{3}x + \dfrac{7}{3}$

$m = -\dfrac{2}{3}$

Since the new line is perpendicular with the given line, the slope of the new line is

$m_\perp = -\dfrac{1}{m}$, where m is the slope of the

given line. $m_\perp = -\dfrac{1}{m} = -\dfrac{1}{\left(-\dfrac{2}{3}\right)} = \dfrac{3}{2}$.

Equation: $y - y_1 = m(x - x_1)$

$$y - 7 = \frac{3}{2}(x - (-3))$$

$$y - 7 = \frac{3}{2}x + \frac{9}{2}$$

$$y - 7 + 7 = \frac{3}{2}x + \frac{9}{2} + 7$$

$$y = \frac{3}{2}x + \frac{23}{2}$$

19. Slope: $m = \dfrac{y_2 - y_1}{x_2 - x_1} = \dfrac{13 - (-5)}{4 - (-2)} = \dfrac{18}{6} = 3$

Equation: $y - y_1 = m(x - x_1)$

$$y - 13 = 3(x - 4)$$

$$y - 13 = 3x - 12$$

$$y = 3x + 1$$

21. Slope: $m = \dfrac{y_2 - y_1}{x_2 - x_1} = \dfrac{(-5) - (-5)}{1 - 0} = \dfrac{0}{1} = 0$

Equation: $y - y_1 = m(x - x_1)$

$$y - (-5) = 0(x - 1)$$

$$y + 5 = 0$$

$$y = -5$$

23. For a linear function, the rate of change is equal to the slope. Therefore, $m = -15$. The equation is

$$y - y_1 = m(x - x_1)$$

$$y - 12 = -15(x - 0)$$

$$y - 12 = -15x$$

$$y = -15x + 12.$$

25. For a linear function, the rate of change is equal to the slope. Therefore, $m = \frac{2}{3}$. The equation is

$$y - y_1 = m(x - x_1)$$

$$y - 9 = \tfrac{2}{3}(x - 3)$$

$$y - 9 = \tfrac{2}{3}x - 2$$

$$y = \tfrac{2}{3}x + 7.$$

27. $\dfrac{f(b) - f(a)}{b - a} = \dfrac{f(2) - f(-1)}{2 - (-1)}$

$$= \dfrac{(2)^2 - (-1)^2}{3}$$

$$= \dfrac{4 - 1}{3}$$

$$= \dfrac{3}{3}$$

$$= 1$$

The average rate of change between the two points is 1.

29. $\dfrac{f(b) - f(a)}{b - a} = \dfrac{f(1) - f(-2)}{1 - (-2)}$

$$= \dfrac{-2 - 7}{3}$$

$$= \dfrac{-9}{3}$$

$$= -3$$

31. $y - y_1 = m(x - x_1)$

$$y - (-10) = -3(x - 1)$$

$$y + 10 = -3x + 3$$

$$y = -3x - 7.$$

The y –intercept is $b = -7$.

33. $f(x+h) = 45 - 15(x+h)$
$\qquad = 45 - 15x - 15h$

$f(x+h) - f(x)$
$= 45 - 15x - 15h - [45 - 15x]$
$= 45 - 15x - 15h - 45 + 15x$
$= -15h$

$$\frac{f(x+h) - f(x)}{h} = \frac{-15h}{h} = -15$$

35. $f(x+h) = 2(x+h)^2 + 4$
$\qquad = 2(x^2 + 2xh + h^2) + 4$
$\qquad = 2x^2 + 4xh + 2h^2 + 4$

$f(x+h) - f(x)$
$= 2x^2 + 4xh + 2h^2 + 4 - [2x^2 + 4]$
$= 2x^2 + 4xh + 2h^2 + 4 - 2x^2 - 4$
$= 4xh + 2h^2$

$$\frac{f(x+h) - f(x)}{h} = \frac{4xh + 2h^2}{h}$$
$$= \frac{h(4x + 2h)}{h}$$
$$= 4x + 2h$$

37. a. The difference in the y-coordinates is consistently 30, while the difference in the x-coordinates is consistently 10. Considering the scatter plot below, a line fits the data exactly.

[0, 60] by [500, 800]

b. Slope: $m = \dfrac{y_2 - y_1}{x_2 - x_1}$

$\qquad = \dfrac{615 - 585}{20 - 10}$

$\qquad = \dfrac{30}{10}$

$\qquad = 3$

Equation: $y - y_1 = m(x - x_1)$
$\qquad\qquad y - 585 = 3(x - 10)$
$\qquad\qquad y - 585 = 3x - 30$
$\qquad\qquad\quad y = 3x + 555$

Section 1.4 Exercises

39. Let $x = $ kWh used and let $y = $ monthly charge in dollars. Then the equation is $y = 0.1034x + 12.00$.

41. Let $t = $ number of years, and let $y = $ value of the machinery after t years. Then the equation is $y = 36,000 - 3,600t$.

43. a. Let $t = $ age in years, and let $H = $ heart rate. Using the information given:

The resulting slope is $\dfrac{6.5}{-10} = -0.65$

Then using the point-slope form of a linear equation and the ordered pair (60, 104):

$H - 104 = -0.65(t - 60)$
$H - 104 = -0.65t + 39$
$H = -0.65t + 143$

b. Let $t = 40$.
$H = -0.65(40) + 143 = -26 + 143 = 117$.
The desired heart rate for a 40 year old person is 117 beats per minute.

45. a. From year 0 to year 5, the automobile depreciates from a value of $26,000 to a value of $1,000. Therefore, the total depreciation is 26,000–1000 or $25,000.

b. Since the automobile depreciates for 5 years in a straight-line (linear) fashion, each year the value declines by
$$\frac{25,000}{5} = \$5,000 \, .$$

c. Let t = the number of years, and let s = the value of the automobile at the end of t years. Then, based on parts a) and b) the linear equation modeling the value is $s = -5000t + 26,000$.

47. Notice that the x and y values always match. The number of deputies always equals the number of patrol cars. Therefore the equation is $y = x$, where x represents the number of deputies, and y represents the number of patrol cars.

49. $m = \dfrac{y_2 - y_1}{x_2 - x_1}$

$= \dfrac{9000 - 4650}{375 - 300}$

$= \dfrac{4350}{75} = 58$

Equation:
$$P - p_1 = m(x - x_1)$$
$$P - 4650 = 58(x - 300)$$
$$P - 4650 = 58x - 17,400$$
$$P = 58x - 12,750$$

51. $m = \dfrac{y_2 - y_1}{x_2 - x_1}$

$= \dfrac{700,000 - 1,310,000}{20 - 10}$

$= \dfrac{-610,000}{10}$

$= -61,000$

Equation:
$$y - y_1 = m(x - x_1)$$
$$y - 700,000 = -61,000(x - 20)$$
$$y - 700,000 = -61,000x + 1,220,000$$
$$y = -61,000x + 1,920,000$$
$$V = -61,000x + 1,920,000$$

53. a. The slope of the equation will be $m = 160$ and the point used will be (66, 2000):

Equation:
$$y - y_1 = m(x - x_1)$$
$$y - 2000 = 160(x - 66)$$
$$y - 2000 = 160x - 10560$$
$$y = 160x - 8560$$
where x = the age of the person between 66 and 70 inclusive, and y = the monthly benefit received.

b. Let $x = 70$.
$$y = 160(70) - 8560$$
$$= 11200 - 8560$$
$$= 2640$$
The monthly benefit for a recipient starting their benefits at age 70 would be $2640.

55. a. Notice that the change in the x-values is consistently 1 while the change in the y-values is consistently 0.045 percentage points per drink. Therefore the table represents a linear function. The rate of change is the slope of the linear function.
$$m = \frac{\text{vertical change}}{\text{horizontal change}} = \frac{0.045}{1} = 0.045$$

b. Let x = the number of drinks, and let y = the blood alcohol percent. Using the slope 0.045 and one of points (5, 0.225), the equation is:

$$y - y_1 = m(x - x_1)$$
$$y - 0.225 = 0.045(x - 5)$$
$$y - 0.225 = 0.045x - 0.225$$
$$y = 0.045x$$

57. a. Let t = the year at the beginning of the decade, and let g = average number of men in the workforce during the decade. Using points (1950, 43.8) and (2050, 100.3) to calculate the slope yields:

$$m = \frac{g_2 - g_1}{t_2 - t_1}$$
$$= \frac{100.3 - 43.8}{2050 - 1950}$$
$$= \frac{56.5}{100} = 0.565$$

Equation:
$$g - g_1 = m(t - t_1)$$
$$g - 43.8 = 0.565(t - 1950)$$
$$g - 43.8 = 0.565t - 1101.75$$
$$g = 0.565t - 1057.95$$

b. Yes. Consider the following table of values based on the equation in comparison to the actual data points.

t	g = Equation Values	Actual Values
1950	43.8	43.8
1960	49.45	46.4
1970	55.1	51.2
1980	60.75	61.5
1990	66.4	69
2000	72.05	75.2
2010	77.7	82.2
2015	80.525	84.2
2020	83.35	85.4
2030	89	88.5
2040	94.65	94
2050	100.3	100.3

c. They are the same since the points (1950, 43.8) and (2050, 100.3) were used to calculate the slope of the linear model.

59. a. Let x = the year, and let p = the percent of Americans considered obese. Using points (2000, 22.1) and (2030, 42.2) to calculate the slope yields:

$$m = \frac{y_2 - y_1}{x_2 - x_1}$$
$$= \frac{42.2 - 22.1}{2030 - 2000}$$
$$= \frac{20.1}{30}$$
$$= 0.67$$

Equation:
$$y - y_1 = m(x - x_1)$$
$$p - 22.1 = 0.67(x - 2000)$$
$$p - 22.1 = 0.67x - 1340$$
$$p = 0.67x - 1317.9$$

b. Let $x = 2020$.
$$p = 0.67(2020) - 1317.9$$
$$= 1353.4 - 1317.9$$
$$= 35.5$$
The percent of Americans that will be considered obese in the year 2020 is projected to be 35.5%.

61. a. $m = \dfrac{y_2 - y_1}{x_2 - x_1}$
$$= \frac{81.6 - 18.4}{80 - 0}$$
$$= \frac{63.2}{80} = 0.79$$

b. It is the same as part a). The average rate of change in the number of women in the workforce between 1950 and 2030 is increasing at a rate of 0.79 million per year.

c.
$$y - y_1 = m(x - x_1)$$
$$y - 18.4 = 0.79(x - 0)$$
$$y - 18.4 = 0.79x$$
$$y = 0.79x + 18.4$$

63. a.
$$\frac{f(b) - f(a)}{b - a} = \frac{f(2012) - f(1960)}{2009 - 1960}$$
$$= \frac{1,570,397 - 212,953}{2012 - 1960}$$
$$= \frac{1,357,444}{52}$$
$$= 26,104.69231$$
$$\approx 26,105 / \text{year}$$

b. The slope of the line connecting the two points is the same as the average rate of change between the two points. Based on part a), $m = 26,105$.

c. The equation of the secant line, using the rounded slope from part a), is given by:
$$y - y_1 = m(x - x_1)$$
$$y - 1,570,397 = 26,105(x - 2012)$$
$$y - 1,570,397 = 26,105x - 52,523,260$$
$$y = 26,105x - 50,952,863$$

d. No. The points on the scatter plot do not approximate a linear pattern.

65. a. No.

b. Yes. The points seem to follow a straight line pattern for years between 2010 and 2030.

c.
$$\frac{f(b) - f(a)}{b - a} = \frac{f(2030) - f(2010)}{2030 - 2010}$$
$$= \frac{2.2 - 3.9}{2030 - 2010}$$
$$= \frac{-1.7}{20}$$
$$= -0.085$$

The average annual rate of change of the data over this period of time is -0.085 points per year.

d.
$$y - y_1 = m(x - x_1)$$
$$y - 3.9 = -0.085(x - 2010)$$
$$y - 3.9 = -0.085x + 170.85$$
$$y = -0.085x + 174.75$$

67. a. Let $x = $ the year, and let $y = $ the number of White non-Hispanic individuals in the U.S. civilian non-institutional population 16 years and older. Then, the average rate of change between 2000 and 2050 is given by:
$$\frac{y_2 - y_1}{x_2 - x_1} = \frac{169.4 - 153.1}{2050 - 2000}$$
$$= \frac{16.3}{50}$$
$$= 0.326$$

The annual average increase in the number of White non-Hispanic individuals in the U.S. civilian non-institutional population 16 years and older is 0.326 million per year.

b.
$$y - y_1 = m(x - x_1)$$
$$y - 153.1 = 0.326(x - 2000)$$
$$y - 153.1 = 0.326x - 652$$
$$y = 0.326x - 498.9$$

c. 2020 corresponds to $x = 2020$.
$$y = 0.326x - 498.9$$
$$y = 0.326(2020) - 498.9$$
$$y = 658.52 - 498.9$$
$$y = 159.62$$

The number of White non-Hispanic individuals in the U.S. civilian non-institutional population 16 years and older in 2020 is projected to be 159.62 million people.

69. $\dfrac{f(b)-f(a)}{b-a} = \dfrac{f(4)-f(2)}{4-2}$

$\qquad = \dfrac{256-192}{4-2}$

$\qquad = \dfrac{64}{2}$

$\qquad = 32$

The average velocity over this period of time is 32 ft/sec.

71. First calculate $f(2+h)$.

$f(2+h) = -16(2+h)^2 + 128(2+h)$

$\qquad = -16(h^2 + 4h + 4) + 128(2+h)$

$\qquad = -16h^2 - 64h - 64 + 256 + 128h$

$\qquad = -16h^2 + 64h + 192$

Now Substitute into the difference quotient:

$\dfrac{f(x+h)-f(x)}{h} = \dfrac{f(2+h)-f(2)}{h}$

$\qquad = \dfrac{\left(-16h^2 + 64h + 192\right) - \left(-16(2)^2 + 128(2)\right)}{h}$

$\qquad = \dfrac{-16h^2 + 64h + 192 + 64 - 256}{h}$

$\qquad = \dfrac{-16h^2 + 64h}{h}$

$\qquad = -16h + 64$

Chapter 1 Skills Check

1. The table represents a function because each x matches with exactly one y.

3. $f(3) = 0$

5. a. $C(3) = 16 - 2(3)^2$

$\qquad = 16 - 2(9) = 16 - 18 = -2$

b. $C(-2) = 16 - 2(-2)^2$

$\qquad = 16 - 2(4) = 16 - 8 = 8$

c. $C(-1) = 16 - 2(-1)^2$

$\qquad = 16 - 2(1) = 16 - 2 = 14$

7.

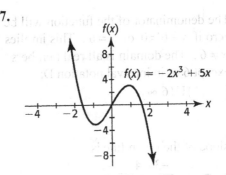

9. $y = -10x^2 + 400x + 10$

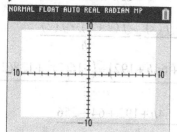

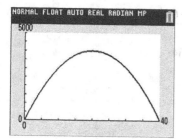

The second view is better.

11. a. Since $y = \sqrt{2x-8}$ will not be a real number if $2x - 8 < 0$, the only values of x that yield real outputs to the function are values that satisfy $2x - 8 \geq 0$. Isolating x yields:

$$2x - 8 + 8 \geq 0 + 8$$
$$\frac{2x}{2} \geq \frac{8}{2}$$
$$x \geq 4$$

Therefore the domain is D: $[4, \infty)$.

b. The denominator of the function will be zero if $x - 6 = 0$ or $x = 6$. This implies $x \neq 6$. The domain is all real numbers except 6 or in interval notation D: $(-\infty, 6) \cup (6, \infty)$.

13. The slope of the given line is
$$m = \frac{y_2 - y_1}{x_2 - x_1} = \frac{-3 - 4}{5 - (-1)} = \frac{-7}{6} = -\frac{7}{6}.$$

Since two parallel lines have the same slope, the slope of any line parallel to this one is also $m = -\frac{7}{6}$.

Since the slopes of perpendicular lines are negative reciprocals of one another,
$$m_{\perp} = -\frac{1}{m} = -\frac{1}{\left(-\frac{7}{6}\right)} = -\left(-\frac{6}{7}\right) = \frac{6}{7}$$

15. a. x-intercept: Let $y = 0$ and solve for x.
$$2x - 3(0) = 12$$
$$2x = 12$$
$$x = 6$$
y-intercept: Let $x = 0$ and solve for y.
$$2(0) - 3y = 12$$
$$-3y = 12$$
$$y = -4$$
x-intercept: $(6, 0)$, y-intercept: $(0, -4)$

b.

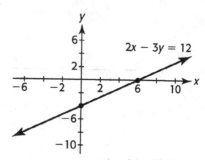

17. The slope is $m = -6$ The y-intercept is $(0, 3)$.

19. $y = mx + b$
$$y = \frac{1}{3}x + 3$$

21. The slope is
$$m = \frac{y_2 - y_1}{x_2 - x_1} = \frac{6-3}{2-(-1)} = \frac{3}{3} = 1.$$

Solving for the equation:
$$y - y_1 = m(x - x_1)$$
$$y - 6 = 1(x - 2)$$
$$y - 6 = x - 2$$
$$y = x + 4$$

23. a. $f(x+h) = 5 - 4(x+h)$
$$= 5 - 4x - 4h$$

b. $f(x+h) - f(x)$
$$= [5 - 4(x+h)] - [5 - 4x]$$
$$= 5 - 4x - 4h - 5 + 4x$$
$$= -4h$$

c. $\dfrac{f(x+h) - f(x)}{h}$
$$= \frac{-4h}{h}$$
$$= -4$$

Chapter 1 Review Exercises

25. a. Yes. If only the Democratic percentages are considered, each year matches with exactly one black voter percentage.

b. $f(1992) = 82$. The table indicates that in 1992, 82% of black voters supported a Democratic candidate for president.

c. When $f(y) = 94$, $y = 1964$. The table indicates that in 1964, 94% of black voters supported a Democratic candidate for president.

26. a. The domain is
$\{1960, 1964, 1968, 1972, 1976, 1980,$
$1984, 1992, 1996, 2000, 2004, 2008, 2012\}$

b. No. 1982 was not a presidential election year.

27.

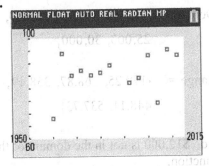

28. a. For 1972 to 2012:
$$m = \frac{y_2 - y_1}{x_2 - x_1}$$
$$= \frac{93 - 87}{2012 - 1972}$$
$$= \frac{6}{40} = 0.15$$

b. The average annual rate of change is the same as the slope in part a), therefore = 0.15 percentage points per year.

c. For 1972 to 1984:
$$\frac{f(b) - f(a)}{b - a} = \frac{89 - 87}{1984 - 1972}$$
$$= \frac{2}{12}$$
$$\approx 0.1667$$
No, the rates of change are not equal.

29. a. Each amount borrowed matches with exactly one monthly payment. The change in y is 89.62 for each change in x of 5000.

b. $f(25,000) = 448.11$. Therefore, borrowing \$25,000 to buy a car from the dealership results in a monthly payment of \$448.11.

c. If $f(A) = \$358.49$, then $A = \$20,000$.

30. a. Domain = $\{10,000, \ 15,000, \ 20,000,$
$25,000, \ 30,000\}$

Range = $\{179.25, \ 268.87, \ 358.49,$
$448.11, \ 537.73\}$

b. No. \$12,000 is not in the domain of the function.

31. a. $f(28,000) = 0.017924(28,000) + 0.01$
$= 501.872 + 0.01$
$= 501.882$

The predicted monthly payment on a car loan of \$28,000 is \$501.88.

b. Any positive input could be used for A. Borrowing a negative amount of money does not make sense in the context of the problem.

32. a. $f(1960) = 15.9$. A 65-year old woman in 1960 is expected to live 15.9 more years. Her overall life expectancy is 80.9 years.

b. $f(2010) = 19.4$. A 65-year old woman in 2010 has a life expectancy of 84.4 years.

c. Since $f(1990) = 19$, the average woman is expected to live 19 years past age 65 in 1990.

33. a. $g(2020) = 16.9$. A 65-year old man in 2020 is expected to live 16.9 more years. His overall life expectancy is 81.9 years.

b. Since $g(1950) = 12.8$, a 65-year old man in 1950 has a life expectancy of 77.8 years.

c. Since $g(1990) = 15$, the average man is expected to live 15 years past age 65 (or to age 80) in 1990. $g(1990) = 15$.

34. No; The function cannot be written in linear form, $y = mx + b$.

35. a. Let $t = $ age, and let $h = $ heart rate in beats per minute. Using points (20, 160) and (90, 104) to calculate the slope yields:

$$m = \frac{h_2 - h_1}{t_2 - t_1}$$
$$= \frac{104 - 160}{90 - 20}$$
$$= \frac{-56}{70} = -0.8$$

Equation:
$$h - h_1 = m(t - t_1)$$
$$h - 160 = -0.8(t - 20)$$
$$h - 160 = -0.8t + 16$$
$$h = -0.8t + 176$$

b. Yes. Consider the following table of values based on the equation in comparison to the actual data points.

t	h = Equation Values	Actual Values
20	160	160
30	152	152
40	144	144
50	136	136
60	128	128
70	120	120
80	112	112
90	104	104

c. They are the same since the points $(20,160)$ and $(90,104)$ were used to calculate the slope of the linear model.

36. a. Using points (2010, 1.94) and (2030, 2.26) to calculate the slope yields:

$$m = \frac{y_2 - y_1}{x_2 - x_1}$$

$$= \frac{2.26 - 1.94}{2030 - 2010}$$

$$= \frac{0.32}{20} = 0.016$$

b. Let $x =$ the year, and let $y =$ crude oil production in billions of barrels. The equation will be:

$$y - y_1 = m(x - x_1)$$

$$y - 1.94 = 0.016(x - 2010)$$

$$y - 1.94 = 0.016x - 32.16$$

$$y = 0.016x - 30.22$$

37. $f(x) = 4500$

38. a. Let $x =$ the number of months past May 2016, and $f(x) =$ the average weekly hours worked. The function is $f(x) = 33.8$.

b. This is a constant function since the rate of change is zero.

39. a. $R(120) = 564(120) = 67,680$. The revenue when 120 units are produced is $67,680.

b. $C(120) = 40,000 + 64(120) = 47,680$. The cost when 120 units are produced is $47,680.

c. Marginal Cost $= MC = 64 = \$64$. Note that MC is the slope of the cost function.

Marginal Revenue $= MR = 564 = \$564$. Note that MR is the slope of the revenue function.

d. $m = 64$

e.

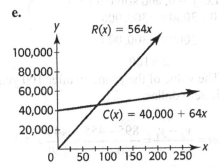

40. a. $P(x) = 564x - (40,000 + 64x)$

$$= 564x - 40,000 - 64x$$

$$= 500x - 40,000$$

b. $P(120) = 500(120) - 40,000$

$$= 60,000 - 40,000$$

$$= \$20,000$$

c. Break-even occurs when $R(x) = C(x)$ or alternately $P(x) = R(x) - C(x) = 0$.

$$500x - 40,000 = 0$$

$$500x = 40,000$$

$$x = \frac{40,000}{500}$$

$$x = 80$$

Eighty units represent break-even for the company.

d. $MP =$ the slope of $P(x) = 500$

e. $MP = MR - MC$

41. a. Let $x = 0$, and solve for y.
$$y + 3000(0) = 300,000$$
$$y = 300,000$$
The initial value of the property is $300,000.

b. Let $y = 0$, and solve for x.
$$0 + 3000x = 300,000$$
$$3000x = 300,000$$
$$x = 100$$
The value of the property after 100 years is zero dollars.

42. a. $m = \dfrac{y_2 - y_1}{x_2 - x_1} = \dfrac{895 - 455}{250 - 150} = \dfrac{440}{100} = 4.4$.

The average rate of change is $4.40 per unit.

b. For a linear function, the slope is the average rate of change. Referring to part a), the slope is 4.4.

c. $y - y_1 = m(x - x_1)$
$$y - 455 = 4.4(x - 150)$$
$$y - 455 = 4.4x - 660$$
$$y = 4.4x - 205$$
$$P(x) = 4.4x - 205$$

d. MP = the slope of $P(x) = 4.4$ or $4.40 per unit.

e. Break-even occurs when $R(x) = C(x)$ or alternately $P(x) = R(x) - C(x) = 0$.

$$4.4x - 205 = 0$$
$$4.4x = 205$$
$$x = \frac{205}{4.4}$$
$$x = 46.5909 \approx 47$$

The company will break even selling approximately 47 units.

Group Activities/Extended Applications

1. Body Mass Index

1. A person uses the table to determine his or her BMI by locating the entry in the table that corresponds to his or her height and weight. The entry in the table is the person's BMI.

2. If a person's BMI is 30 or higher, the person is considered obese and at risk for health problems.

3. a. Determine the heights and weights that produce a BMI of exactly 30 based on the table.

Height (inches)	Weight (pounds)
61	160
63	170
65	180
67	190
68	200
69	200
72	220
73	230

b.

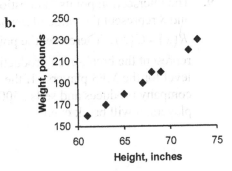

A linear model is reasonable, but not exact.

c. $m = \dfrac{y_2 - y_1}{x_2 - x_1}$

$= \dfrac{230 - 160}{73 - 61}$

$= \dfrac{70}{12}$

$= 5.8\overline{3}$

$y - y_1 = m(x - x_1)$

$y - 160 = \dfrac{70}{12}(x - 61)$

$y - 160 = \dfrac{70}{12}x - \dfrac{4270}{12}$

$y = \dfrac{70}{12}x - \dfrac{4270}{12} + \dfrac{1920}{12}$

$y = \dfrac{70}{12}x - \dfrac{2350}{12}$

$y = 5.8\overline{3}x - 195.8\overline{3}$

d.

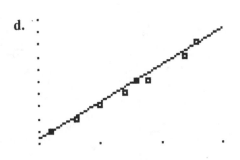

[60, 75] by [150, 250]

The line fits the data points well, but not perfectly.

e.   Any data point that lies exactly along the line generated from the model will yield a BMI of 30. If a height is substituted into the model, the output weight would generate a BMI of 30. That weight or any higher weight for the given height would place a person at risk for health problems.

2.   Total Revenue, Total Cost, and Profit

Let x represent the number of units produced and sold.

1.   The revenue function is $R(x) = 98x$.

2.   Marginal Revenue $= MR = 98 = \$98$. Note that MR is the slope of the revenue function.

3. $C(x) = 23x + 262{,}500$

4.   Marginal Cost $= MC = 23 = \$23$. Note that MC is the slope of the cost function.

Neither. $MC \cdot x$ is the variable cost.

5. $P(x) = R(x) - C(x)$

$= [98x] - [23x - 262{,}500]$

$= 75x - 262{,}500$

$MP =$ the slope of $P(x) = 75$

6.   If $x = 0$, then

$R(0) = 98(0) = 0$

$C(0) = 23(0) + 262{,}500 = 262{,}500$

$P(0) = 75(0) - 262{,}500 = -262{,}500$

7.

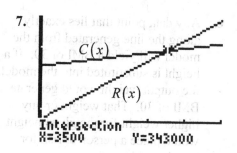

[0, 5000] by [0, 500,000]

The intersection point is approximately $x = 3500$ units.

8.

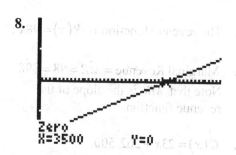

[0, 5000] by [–300,000, 300,000]

$x = 3500$ is the intersection point

9. The intersection points in questions 7 and 8 represent the value of x where $R(x) = C(x)$. Therefore, the points represent the break-even production level for the MP3 players. If the company produces and sells 3500 MP3 players, it will break even.

CHAPTER 2
Linear Models, Equations, and Inequalities

Toolbox Exercises

1. $3x = 6$
    Division Property
    $$\frac{3x}{3} = \frac{6}{3}$$
    $x = 2$

2. $x - 7 = 11$
    Addition Property
    $x - 7 = 11$
    $x - 7 + 7 = 11 + 7$
    $x = 18$

3. $x + 3 = 8$
    Subtraction Property
    $x + 3 = 8$
    $x + 3 - 3 = 8 - 3$
    $x = 5$

4. $x - 5 = -2$
    Addition Property
    $x - 5 = -2$
    $x - 5 + 5 = -2 + 5$
    $x = 3$

5. $\frac{x}{3} = 6$
    Multiplication Property
    $$\frac{x}{3} = 6$$
    $$3\left(\frac{x}{3}\right) = 3(6)$$
    $x = 18$

6. $-5x = 10$
    Division Property
    $$\frac{-5x}{-5} = \frac{10}{-5}$$
    $x = -2$

7. $2x + 8 = -12$
    Subtraction Property and Division Property
    $2x + 8 = -12$
    $2x + 8 - 8 = -12 - 8$
    $2x = -20$
    $$\frac{2x}{2} = \frac{-20}{2}$$
    $x = -10$

8. $\frac{x}{4} - 3 = 5$
    Addition Property and Multiplication Property
    $$\frac{x}{4} - 3 = 5$$
    $$\frac{x}{4} - 3 + 3 = 5 + 3$$
    $$\frac{x}{4} = 8$$
    $$4\left(\frac{x}{4}\right) = 4(8)$$
    $x = 32$

9. $4x - 3 = 6 + x$
    $4x - x - 3 = 6 + x - x$
    $3x - 3 = 6$
    $3x - 3 + 3 = 6 + 3$
    $3x = 9$
    $$\frac{3x}{3} = \frac{9}{3}$$
    $x = 3$

10. $3x - 2 = 4 - 7x$

$3x + 7x - 2 = 4 - 7x + 7x$

$10x - 2 = 4$

$10x - 2 + 2 = 4 + 2$

$10x = 6$

$\dfrac{10x}{10} = \dfrac{6}{10}$

$x = \dfrac{6}{10}$

$x = \dfrac{3}{5}$

11. $\dfrac{3x}{4} = 12$

$4\left(\dfrac{3x}{4}\right) = 4(12)$

$3x = 48$

$\dfrac{3x}{3} = \dfrac{48}{3}$

$x = 16$

12. $\dfrac{5x}{2} = -10$

$2\left(\dfrac{5x}{2}\right) = 2(-10)$

$5x = -20$

$\dfrac{5x}{5} = \dfrac{-20}{5}$

$x = -4$

13.

$3(x - 5) = -2x - 5$

$3x - 15 = -2x - 5$

$3x + 2x - 15 = -2x + 2x - 5$

$5x - 15 = -5$

$5x - 15 + 15 = -5 + 15$

$5x = 10$

$\dfrac{5x}{5} = \dfrac{10}{5}$

$x = 2$

14.

$-2(3x - 1) = 4x - 8$

$-6x + 2 = 4x - 8$

$-6x - 4x + 2 = 4x - 4x - 8$

$-10x + 2 = -8$

$-10x + 2 - 2 = -8 - 2$

$-10x = -10$

$\dfrac{-10x}{-10} = \dfrac{-10}{-10}$

$x = 1$

15.

$2x - 7 = -4\left(4x - \dfrac{1}{2}\right)$

$2x - 7 = -16x + 2$

$2x + 16x - 7 = -16x + 16x + 2$

$18x - 7 = 2$

$18x - 7 + 7 = 2 + 7$

$18x = 9$

$\dfrac{18x}{18} = \dfrac{9}{18}$

$x = \dfrac{1}{2}$

16.

$-2(2x - 6) = 3\left(3x - \dfrac{1}{3}\right)$

$-4x + 12 = 9x - 1$

$-4x - 9x + 12 = 9x - 9x - 1$

$-13x + 12 = -1$

$-13x + 12 - 12 = -1 - 12$

$-13x = -13$

$\dfrac{-13x}{-13} = \dfrac{-13}{-13}$

$x = 1$

17. $y = 2x$ and $x + y = 12$

$x + (2x) = 12$

$3x = 12$

$x = 4$

18. $y = 4x$ and $x + y = 25$

$x + (4x) = 25$

$5x = 25$

$x = 5$

19. $y = 3x$ and $2x + 4y = 42$

$2x + 4(3x) = 42$

$2x + 12x = 42$

$14x = 42$

$x = 3$

20. $y = 6x$ and $3x + 2y = 75$

$3x + 2(6x) = 75$

$3x + 12x = 75$

$15x = 75$

$x = 5$

21. $3x - 5x = 2x + 7$

$-2x = 2x + 7$

$-4x = 7$

$x = \dfrac{7}{-4}$

This is a conditional equation.

22. $3(x + 1) = 3x - 7$

$3x + 3 = 3x - 7$

$3 = -7$

This is a contradiction equation.

23. $9x - 2(x - 5) = 3x + 10 + 4x$

$9x - 2x + 10 = 7x + 10$

$7x + 10 = 7x + 10$

This is an identity equation.

24. $\dfrac{x}{2} - 5 = \dfrac{x}{4} + 2$

$2x - 20 = x + 8$

$x = 28$

This is a conditional equation.

25. $5x + 1 > -5$

$5x > -6$

$x > -\dfrac{6}{5}$

26. $1 - 3x \geq 7$

$-3x \geq 6$

$x \leq -2$

27. $\dfrac{x}{4} > -3$

$x > -12$

28. $\dfrac{x}{6} > -2$

$x > -12$

29. $\dfrac{x}{4} - 2 > 5x$

$\dfrac{x}{4} > 5x + 2$

$x > 20x + 8$

$-19x > 8$

$x < -\dfrac{8}{19}$

30. $\dfrac{x}{2}+3>6x$

$x+6>12x$

$-11x+6>0$

$-11x>-6$

$x<\dfrac{6}{11}$

31. $-3(x-5)<-4$

$-3x+15<-4$

$-3x<-19$

$x>\dfrac{19}{3}$

32. $-\dfrac{1}{2}(x+4)<6$

$-\dfrac{1}{2}x-2<6$

$-\dfrac{1}{2}x<8$

$x>-16$

33. There are 6 significant digits.

34. There are 4 significant digits.

35. There are 2 significant digits.

36. There are 5 significant digits.

37. There are 5 significant digits.

38. There are 2 significant digits.

39. a. $f(x)=0.008x^2-632.578x+480.650$

  b. $f(x)=0.00754x^2-633x+481$

  c. $f(x)=0.007453x^2-632.6x+480.7$

Section 2.1 Skills Check

1.

$5x-14=23+7x$

$5x-7x-14=23+7x-7x$

$-2x-14=23$

$-2x-14+14=23+14$

$-2x=37$

$\dfrac{-2x}{-2}=\dfrac{37}{-2}$

$x=-\dfrac{37}{2}$

$x=-18.5$

Applying the intersections of graphs method, graph $y=5x-14$ and $y=23+7x$. Determine the intersection point from the graph:

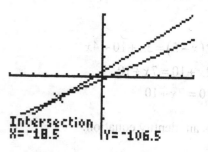

Intersection
X=-18.5   Y=-106.5

[–40, 40] by [–300, 300]

3.
$$3(x-7)=19-x$$
$$\boxed{3x-21}=19-x$$
$$3x+x-21=19-x+x$$
$$4x-21=19$$
$$4x-21+21=19+21$$
$$4x=40$$
$$\frac{4x}{4}=\frac{40}{4}$$
$$x=10$$

Applying the intersections of graphs method yields:

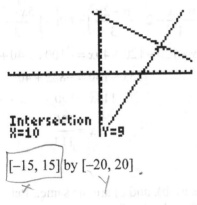

Intersection
X=10   Y=9

[−15, 15] by [−20, 20]

5.
$$x-\frac{5}{6}=3x+\frac{1}{4}$$
$$LCD:12$$
$$12\left(x-\frac{5}{6}\right)=12\left(3x+\frac{1}{4}\right)$$
$$12x-10=36x+3$$
$$12x-36x-10=36x-36x+3$$
$$-24x-10=3$$
$$-24x-10+10=3+10$$
$$-24x=13$$
$$\frac{-24x}{-24}=\frac{13}{-24}$$
$$x=-\frac{13}{24}\approx-0.5416667$$

Applying the intersections of graphs method yields:

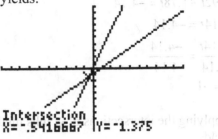

Intersection
X=-.5416667   Y=-1.375

[−10, 10] by [−10, 10]

7.
$$\frac{5(x-3)}{6}-x=1-\frac{x}{9}$$
$$LCD:18$$
$$18\left(\frac{5(x-3)}{6}-x\right)=18\left(1-\frac{x}{9}\right)$$
$$15(x-3)-18x=18-2x$$
$$15x-45-18x=18-2x$$
$$-3x-45=18-2x$$
$$-1x-45=18$$
$$-1x=63$$
$$x=-63$$

Applying the intersections of graphs method yields:

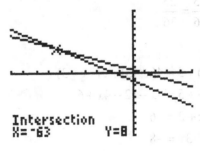

Intersection
X=-63   Y=8

[−100, 50] by [−20, 20]

9. $5.92t = 1.78t - 4.14$

$5.92t - 1.78t = -4.14$

$4.14t = -4.14$

$\dfrac{4.14t}{4.14} = \dfrac{-4.14}{4.14}$

$t = -1$

Applying the intersections of graphs method yields:

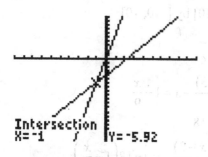

Intersection
X=-1   Y=-5.92

[–10, 10] by [–20, 10]

11. $\dfrac{3}{4} + \dfrac{1}{5}x - \dfrac{1}{3} = \dfrac{4}{5}x$

$LCD = 60$

$60\left(\dfrac{3}{4} + \dfrac{1}{5}x - \dfrac{1}{3}\right) = 60\left(\dfrac{4}{5}x\right)$

$45 + 12x - 20 = 48x$

$-36x = -25$

$x = \dfrac{-25}{-36} = \dfrac{25}{36}$

13. $3(x-1) + 5 = 4(x-3) - 2(2x-3)$

$3x - 3 + 5 = 4x - 12 - 4x + 6$

$3x + 2 = -6$

$3x = -8$

$x = -\dfrac{8}{3}$

15. $2[3(x-10) - 3(2x+1)] = 3[-(x-6) - 5(2x-4)]$

$2[3x - 30 - 6x - 3] = 3[-x + 6 - 10x + 20]$

$6x - 60 - 12x - 6 = -3x + 18 - 30x + 60$

$-6x - 66 = -33x + 78$

$27x = 144$

$x = \dfrac{144}{27} = \dfrac{16}{3}$

17. $\dfrac{6-x}{4} - 2 - \dfrac{6-2x}{3} = -\left[\dfrac{5x-2}{3} - \dfrac{x}{5}\right]$

$LCD = 60$

$60\left(\dfrac{6-x}{4} - 2 - \dfrac{6-2x}{3}\right) = 60\left(-\left[\dfrac{5x-2}{3} - \dfrac{x}{5}\right]\right)$

$90 - 15x - 120 - 120 + 40x = -100x + 40 + 12x$

$25x - 150 = -88x + 40$

$113x = 190$

$x = \dfrac{190}{113}$

19. Answers a), b), and c) are the same. Let $f(x) = 0$ and solve for x.

$32 + 1.6x = 0$

$1.6x = -32$

$x = -\dfrac{32}{1.6}$

$x = -20$

The solution to $f(x) = 0$, the x-intercept of the function, and the zero of the function are all –20.

21. Answers a), b), and c) are the same. Let $f(x) = 0$ and solve for x.

$$\frac{3}{2}x - 6 = 0$$

$LCD : 2$

$$2\left(\frac{3}{2}x - 6\right) = 2(0)$$

$$3x - 12 = 0$$

$$3x = 12$$

$$x = 4$$

The solution to $f(x) = 0$, the x-intercept of the function, and the zero of the function are all 4.

23. a. The x-intercept is 2, since an input of 2 creates an output of 0 in the function.

b. The y-intercept is -34, since the output of -34 corresponds with an input of 0.

c. The solution to $f(x) = 0$ is equal to the x-intercept for the function. Therefore, the solution to $f(x) = 0$ is 2.

25. The answers to a) and b) are the same. The graph crosses the x-axis at $x = 40$.

27. Using the first screen, the equation that would need to be solved is $2x - 5 = 3(x - 2)$. Using the second screen, the y values are the same when $x = 1$.

29. Answers a), b), and c) are the same. Let $f(x) = 0$ and solve for x.

$$4x - 100 = 0$$

$$4x = 100$$

$$x = 25$$

The zero of the function, the x-intercept of the graph of the function, and the solution to $f(x) = 0$ are all 25.

31. Answers a), b), and c) are the same. Let $f(x) = 0$ and solve for x.

$$330 + 40x = 0$$

$$40x = -330$$

$$x = -8.25$$

The zero of the function, the x-intercept of the graph of the function, and the solution to $f(x) = 0$ are all -8.25.

33. Applying the intersections of graphs method yields:

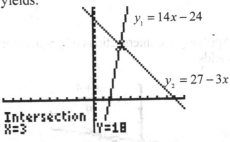

$[-10, 10]$ by $[-10. 30]$

The solution is the x-coordinate of the intersection point or $x = 3$.

35. Applying the intersections of graphs method yields:

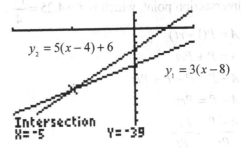

$[-10, 5]$ by $[-70, 10]$

The solution is the x-coordinate of the intersection point or $s = -5$.

37. Applying the intersections of graphs method yields:

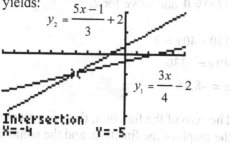

$$y_2 = \frac{5x-1}{3} + 2$$

$$y_1 = \frac{3x}{4} - 2$$

Intersection
X=-4   Y=-5

[−10, 5] by [−20, 10]

The solution is the x-coordinate of the intersection point. $t = -4$.

39. Applying the intersections of graphs method yields:

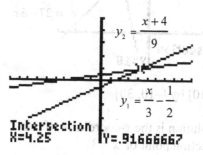

$$y_2 = \frac{x+4}{9}$$

$$y_1 = \frac{x}{3} - \frac{1}{2}$$

Intersection
X=4.25   Y=.91666667

[−10, 10] by [−5, 5]

The solution is the x-coordinate of the intersection point, which is $t = 4.25 = \frac{17}{4}$.

41. $A = P(1+rt)$

$A = P + Prt$

$A - P = P - P + Prt$

$A - P = Prt$

$\dfrac{A-P}{Pt} = \dfrac{Prt}{Pt}$

$\dfrac{A-P}{Pt} = r$ or $r = \dfrac{A-P}{Pt}$

43. $5F - 9C = 160$

$5F - 9C + 9C = 160 + 9C$

$5F = 160 + 9C$

$\dfrac{5F}{5} = \dfrac{160 + 9C}{5}$

$F = \dfrac{9}{5}C + \dfrac{160}{5}$

$F = \dfrac{9}{5}C + 32$

45.

$$\frac{P}{2} + A = 5m - 2n$$

$LCD : 2$

$$2\left(\frac{P}{2} + A\right) = 2(5m - 2n)$$

$P + 2A = 10m - 4n$

$P + 2A - 10m = 10m - 4n - 10m$

$\dfrac{P + 2A - 10m}{-4} = \dfrac{-4n}{-4}$

$\dfrac{P + 2A - 10m}{-4} = n$

$n = \dfrac{P}{-4} + \dfrac{2A}{-4} - \dfrac{10m}{-4}$

$n = \dfrac{5m}{2} - \dfrac{P}{4} - \dfrac{A}{2}$

47. $5x - 3y = 5$

$-3y = -5x + 5$

$y = \dfrac{-5x + 5}{-3}$

$y = \dfrac{5}{3}x - \dfrac{5}{3}$

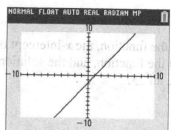

49. $x^2 + 2y = 6$

$$2y = 6 - x^2$$

$$y = \frac{6 - x^2}{2}$$

$$y = 3 - \frac{1}{2}x^2 \quad \text{or}$$

$$y = -\frac{1}{2}x^2 + 3$$

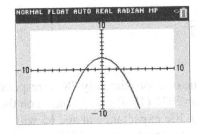

Section 2.1 Exercises

51. Let $y = 690{,}000$ and solve for x.

$$690{,}000 = 828{,}000 - 2300x$$

$$-138{,}000 = -2300x$$

$$x = \frac{-138{,}000}{-2300}$$

$$x = 60$$

After 60 months or 5 years the value of the building will be $690,000.

53. $S = P(1 + rt)$

$$9000 = P(1 + (0.10)(5))$$

$$9000 = P(1 + 0.50)$$

$$9000 = 1.5P$$

$$P = \frac{9000}{1.5} = 6000$$

$6000 must be invested as the principal.

55. $M = 0.819W - 3.214$

$$41.52 = 0.819W - 3.214$$

$$0.819W = 44.734$$

$$W = \frac{44.734}{0.819} = 54.620$$

The median annual salary for whites is approximately $54,620.

57. Let $f(x) = 8351.64$, and calculate x.

$$8351.64 = 486.48x + 3486.84$$

$$486.48x = 4864.8$$

$$x = \frac{4864.8}{486.48} = 10$$

In the year 2020, 10 years after 2010, the federal income tax per capita will be $8,351.64.

59. Let $y = 3236$, and solve for x.

$$3236 = -286.8x + 8972$$

$$-5736 = -286.8x$$

$$x = \frac{-5736}{-286.8}$$

$$x = 20$$

An x-value of 20 corresponds to the year 2026 (2006 + 20). The number of banks will be 3236 in 2026.

61. Let $y = 4.29$, and calculate x.

$$132x + 1000(4.29) = 9570$$

$$132x + 4290 = 9570$$

$$132x = 5280$$

$$x = \frac{5280}{132} = 40$$

In the year 2020 (1980 + 40), the marriage rate per 1000 population will be 4.29.

63. Note that p is in millions. A population of 320,000,000 corresponds to a p-value of 320. Let $p = 320$ and solve for x.

$$320 = 2.6x + 177$$
$$320 - 177 = 2.6x$$
$$143 = 2.6x$$
$$x = \frac{143}{2.6}$$
$$x = 55$$

An x-value of 55 corresponds to the year 2015 (1960 + 55). Based on the model, in 2015, the population is estimated to be 320,000,000.

65. a. Let $H = 130$, and calculate t.

$$130 = -0.65x + 143$$
$$-13 = -0.65x$$
$$x = \frac{-13}{-0.65} = 20$$

A person 20 years old should have the desired heart rate for weight loss of 130.

b. Let $H = 104$, and calculate t.

$$104 = -0.65x + 143$$
$$-39 = -0.65x$$
$$x = \frac{-39}{-0.65} = 60$$

A person 60 years old should have the desired heart rate for weight loss of 104.

67. Let $y = 14.8$, and solve for x.

$$14.8 = 0.876x + 6.084$$
$$14.8 - 6.084 = 0.876x$$
$$8.716 = 0.876x$$
$$x = \frac{8.716}{0.876}$$
$$x = 9.94977 \approx 10$$

An x-value of 10 corresponds to the year 1990 (1980 + 10). The model shows that it was in 1990 that the number of Hispanics in the U.S. civilian population was 14.8 million.

69. Let $y = 20$, and solve for x.

$$20 = 0.077x + 14$$
$$6 = 0.077x$$
$$x = \frac{6}{0.077}$$
$$x \approx 78$$

An x-value of approximately 78 corresponds to the year 2028 (1950 + 78). Based on the model, the life expectancy at age 65 will reach 20 additional years in 2028.

71. Let x represent the score on the fifth exam.

$$90 = \frac{92 + 86 + 79 + 96 + x}{5}$$

LCD: 5

$$5(90) = 5\left(\frac{92 + 86 + 79 + 96 + x}{5}\right)$$
$$450 = 353 + x$$
$$x = 97$$

The student must score 97 on the fifth exam to earn a 90 in the course.

73. Let x = the company's 1999 revenue in billions of dollars.

$$0.94x = 74$$
$$x = \frac{74}{0.94}$$
$$x \approx 78.723$$

The company's 1999 revenue was approximately \$78.723 billion.

75. Commission Reduction

$$= (20\%)(50,000)$$
$$= 10,000$$

New Commission

$$= 50,000 - 10,000$$
$$= 40,000$$

To return to a $50,000 commission, the commission must be increased $10,000. The percentage increase is now based on the $40,000 commission.

Let x represent the percent increase from the second year.

$$40,000x = 10,000$$
$$x = 0.25 = 25\%$$

77. Total cost = Original price + Sales tax

Let x = original price.

$$29,998 = x + 6\%x$$
$$29,998 = x + 0.06x$$
$$29,998 = 1.06x$$
$$x = \frac{29,998}{1.06} = 28,300$$

Sales tax = $29,998 - 28,300 = \$1698$

79. $A = P + Prt$

$$A - P = P - P + Prt$$
$$A - P = Prt$$
$$\frac{A - P}{Pr} = \frac{Prt}{Pr}$$
$$\frac{A - P}{Pr} = t \text{ or } t = \frac{A - P}{Pr}$$

81.

$$A = P(1 + rt)$$
$$A = P + Prt$$
$$A - P = P - P + Prt$$
$$A - P = Prt$$
$$\frac{A - P}{Pt} = \frac{Prt}{Pt}$$
$$\frac{A - P}{Pt} = r$$
$$\frac{3200 - 2000}{2000(6)} = r$$
$$\frac{1200}{12000} = \frac{1}{10} = r$$
$$r = .1 = 10\%$$

83. $\dfrac{I_1}{r_1} = \dfrac{I_2}{r_2}$

$$\frac{920}{.12} = \frac{x}{.08}$$
$$x = \frac{920(.08)}{.12} = \$613.33$$

85. Yes, the circumference of a circle varies directly with the radius of a circle since the relationship fits the direct variation format of $y = kx$, or in the case of a circle, $C = 2\pi r$, where the constant of variation is 2π.

87. a. Since $y = kx$ is the direct variation format, let $y = B$ (BMI) and $x = w$ (weight of person in kilograms).
$$B = kw$$

$$20 = k(45), \text{ so } k = \frac{4}{9}$$

Thus the constant of variation is 4/9.

b. Then $32 = \dfrac{4}{9}A$

$$\frac{9}{4}(32) = A$$
$$A = 72$$

The woman would weigh 72 kilograms.

Section 2.2 Skills Check

1. No. The data points do not lie close to a
   straight line.

3. Can be modeled approximately since not all
   points lie exactly on a line.

5.

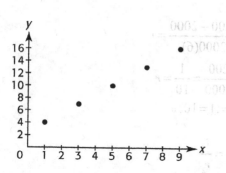

7. Exactly. The differences of the inputs and
   outputs are constant for the first three.

9. Using a spreadsheet program yields

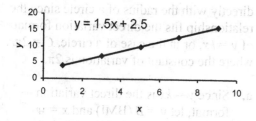

11.

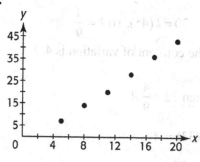

13. Using a spreadsheet program yields

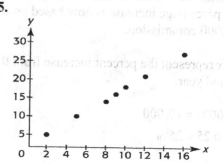

15.

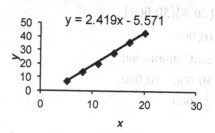

17.

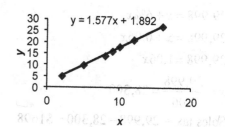

19. Y1=-2X+8

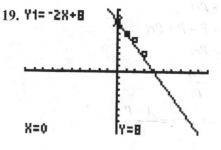

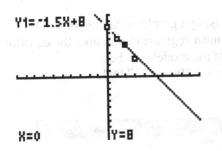

`Y1=-1.5X+8`

`X=0`   `Y=8`

The second equation, $y = -1.5x + 8$, is a better fit to the data points.

21. a. Exactly linear. The first differences are constant.

b. Nonlinear. The first differences vary, but don't grow at an approximately constant rate.

c. Approximately linear. The first differences increase at a constant rate.

Section 2.2 Exercises

23. a. Discrete. There are gaps between the years.

b. Continuous. Gaps between the years no longer exist.

c. No. A line would not fit the points on the scatter plot. A non-linear function is better.

25. a. No; increases from one year to the next are not constant.

b. Yes; increases from one year to the next are constant.

c. $m = \dfrac{1320 - 1000}{70 - 66} = \dfrac{320}{4} = 80$

$S - s_1 = m(x - x_1)$

$S - 1000 = 80(x - 66)$

$S = 80x - 4280$

27. a. Yes, the first differences are constant for uniform inputs.

b. Two; two point are needed to find the equation of any linear equation.

c. $m = \dfrac{45 - 40}{99 - 88} = \dfrac{5}{11}$

$D - d_1 = m(w - w_1)$

$D - 40 = \dfrac{5}{11}(w - 88)$

$D = \dfrac{5}{11}w$

d. For $w = 209$,

$D = \dfrac{5}{11}w$

$D = \dfrac{5}{11} \cdot 209$

$D = 95$

A dosage of 95 mg would be needed for a 209 pound patient.

29. a. No, the first differences are not constant for uniform inputs.

b. Using a graphing calculator and the linear regression function, the equation of the model will be $y = 0.328x + 6.316$.

c. Let $x = 23$,

$y = 0.328x + 6.316$

$y = 0.328(23) + 6.316$

$y = 7.544 + 6.316$

$y = 13.86$

The total disposable income in 2023 in projected to be $13.86 billion.

31. a. Using a graphing calculator yields

b. Using a graphing calculator and the linear regression function, the equation of the model will be
$y = 2375x + 39,630.$

c. The slope is 2375; the average annual wage is expected to increase $2375 per year.

33. a. Using a graphing calculator and the linear regression function, the equation of the model will be
$y = 0.465x + 16.696.$

b. Let $x = 17$,
$y = 0.465x + 16.696$
$y = 0.465(17) + 16.696$
$y = 7.905 + 16.696$
$y = 24.601$
The percent of U.S. adults with diabetes in 2027 in projected to be 24.6%.

c.

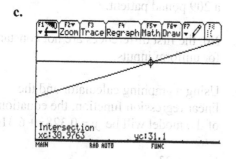

[0, 50] by [0, 40]

According to the graph, in the year 2041, the percent of U.S. adults with diabetes will be 31.1%.

35. a. Using a graphing calculator yields

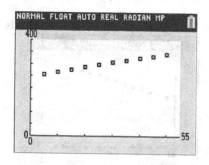

b. Using a graphing calculator and the linear regression function, the equation of the model will be
$y = 1.890x + 247.994.$

c.

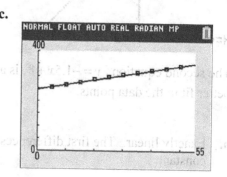

The model is a good fit.

d. Let $x = 32$,
$y = 1.890x + 247.994$
$y = 1.890(32) + 247.994$
$y = 60.48 + 247.994$
$y = 308.474$
The U.S. population for residents over age 16 in 2042 in projected to be 308.474 million.

e. Let $y = 336.827$,
$y = 1.890x + 247.994$
$336.827 = 1.890x + 247.994$
$88.833 = 1.890x$
$x = 47$
The U.S. population for residents over age 16 will reach 336.827 million in the year 2057 (2010 + 57).

37. a. Using a graphing calculator yields

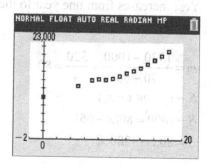

b. $y = 579x + 9410$

c.

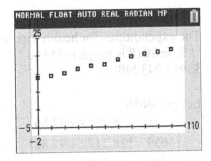

It appears to not be a good fit.

39. a. Using a graphing calculator yields

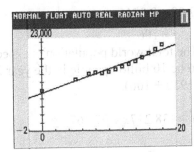

b. $y = 0.077x + 13.827$

c.

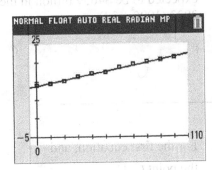

It appears to be a good fit.

d. Let $x = 72$,

$y = 0.077x + 13.827$

$y = 0.077(72) + 13.827$

$y = 5.544 + 13.827$

$y = 19.371$

The additional years of life expectancy at age 65 in 2022 is projected to be 19.4 years.

e.

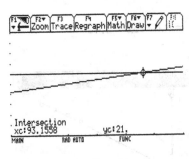

[70, 100] by [15, 35]

According to the graph, in the year 2043 (when $x = 93$), the additional life expectancy at age 65 will be 21 years.

41. a. $y = 0.297x + 2.043$

b. Using a graphing calculator yields

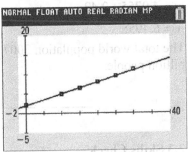

It appears to be a good fit.

c. The slope of the equation is approximately 0.3 percentage points per year.

43. a. Using a graphing calculator yields

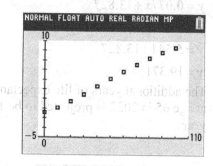

b. $y = 0.0715x + 2.42$

c.

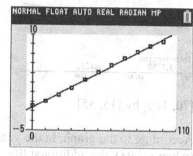

It appears to be a good fit.

d. Let $x = 57$,
$$y = 0.0715x + 2.42$$
$$y = 0.0715(57) + 2.42$$
$$y = 4.0755 + 2.42$$
$$y = 6.4955$$
The total world population 2007 was 6.5 billion people.

e. Let $y = 10$,
$$y = 0.0715x + 2.42$$
$$10 = 0.0715x + 2.42$$
$$7.58 = 0.0715x$$
$$x = 106$$
The total world population is expected to be 10 billion people in the year 2056 (1950 + 106).

45. a. $y = 138.217x + 97.867$

b. Let $x = 28$,
$$y = 138.217x + 97.867$$
$$y = 138.217(28) + 97.867$$
$$y = 3870.076 + 97.867$$
$$y = 3967.943$$
The expenditures for health care in the U.S. in 2018 is projected to be $3967.943 billion.

c. Let $y = 4659$
$$y = 138.217x + 97.867$$
$$4659 = 138.217x + 97.867$$
$$4561.133 = 138.217x$$
$$x = 33$$
The expenditures for health care is expected to be $4659 billion in the year 2023 (1990 + 33).

Section 2.3 Skills Check

1. To determine if an ordered pair is a solution of the system of equations, substitute each ordered pair into both equations.

a. For the first equation, and the point $(2,1)$
$$2x + 3y = -1$$
$$2(2) + 3(1) = 4 + 3 = 7$$
For the second equation, and the point $(2,1)$
$$x - 4y = -6$$
$$(2) - 4(1) = 2 - 4 = -2$$

The point (2, 1) works in neither equation, thus is not a solution of the system.

b. For the first equation, and the point $(-2, 1)$

$2x + 3y = -1$

$2(-2) + 3(1) = -4 + 3 = -1$

For the second equation, and the point $(-2, 1)$

$x - 4y = -6$

$(-2) - 4(1) = -2 - 4 = -6$

The point $(-2, 1)$ works in both of the equations, thus is a solution of the system.

3. Applying the substitution method
$y = 3x - 2$ and $y = 3 - 2x$

$3 - 2x = 3x - 2$

$3 - 2x - 3x = 3x - 3x - 2$

$3 - 5x = -2$

$3 - 3 - 5x = -2 - 3$

$-5x = -5$

$x = 1$

Substituting to find y

$y = 3(1) - 2 = 1$

$(1, 1)$ is the intersection point.

5. Applying the intersection of graphs method, for $y = 3x - 12$ and $y = 4x + 2$

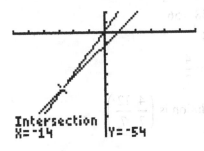

[−30, 30] by [−100, 20]
The solution for #5 is $(-14, -54)$.

7. Solving the equations for y

$4x - 3y = -4$

$-3y = -4x - 4$

$y = \dfrac{-4x - 4}{-3}$

$y = \dfrac{4}{3}x + \dfrac{4}{3}$

and

$2x - 5y = -4$

$-5y = -2x - 4$

$y = \dfrac{-2x - 4}{-5}$

$y = \dfrac{2}{5}x + \dfrac{4}{5}$

Applying the intersection of graphs method

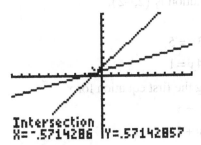

[−10, 10] by [−10, 10]

The solution is $(-0.5714, 0.5714)$ or $\left(-\dfrac{4}{7}, \dfrac{4}{7}\right)$.

9. $\begin{cases} 2x + 5y = 6 & (Eq\ 1) \\ x + 2.5y = 3 & (Eq\ 2) \end{cases}$

$\begin{cases} 2x + 5y = 6 & (Eq\ 1) \\ -2x - 5y = -6 & -2 \times (Eq\ 2) \end{cases}$

$0 = 0$

There are infinitely many solutions to the system. The graphs of both equations represent the same line.

11.
$$\begin{cases} x = 5y + 12 \\ 3x + 4y = -2 \end{cases}$$

Substituting the first equation into the second equation

$$3(5y + 12) + 4y = -2$$
$$15y + 36 + 4y = -2$$
$$19y + 36 = -2$$
$$19y = -38$$
$$y = -2$$

Substituting to find x

$$x = 5(-2) + 12$$
$$x = -10 + 12$$
$$x = 2$$

The solution is $(2, -2)$.

13.
$$\begin{cases} 2x - 3y = 5 \\ 5x + 4y = 1 \end{cases}$$

Solving the first equation for x

$$2x - 3y = 5$$
$$2x = 3y + 5$$
$$x = \frac{3y + 5}{2}$$

Substituting into the second equation

$$5\left(\frac{3y + 5}{2}\right) + 4y = 1$$
$$2\left[5\left(\frac{3y + 5}{2}\right) + 4y\right] = 2[1]$$
$$15y + 25 + 8y = 2$$
$$23y + 25 = 2$$
$$23y = -23$$
$$y = -1$$

Substituting to find x

$$2x - 3(-1) = 5$$
$$2x + 3 = 5$$
$$2x = 2$$
$$x = 1$$

The solution is $(1, -1)$.

15.
$$\begin{cases} x + 3y = 5 & (Eq\,1) \\ 2x + 4y = 8 & (Eq\,2) \end{cases}$$

$$\begin{cases} -2x - 6y = -10 & -2 \times (Eq\ 1) \\ 2x + 4y = 8 & (Eq\,2) \end{cases}$$

$$-2y = -2$$
$$y = 1$$

Substituting to find x

$$x + 3(1) = 5$$
$$x + 3 = 5$$
$$x = 2$$

The solution is $(2, 1)$.

17.
$$\begin{cases} 5x + 3y = 8 & (Eq\,1) \\ 2x + 4y = 8 & (Eq\,2) \end{cases}$$

$$\begin{cases} -10x - 6y = -16 & -2 \times (Eq\,1) \\ 10x + 20y = 40 & 5 \times (Eq\,2) \end{cases}$$

$$14y = 24$$
$$y = \frac{24}{14} = \frac{12}{7}$$

Substituting to find x

$$2x + 4\left(\frac{12}{7}\right) = 8$$
$$7\left[2x + \left(\frac{48}{7}\right)\right] = 7[8]$$
$$14x + 48 = 56$$
$$14x = 8$$
$$x = \frac{8}{14} = \frac{4}{7}$$

The solution is $\left(\frac{4}{7}, \frac{12}{7}\right)$.

19.

$$\begin{cases} 0.3x + 0.4y = 2.4 & (Eq\,1) \\ 5x - 3y = 11 & (Eq\,2) \end{cases}$$

$$\begin{cases} 9x + 12y = 72 & 30 \times (Eq\,1) \\ 20x - 12y = 44 & 4 \times (Eq\,2) \end{cases}$$

$$29x = 116$$

$$x = \frac{116}{29} = 4$$

Substituting to find y

$$5(4) - 3y = 11$$

$$20 - 3y = 11$$

$$-3y = -9$$

$$y = 3$$

The solution is $(4,3)$.

21.

$$\begin{cases} 3x + 6y = 12 & (Eq\,1) \\ 4y - 8 = -2x & (Eq\,2) \end{cases}$$

$$\begin{cases} 3x + 6y = 12 & (Eq\,1) \\ 2x + 4y = 8 & (Eq\,2) \end{cases}$$

$$\begin{cases} -6x - 12y = -24 & -2 \times (Eq\ 1) \\ 6x + 12y = 24 & 3 \times (Eq\ 2) \end{cases}$$

$$0 = 0$$

Infinitely many solutions. The lines are the same. This is a dependent system.

23.

$$\begin{cases} 6x - 9y = 12 & (Eq\,1) \\ 3x - 4.5y = -6 & (Eq\,2) \end{cases}$$

$$\begin{cases} 6x - 9y = 12 & (Eq\ 1) \\ -6x + 9y = 12 & -2 \times (Eq\ 2) \end{cases}$$

$$0 = 24$$

No solution. Lines are parallel. This is an inconsistent system.

25.

$$\begin{cases} y = 3x - 2 \\ y = 5x - 6 \end{cases}$$

Substituting the first equation into the second equation

$$3x - 2 = 5x - 6$$

$$-2x - 2 = -6$$

$$-2x = -4$$

$$x = 2$$

Substituting to find y

$$y = 3(2) - 2 = 6 - 2 = 4$$

The solution is $(2,4)$.

27.

$$\begin{cases} 4x + 6y = 4 \\ x = 4y + 8 \end{cases}$$

Substituting the second equation into the first equation

$$4(4y + 8) + 6y = 4$$

$$16y + 32 + 6y = 4$$

$$22y + 32 = 4$$

$$22y = -28$$

$$y = \frac{-28}{22} = -\frac{14}{11}$$

Substituting to find x

$$x = 4\left(-\frac{14}{11}\right) + 8$$

$$x = -\frac{56}{11} + \frac{88}{11} = \frac{32}{11}$$

The solution is $\left(\frac{32}{11}, -\frac{14}{11}\right)$.

29.

$$\begin{cases} 2x - 5y = 16 & (Eq\,1) \\ 6x - 8y = 34 & (Eq\,2) \end{cases}$$

$$\begin{cases} -6x + 15y = -48 & -3 \times (Eq\ 1) \\ 6x - 8y = 34 & (Eq\ 2) \end{cases}$$

$7y = -14$

$y = -2$

Substituting to find x

$2x - 5(-2) = 16$

$2x + 10 = 16$

$2x = 6$

$x = 3$

The solution is $(3, -2)$.

31.

$$\begin{cases} 3x = 7y - 1 & (Eq\,1) \\ 4x + 3y = 11 & (Eq\,2) \end{cases}$$

$$\begin{cases} 3x - 7y = -1 & (Eq\,1) \\ 4x + 3y = 11 & (Eq\,2) \end{cases}$$

$$\begin{cases} -12x + 28y = 4 & -4 \times (Eq\,1) \\ 12x + 9y = 33 & 3 \times (Eq\,2) \end{cases}$$

$37y = 37$

$y = \dfrac{37}{37} = 1$

Substituting to find x

$3x - 7(1) = -1$

$3x - 7 = -1$

$3x = 6$

$x = 2$

The solution is $(2,1)$.

33.

$$\begin{cases} 4x - 3y = 9 & (Eq\,1) \\ 8x - 6y = 16 & (Eq\,2) \end{cases}$$

$$\begin{cases} -8x + 6y = -18 & -2 \times (Eq\ 1) \\ 8x - 6y = 16 & (Eq\ 2) \end{cases}$$

$0 = -2$

No solution. Lines are parallel.

Section 2.3 Exercises

35.

$R = C$

$76.50x = 2970 + 27x$

$49.50x = 2970$

$x = \dfrac{2970}{49.50}$

$x = 60$

Applying the intersections of graphs method yields $x = 60$.

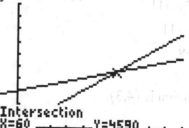

Intersection
X=60   Y=4590

$[-10, 100]$ by $[-10, 10{,}000]$

Break-even occurs when the number of units produced and sold is 60.

37.

$R = C$

$15.80x = 8593.20 + 3.20x$

$12.60x = 8593.20$

$x = \dfrac{8593.20}{12.60}$

$x = 682$

Break-even occurs when the number of units produced and sold is 682.

39. a.

$$\begin{cases} p + 2q = 320 & (Eq\,1) \\ p - 8q = 20 & (Eq\,2) \end{cases}$$

$$\begin{cases} 4p + 8q = 1280 & 4 \times (Eq\ 1) \\ p - 8q = 20 & (Eq\ 2) \end{cases}$$

$5p = 1300$

$p = \dfrac{1300}{5} = 260$

The equilibrium price is \$260.

b. Substituting to find q

$$260 - 8q = 20$$

$$-8q = -240$$

$$q = \frac{-240}{-8} = 30$$

The equilibrium quantity occurs when 30 units are demanded and supplied.

41. a. For the drug Concerta, let

$x_1 = 0, y_1 = 2.4$, and $x_2 = 11, y_2 = 10$.

Then, $\dfrac{y_2 - y_1}{x_2 - x_1} = \dfrac{10 - 2.4}{11 - 0} = 0.69$

Since $x_1 = 0, y_1$ is the y-intercept.

Thus the equation is $y = 0.69x + 2.4$, representing the market share for Concerta as a linear function of time.

b. For the drug Ritalin, let

$x_1 = 0, y_1 = 7.7$, and $x_2 = 11, y_2 = 6.9$.

Then, $\dfrac{y_2 - y_1}{x_2 - x_1} = \dfrac{6.9 - 7.7}{11 - 0} = -0.073$

Since $x_1 = 0, y_1$ is the y-intercept.

Thus the equation is $y = -0.073x + 7.7$, representing the market share for Ritalin as a linear function of time.

c. To determine the number of weeks past the release date when the two market shares are equal, set the two linear equations equal, and solve for x. Thus,

$$0.69x + 2.4 = -0.073x + 7.7$$

$$0.763x = 5.3$$

$$x = 6.946 \approx 7 \text{ weeks}$$

43. The populations will be equal when the two equations are equal,

$$0.293x + 157.8454 = 1.73x + 54.094$$

$$103.7514 = 1.437x$$

$$x = 72$$

In the year 2072 (2000 + 72) is when the White, non-Hispanic population is projected to be the same as the remainder of the population.

45. $y_H = 0.224x + 9.0$ (the % of Hispanics)

$y_B = 0.057x + 12.3$ (the % of blacks)

Setting the first equation equal to the second equation, solve for x:

$$0.224x + 9.0 = 0.057x + 12.3$$

$$.167x = 3.3$$

$$x = \frac{3.3}{.167} = 19.76 \approx 20$$

Thus, 20 years after 1990, in the year 2010, the percent of Hispanics in the U.S. civilian non-institutional population equaled the percent of blacks.

47. Let $h =$ the 2016 revenue, and let $l =$ the 2013 revenue.

$$\begin{cases} h + 2l = 2144.9 & (Eq\,1) \\ h - l = 135.5 & (Eq\,2) \end{cases}$$

$$\begin{cases} h + 2l = 2144.9 & (Eq\ 1) \\ 2h - 2l = 271 & 2 \times (Eq\ 2) \end{cases}$$

$$3h = 2415.9$$

$$h = \frac{2415.9}{3} = 805.3$$

Substituting to calculate l

$$805.3 - l = 135.5$$

$$-l = -669.8$$

$$l = 669.8$$

The 2013 revenue is $669.8 million, while the 2016 revenue is $805.3 million.

49. a. $x+y=2400$

b. $30x$

c. $45y$

d. $30x+45y=84,000$

e.

$$\begin{cases} x+y=2400 & (Eq\ 1) \\ 30x+45y=84,000 & (Eq\ 2) \end{cases}$$

$$\begin{cases} -30x-30y=-72,000 & -30\times(Eq\ 1) \\ 30x+45y=84,000 & (Eq\ 2) \end{cases}$$

$$15y=12,000$$

$$y=\frac{12,000}{15}=800$$

Substituting to calculate x

$$x+800=2400$$

$$x=1600$$

The promoter needs to sell 1600 tickets at $30 per ticket and 800 tickets at $45 per ticket.

51. a.

Let $x=$ the amount in the safer account, and let $y=$ the amount in the riskier account.

$$\begin{cases} x+y=100,000 & (Eq\ 1) \\ 0.08x+0.12y=9,000 & (Eq\ 2) \end{cases}$$

$$\begin{cases} -0.08x-0.08y=-8,000 & -0.08\times(Eq\ 1) \\ 0.08x+0.12y=9,000 & (Eq\ 2) \end{cases}$$

$$0.04y=1000$$

$$y=\frac{1000}{0.04}=25,000$$

Substituting to calculate x

$$x+25,000=100,000$$

$$x=75,000$$

b. $75,000 is invested in the 8% account, and $25,000 is invested in the 12% account. Using two accounts minimizes investment risk.

53.

Let $x=$ the amount in the money market fund, and let $y=$ the amount in the mutual fund.

$$\begin{cases} x+y=250,000 & (Eq\ 1) \\ 0.066x+0.086y=.07(x+y) & (Eq\ 2) \end{cases}$$

$$66x+86y=70x+70y \qquad 1000\times(Eq\ 2)$$

$$16y=4x$$

$$4y=x$$

Then substituting $4y$ for x in $Eq\ 1$, gives

$$4y+y=250,000$$

$$5y=250,000$$

$$y=50,000$$

Substituting to calculate x

$$x+50,000=250,000$$

$$x=200,000$$

Thus, $200,000 is invested in the money market fund, and $50,000 is invested in the mutual fund.

55.

Let $x=$ the amount (cc's) of the 10% solution, and let $y=$ the amount (cc's) of the 5% solution.

$$\begin{cases} x+y=100 \text{ (cc's of the final solution)} \\ 0.10x+0.05y=0.08(100) \end{cases}$$

$$\begin{cases} 10x+5y=800 \\ -5x-5y=-500 \end{cases}$$

Then using the elimination method

$$5x=300$$

$$x=60 \text{ cc's of the 10% solution}$$

Substituting to calculate y

$$y=40 \text{ cc's of the 5% solution}$$

Thus, 60 cc's of the 10% solution should be mixed with 40 cc's of the 5% solution to get 100 cc's of the 8% solution.

57.

Let x = the number of glasses of milk, and let y = the number of 1/4-pound servings of meat.

Then the protein equation is:

$8.5x + 22y = 69.5$ $(Eq\ 1)$

and the iron equation is:

$0.1x + 3.4y = 7.1$ $(Eq\ 2)$

$\begin{cases} 85x + 220y = 695 & 10 \times (Eq\ 1) \\ 1x + 34y = 71 & 10 \times (Eq\ 2) \end{cases}$

$-85x - 2890y = -6035$ $-85 \times (Eq\ 2)$

$-2670y = -5340$

$y = \dfrac{-5340}{-2670} = 2$

Substituting to calculate x

$8.5x + 22(2) = 69.5$

$8.5x + 44 = 69.5$

$8.5x = 25.5$

$x = \dfrac{25.5}{8.5} = 3$

Thus, 3 glasses of milk and 2 servings of meat provide the desired quantities for the diet

59. To find when the percent of those enrolled equals that of those not enrolled, set the equations equal to each other and solve for x. Then find the percent by substituting into either one of the original equations.

$-0.282x + 19.553 = -0.086x + 13.643$

$-0.196x = -5.91$

$x = 30.153$, the number of years after the year 2000, during the year 2031.

Substituting to calculate y:

$y = -0.282(30.153) + 19.553$

$y = 11\%$

It is estimated that in the year 2031, the percent of young adults aged 18 to 22 who use alcohol will be the same for those enrolled in college as for those not enrolled in college. That percent will be 11%.

61.

L1	L2	L3	1
50	210	0	
60	190	40	
70	170	80	
80	150	120	
100	110	200	

L1(1)=50

a. Demand function: Finding a linear model using L_2 as input and L_1 as output yields $p = -\frac{1}{2}q + 155$.

LinReg
y=ax+b
a=-.5
b=155

b. Supply function: Finding a linear model using L_3 as input and L_1 as output yields $p = \frac{1}{4}q + 50$.

LinReg
y=ax+b
a=.25
b=50

c. Applying the intersection of graphs method

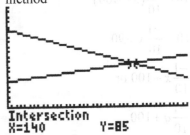

[0, 200] by [−50, 200]

When the price is $85, 140 units are both supplied and demanded. Therefore, equilibrium occurs when the price is $85 per unit.

63. Let $x =$ the number of years after 62.

$$750x = 1000(x-4)$$
$$750x = 1000x - 4000$$
$$-250x = -4000$$
$$x = 16$$

When the person is 78 (62 +16) is the age when the social security benefits paid would be the same.

65. a. $300x + 200y = 100,000$

b. $x = 2y$
$$300(2y) + 200y = 100,000$$
$$800y = 100,000$$
$$y = \frac{100,000}{800} = 125$$

Substituting to calculate x
$$x = 2(125) = 250$$

There are 250 clients in the first group and 125 clients in the second group.

67. The slope of the demand function is
$$m = \frac{y_2 - y_1}{x_2 - x_1} = \frac{10-60}{900-400} = \frac{-50}{500} = \frac{-1}{10}.$$

Calculating the demand equation:
$$y - y_1 = m(x - x_1)$$
$$y - 10 = \frac{-1}{10}(x - 900)$$
$$y - 10 = \frac{-1}{10}x + 90$$
$$y = \frac{-1}{10}x + 100 \text{ or}$$
$$p = \frac{-1}{10}q + 100$$

Likewise, the slope of the supply function is
$$m = \frac{y_2 - y_1}{x_2 - x_1} = \frac{30-50}{700-1400} = \frac{-20}{-700} = \frac{2}{70}.$$

Calculating the supply equation:
$$y - y_1 = m(x - x_1)$$
$$y - 30 = \frac{2}{70}(x - 700)$$
$$y - 30 = \frac{2}{70}x - 20$$
$$y = \frac{2}{70}x + 10 \text{ or}$$
$$p = \frac{2}{70}q + 10$$

To find the quantity, q, that produces market equilibrium, set the equations equal.
$$\frac{-1}{10}q + 100 = \frac{2}{70}q + 10$$
$$70\left(\frac{-1}{10}q + 100\right) = 70\left(\frac{2}{70}q + 10\right)$$
$$-7q + 7000 = 2q + 700$$
$$-9q = -6300$$
$$q = \frac{-6300}{-9} = 700$$

To find the price, p, at market equilibrium, solve for p.
$$p = \frac{2}{70}(700) + 10$$
$$p = 2(10) + 10$$
$$p = 30$$

700 units priced at $30 represents the market equilibrium.

Section 2.4 Skills Check

1. Algebraically:

$$6x - 1 \le 11 + 2x$$
$$4x - 1 \le 11$$
$$4x \le 12$$
$$x \le \frac{12}{4}$$
$$x \le 3$$

Graphically:

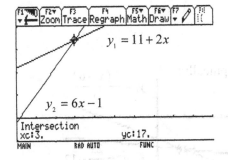

[0, 10] by [−5, 20]

$6x - 1 \le 11 + 2x$ implies that the solution region is $x \le 3$.
The interval notation is $(-\infty, 3]$.

The graph of the solution is:

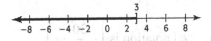

3. Algebraically:

$$4(3x - 2) \le 5x - 9$$
$$12x - 8 \le 5x - 9$$
$$7x \le -1$$
$$\frac{7x}{7} \le \frac{-1}{7}$$
$$x \le -\frac{1}{7}$$

Graphically:

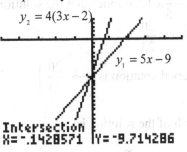

[−5, 5] by [−25, 5]
$4(3x - 2) \le 5x - 9$ implies that the solution region is $x \le -\frac{1}{7}$.

The interval notation is $\left(-\infty, -\frac{1}{7}\right]$.

The graph of the solution is:

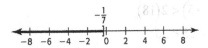

5. Algebraically:

$$4x + 1 < -\frac{3}{5}x + 5$$
$$5(4x + 1) < 5\left(-\frac{3}{5}x + 5\right)$$
$$20x + 5 < -3x + 25$$
$$23x + 5 < 25$$
$$23x < 20$$
$$x < \frac{20}{23}$$

Graphically:

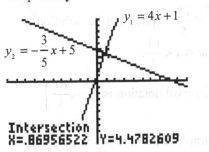

[−10, 10] by [−10, 10]

$4x + 1 < -\dfrac{3}{5}x + 5$ implies that the solution

region is $x < \dfrac{20}{23}$.

The interval notation is $\left(-\infty, \dfrac{20}{23}\right)$.

The graph of the solution is:

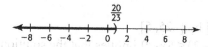

7. Algebraically:

$$\dfrac{x-5}{2} < \dfrac{18}{5}$$

$$10\left(\dfrac{x-5}{2}\right) < 10\left(\dfrac{18}{5}\right)$$

$$5(x-5) < 2(18)$$

$$5x - 25 < 36$$

$$5x < 61$$

$$x < \dfrac{61}{5}$$

Graphically:

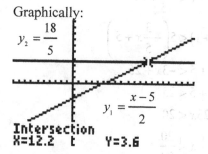

[−10, 20] by [−10, 10]

$\dfrac{x-5}{2} < \dfrac{18}{5}$ implies that the solution

region is $x < \dfrac{61}{5}$.

The interval notation is $\left(-\infty, \dfrac{61}{5}\right)$.

The graph of the solution is

9. Algebraically:

$$\dfrac{3(x-6)}{2} \geq \dfrac{2x}{5} - 12$$

$$10\left(\dfrac{3(x-6)}{2}\right) \geq 10\left(\dfrac{2x}{5} - 12\right)$$

$$5(3(x-6)) \geq 2(2x) - 120$$

$$15(x-6) \geq 4x - 120$$

$$15x - 90 \geq 4x - 120$$

$$11x - 90 \geq -120$$

$$11x \geq -30$$

$$x \geq -\dfrac{30}{11}$$

Graphically:

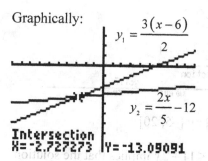

[−10, 10] by [−35, 15]

$\dfrac{3(x-6)}{2} \geq \dfrac{2x}{5} - 12$ implies that the solution

region is $x \geq -\dfrac{30}{11}$.

The interval notation is $\left[-\dfrac{30}{11}, \infty\right)$.

The graph of the solution is:

11. Algebraically:

$$2.2x - 2.6 \geq 6 - 0.8x$$
$$3.0x - 2.6 \geq 6$$
$$3.0x \geq 8.6$$
$$x \geq \frac{8.6}{3.0}$$
$$x \geq 2.8\overline{6}$$

Graphically:

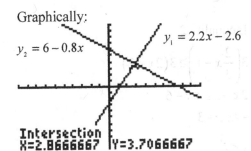

$y_2 = 6 - 0.8x$ $y_1 = 2.2x - 2.6$

Intersection
X=2.8666667 Y=3.7066667

[−10, 10] by [−10, 10]

$2.2x - 2.6 \geq 6 - 0.8x$ implies that the solution region is $x \geq 2.8\overline{6}$.

The interval notation is $\left[2.8\overline{6}, \infty\right)$.

The graph of the solution is:

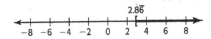

2.8$\overline{6}$

13. Applying the intersection of graphs method yields:

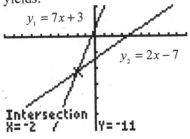

$y_1 = 7x + 3$ $y_2 = 2x - 7$

Intersection
X=-2   Y=-11

[−10, 10] by [−30, 10]

$7x + 3 < 2x - 7$ implies that the solution region is $x < -2$.

The interval notation is $(-\infty, -2)$.

15. To apply the x-intercept method, first rewrite the inequality so that zero is on one side of the inequality.

$$5(2x + 4) \geq 6(x - 2)$$
$$10x + 20 \geq 6x - 12$$
$$4x + 32 \geq 0$$

Let $f(x) = 4x + 32$, and determine graphically where $f(x) \geq 0$.

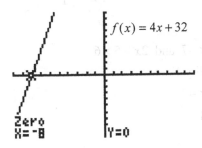

$f(x) = 4x + 32$

Zero
X=-8   Y=0

[−10, 10] by [−10, 10]

$f(x) \geq 0$ implies that the solution region is $x \geq -8$.

The interval notation is $[-8, \infty)$.

17. a. The x-coordinate of the intersection point is the solution. $x = -1$.

b. $(-\infty, -1)$

19.
$$17 \leq 3x - 5 < 31$$
$$17 + 5 \leq 3x - 5 + 5 < 31 + 5$$
$$22 \leq 3x < 36$$
$$\frac{22}{3} \leq \frac{3x}{3} < \frac{36}{3}$$
$$\frac{22}{3} \leq x < 12$$

The interval notation is $\left[\frac{22}{3}, 12\right)$.

21. $2x+1\geq 6$ and $2x+1\leq 21$

$$6\leq 2x+1\leq 21$$

$$5\leq 2x\leq 20$$

$$\frac{5}{2}\leq x\leq 10$$

$$x\geq\frac{5}{2} \text{ and } x\leq 10$$

The interval notation is $\left[\frac{5}{2},10\right]$.

23. $3x+1<-7$ and $2x-5>6$

Inequality 1

$$3x+1<-7$$

$$3x<-8$$

$$x<-\frac{8}{3}$$

Inequality 2

$$2x-5>6$$

$$2x>11$$

$$x>\frac{11}{2}$$

$$x<-\frac{8}{3} \text{ and } x>\frac{11}{2}$$

Since "and" implies that these two inequalities must both be true at the same time, and there is no set of numbers where this occurs, there is no solution to this system of inequalities.

25. $\frac{3}{4}x-2\geq 6-2x$ or $\frac{2}{3}x-1\geq 2x-2$

Inequality 1

$$4\left(\frac{3}{4}x-2\right)\geq 4\left(6-2x\right)$$

$$3x-8\geq 24-8x$$

$$11x\geq 32$$

$$x\geq\frac{32}{11}$$

Inequality 2

$$3\left(\frac{2}{3}x-1\right)\geq 3\left(2x-2\right)$$

$$2x-3\geq 6x-6$$

$$-4x\geq -3$$

$$x\leq\frac{3}{4}$$

$$x\geq\frac{32}{11} \text{ or } x\leq\frac{3}{4}$$

Since "or" implies that one or the other of these inequalities is true, the solution for this system, in interval notation, is

$$\left(-\infty,\frac{3}{4}\right]\cup\left[\frac{32}{11},\infty\right)$$

27.

$$37.002\leq 0.554x-2.886\leq 77.998$$

$$37.002+2.886\leq 0.554x-2.886+2.886\leq 77.998+2.886$$

$$39.888\leq 0.554x\leq 80.884$$

$$\frac{39.888}{0.554}\leq\frac{0.554x}{0.554}\leq\frac{80.884}{0.554}$$

$$72\leq x\leq 146$$

The interval notation is $\left[72,146\right]$.

Section 2.4 Exercises

29. a. $V = 12,000 - 2000t$

b. $12,000 - 2000t < 8000$

c. $12,000 - 2000t \geq 6000$

31. $\qquad F \leq 32$

$$\frac{9}{5}C + 32 \leq 32$$

$$\frac{9}{5}C \leq 0$$

$$C \leq 0$$

A Celsius temperature at or below zero degrees is "freezing."

33.
Position 1 income $= 3100$

Position 2 income $= 2000 + 0.05x$, where x represents the sales within a given month. The income from the second position will exceed the income from the first position when $\rightarrow$ more

$$2000 + 0.05x > 3100$$

$$0.05x > 1100$$

$$x > \frac{1100}{0.05}$$

$$x > 22,000$$

When monthly sales exceed $22,000, the second position is more profitable than the first position.

35.

Let $x =$ Stan's final exam grade.

$$80 \leq \frac{78 + 69 + 92 + 81 + 2x}{6} \leq 89$$

$$6(80) \leq 6\left(\frac{78 + 69 + 92 + 81 + 2x}{6}\right) \leq 6(89)$$

$$480 \leq 320 + 2x \leq 534$$

$$480 - 320 \leq 320 - 320 + 2x \leq 534 - 320$$

$$160 \leq 2x \leq 214$$

$$\frac{160}{2} \leq \frac{2x}{2} \leq \frac{214}{2}$$

$$80 \leq x \leq 107$$

If the final exam does not contain any bonus points, Stan needs to score between 80 and 100 to earn a grade of B for the course.

37. $\qquad p < 27$

$$-1.873t + 60.643 < 27$$

$$1.873t < 33.643$$

$$t < 18$$

From the year 2018 (2000 +18) is when the percent of 12th graders who have ever used cigarettes is projected to be below 27 %.

39. Let $x =$ age after 70.

$$2000(x + 4) < 2640x$$

$$2000x + 8000 < 2640x$$

$$8000 < 640x$$

$$12.5 < x$$

From age 82.5 (70 + 12.5) is when the social security benefits would be more for a person that delayed their benefits until age 70.

41. Let x represent the actual life of the HID headlights.

$$1500 - 10\%(1500) \leq x \leq 1500 + 10\%(1500)$$
$$1500 - 150 \leq x \leq 1500 + 150$$
$$1350 \leq x \leq 1650$$

43. If x is the number of years after 2010, and if $y = 2375x + 39{,}630$, then the average annual wage of U.S. workers will be at least $\$68{,}130$ is given by:

$$2375x + 39{,}630 \geq 68{,}130$$
$$2375x \geq 28{,}500$$
$$x \geq 12 \text{ years after } 2010$$

Therefore, 12 years from 2010 is 2022 and after.

45. a. Since the rate of increase is constant, the equation modeling the value of the home is linear.

$$m = \frac{y_2 - y_1}{x_2 - x_1}$$
$$= \frac{270{,}000 - 190{,}000}{4 - 0}$$
$$= \frac{80{,}000}{4}$$
$$= 20{,}000$$

Solving for the equation:
$$y - y_1 = m(x - x_1)$$
$$y - 190{,}000 = 20{,}000(x - 0)$$
$$y - 190{,}000 = 20{,}000x$$
$$y = 20{,}000x + 190{,}000$$

b.

$$y > 400{,}000$$
$$20{,}000x + 190{,}000 > 400{,}000$$
$$20{,}000x > 210{,}000$$
$$x > \frac{210{,}000}{20{,}000}$$
$$x > 10.5$$

2010 corresponds to
$$x = 2010 - 1996 = 14.$$
Therefore, $11 \leq x \leq 14$.
Or, $y > 400{,}000$
between 2007 and 2010, inclusive.
The value of the home will be greater than $\$400{,}000$ between 2007 and 2010.

c. Since the housing market took such a down turn in the last year or so, it does not seem reasonable that this model remained accurate until the end of 2010.

47.
$$P(x) > 10{,}900$$
$$6.45x - 2000 > 10{,}900$$
$$6.45x > 12{,}900$$
$$x > \frac{12{,}900}{6.45}$$
$$x > 2000$$

A production level above 2000 units will yield a profit greater than $\$10{,}900$.

49.
$$P(x) \geq 0$$
$$6.45x - 9675 \geq 0$$
$$6.45x \geq 9675$$
$$x \geq \frac{9675}{6.45}$$
$$x \geq 1500$$

Sales of 1500 feet or more of PVC pipe will avoid a loss for the hardware store.

51. Recall that Profit = Revenue − Cost.
Let $x =$ the number of boards manufactured and sold.

$P(x) = R(x) - C(x)$

$R(x) = 489x$

$C(x) = 125x + 345,000$

$P(x) = 489x - (125x + 345,000)$

$P(x) = 489x - 125x - 345,000$

$P(x) = 364x - 345,000$

To make a profit, $P(x) > 0$.

$364x - 345,000 > 0$

53. Since at least implies greater than or equal to, $H(x) \geq 14.6$. Thus,

$0.224x + 9.0 \geq 14.6$

$0.224x \geq 5.6$

$x \geq \dfrac{5.6}{0.224}$

$x \geq 25$

Thus, the Hispanic population is at least 14.6%, 25 years after 1990, on or after the year 2015.

55. If x is the number of years after 1990, and if $y = 138.2x + 97.87$, then the expenditures for health care will exceed \$4658.5 billion is given by:

$138.2x + 97.87 > 4658.5$

$138.2x > 4560.63$

$x > 33.0002 \approx 33$ years after 1990

Therefore, 33 years from 1990 is after 2023.

Chapter 2 Skills Check

1. a.

$3x + 22 = 8x - 12$

$3x - 8x + 22 = 8x - 8x - 12$

$-5x + 22 = -12$

$-5x + 22 - 22 = -12 - 22$

$-5x = -34$

$\dfrac{-5x}{-5} = \dfrac{-34}{-5}$

$x = \dfrac{34}{5} = 6.8$

b. Applying the intersection of graphs method yields $x = 6.8$.

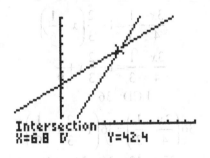

$[-5, 15]$ by $[-10, 60]$

3. **a.**
$$\frac{3(x-2)}{5} - x = 8 - \frac{x}{3}$$

LCD: 15

$$15\left(\frac{3(x-2)}{5} - x\right) = 15\left(8 - \frac{x}{3}\right)$$

$$3(3(x-2)) - 15x = 120 - 5x$$

$$3(3x-6) - 15x = 120 - 5x$$

$$9x - 18 - 15x = 120 - 5x$$

$$-6x - 18 = 120 - 5x$$

$$-1x - 18 = 120$$

$$-1x = 138$$

$$x = -138$$

b. Applying the intersection of graphs method yields $x = -138$.

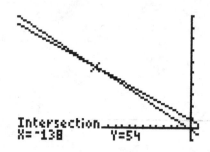

$[-250, 10]$ by $[-10, 100]$

5. **a.**
$$\frac{3x}{4} - \frac{1}{3} = 1 - \frac{2}{3}\left(x \div \frac{1}{6}\right)$$

$$\frac{3x}{4} - \frac{1}{3} = 1 - \frac{2}{3}x + \frac{1}{9}$$

LCD: 36

$$36\left(\frac{3x}{4} - \frac{1}{3}\right) = 36\left(1 - \frac{2}{3}x + \frac{1}{9}\right)$$

$$27x - 12 = 36 - 24x + 4$$

$$51x = 52$$

$$x = \frac{52}{51} \approx 1.0196$$

b. Applying the intersection of graphs method yields $x = 52/51 \approx 1.0196$.

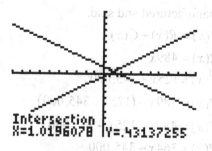

Intersection
X=1.0196078 Y=.43137255

$[-10,10]$ by $[-10, -10]$

7. **a.** $f(x) = 7x - 105$

$$7x - 105 = 0$$

$$7x = 105$$

$$x = 15$$

b. The x-intercepts of the graph are the same as the zeros of the function.

c. Solving the equation $f(x) = 0$ is the same as finding the zeros of the function and the x-intercepts of the graph.

9. Solve for y:

$$4x - 3y = 6$$

$$-3y = -4x + 6$$

$$y = \frac{-4x+6}{-3} = \frac{4x-6}{3}$$

$$y = \frac{4}{3}x - 2$$

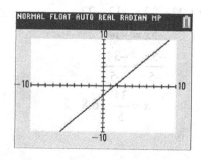

11. $y = 2.8947x - 11.211$

13. No. The data points in the table do not fit the linear model exactly.

15. $\begin{cases} 3x + 2y = -3 & (Eq1) \\ 2x - 3y = 3 & (Eq2) \end{cases}$

$\begin{cases} 9x + 6y = -9 & 3 \times (Eq1) \\ 4x - 6y = 6 & 2 \times (Eq2) \end{cases}$

$13x = -3$

$x = -\dfrac{3}{13}$

Substituting to find y

$3\left(-\dfrac{3}{13}\right) + 2y = -3$

$-\dfrac{9}{13} + 2y = -\dfrac{39}{13}$

$2y = -\dfrac{30}{13}$

$y = -\dfrac{15}{13}$

The solution is $\left(-\dfrac{3}{13}, -\dfrac{15}{13}\right)$.

17. $\begin{cases} -6x + 4y = 10 & (Eq1) \\ 3x - 2y = 5 & (Eq2) \end{cases}$

$\begin{cases} -6x + 4y = 10 & (Eq1) \\ 6x - 4y = 10 & 2 \times (Eq2) \end{cases}$

$0 = 20$

No solution. Lines are parallel.

19.

$\begin{cases} 2x + y = -3 & (Eq1) \\ 4x - 2y = 10 & (Eq2) \end{cases}$

$\begin{cases} 4x + 2y = -6 & 2 \times (Eq1) \\ 4x - 2y = 10 & (Eq2) \end{cases}$

$8x = 4$

$x = \dfrac{1}{2}$

Substituting to find y

$2\left(\dfrac{1}{2}\right) + y = -3$

$1 + y = -3$

$y = -4$

The solution is $\left(\dfrac{1}{2}, -4\right)$.

21. Algebraically:

$$3x - \frac{1}{2} \le \frac{x}{5} + 2$$

$$10\left(3x - \frac{1}{2}\right) \le 10\left(\frac{x}{5} + 2\right)$$

$$30x - 5 \le 2x + 20$$

$$28x - 5 \le 20$$

$$28x \le 25$$

$$x \le \frac{25}{28} \approx 0.893$$

Graphically:

$y_2 = \dfrac{x}{5} + 2$

$y_1 = 3x - \dfrac{1}{2}$

Intersection
X=.89285714   Y=2.1785714

$[-10, 10]$ by $[-10, 10]$

$3x - \dfrac{1}{2} \le \dfrac{x}{5} + 2$ implies that the

solution is $x \le \dfrac{25}{28}$.

The interval notation is $\left(-\infty, \dfrac{25}{28}\right]$.

Chapter 2 Review Exercises

23. a. Yes. As each amount borrowed increases by $5000, the monthly payment increases by $89.62.

b. Yes. Since the first differences are constant, a linear model will fit the data exactly.

c.

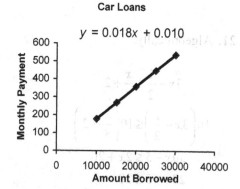

Car Loans

$y = 0.018x + 0.010$

Rounded model:

$P = f(A) = 0.018A + 0.010$

Unrounded model:

$f(A) = 0.017924A + 0.010$

24. a. Using the linear model from part c) in problem 23:

$f(28,000) = 0.018(28,000) + 0.010$

$= 504.01$

The predicted monthly payment on a car loan of $28,000 is $504.01.

b. Yes. Any input could be used for A.

c. $f(A) \le 500$

$0.017924A + 0.010 \le 500$

$0.017924A \le 499.99$

$A \le \dfrac{499.99}{0.017924}$

$A \le 27,895.00$

The loan amount should be less than or equal to $27,895.00.

25. $2.158 = 0.0154x + 1.85$

$0.308 = 0.0154x$

$x = 20$

$x = 20$, and the year is 2030 $(2010 + 20)$.

26. $f(x) = 4500$

27. a. $m = \dfrac{y_2 - y_1}{x_2 - x_1} = \dfrac{895 - 455}{250 - 150} = \dfrac{440}{100} = 4.4$

The average rate of change is $4.40 per unit.

$y - y_1 = m(x - x_1)$

$y - 455 = 4.4(x - 150)$

$y - 455 = 4.4x - 660$

$y = 4.4x - 205$

$P(x) = 4.4x - 205$

b. Profit occurs when $P(x) > 0$.

$4.4x - 205 > 0$

$4.4x > 205$

$x = \dfrac{205}{4.4} = 46.59\overline{09} \approx 47$

The company will make a profit when producing and selling at least 47 units.

28. a. Let x = monthly sales.

$2100 = 1000 + 5\%x$

$2100 = 1000 + 0.05x$

$1100 = 0.05x$

$x = \dfrac{1100}{0.05} = 22,000$

If monthly sales are \$22,000, both positions will yield the same monthly income.

b. Considering the solution from part a), if sales exceed \$22,000 per month, the second position will yield a greater salary.

29.

Original Cost $= 24,000 \times 12 = 288,000$

Desired Profit $= 10\%(288,000) = 28,800$

Revenue from 8 sold cars, with 12% avg profit

$= 8(24,000 + 12\% \times 24,000)$

$= 8(24,000 + 2,880) = 215,040$

Solve for x where x is the selling price of the remaining four cars.

The total Revenue will be $215,040 + 4x$.

The desired Profit = Revenue − Cost

$28,800 = (215,040 + 4x) - 288,000$

$28,800 = 4x - 72,960$

$4x = 101,760$

$x = \dfrac{101,760}{4} = 25,440$

The remaining four cars should be sold for \$25,440 each.

30. Let x = amount invested in the safe account, and let $420,000 - x$ = amount invested in the risky account.

$6\%x + 10\%(420,000 - x) = 30,000$

$0.06x + 42,000 - 0.10x = 30,000$

$-0.04x = -12,000$

$x = \dfrac{-12,000}{-0.04}$

$x = 300,000$

The couple should invest \$300,000 in the safe account and \$120,000 in the risky account.

31. a. $m = \dfrac{13 - 7}{90 - 70} = \dfrac{6}{20} = 0.3$

$y - 7 = 0.3(x - 70)$

$y - 7 = 0.3x - 21$

$y = 0.3x - 14$

b.

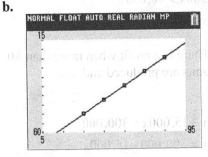

c. Yes; the points do fit exactly.

d. let $x = 78$.

$y = 0.3(78) - 14$

$y = 23.4 - 14$

$y = 9.4$

The annual rate of return on a donation at age 78 will be 9.4%.

e.   let $y = 12.4$.

$12.4 = 0.3x - 14$

$26.4 = 0.3x$

$x = 88$

A person that is 88 years old can expect a 12.4% annual rate of return on a donation.

32. Profit occurs when $R(x) > C(x)$.

$500x > 48,000 + 100x$

$400x > 48,000$

$x > 120$

The company will make a profit when producing and selling more than 120 units.

33. a. $P(x) = 564x - (40,000 + 64x)$

$= 564x - 40,000 - 64x$

$= 500x - 40,000$

b. $500x - 40,000 > 0$

$500x > 40,000$

$x > 80$

c.   There is a profit when more than 80 units are produced and sold.

34. a. $y + 15,000x = 300,000$

$y = 300,000 - 15,000x$

$300,000 - 15,000x < 150,000$

$-15,000x < -150,000$

$x > 10$

b.   After 10 years, the property value will be below $150,000.

35. a. $p = -1.873t + 60.643$

b.   let $t = 20$ (2020 − 2000).

$p = -1.873(20) + 60.643$

$p = -37.46 + 60.643$

$p = 23.183$

The model predicts that in the year 2020, the percent of 12th graders who have ever used cigarettes will be 23.2%. This is extrapolation because it goes beyond the given data.

36. a. $y = 0.0638x + 15.702$

b.

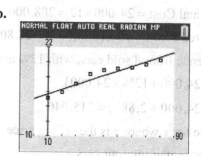

c. $f(99) = 0.0638(99) + 15.702$

$= 22.018$

In 2049 (1950 + 99), the average woman is expected to live 22 years beyond age 65. Her life expectancy is 87 years.

d. $y > 84 - 65 = 19$

$0.0638x + 15.702 > 19$

$0.0638x > 3.298$

$x > \dfrac{3.298}{0.0638}$

$x > 51.69$

In 2002 (1950 + 52) and beyond, the average woman of age 65 is expected to live more than 84 years.

37. a. $y = 0.0655x + 12.324$

b.

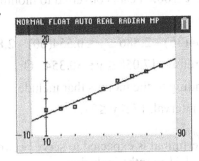

c. $g(130) = 0.0655(130) + 12.324$

$= 8.515 + 12.324$

$= 20.839 \approx 20.8$

In 2080 (1950 + 130), a 65-year old male is expected to live 20.8 more years. The overall life expectancy is 86 years.

d.

A life expectancy of 90 years translates into 90 - 65 = 25 years beyond age 65.

Therefore, let $g(x) = 25$.

$0.0655x + 12.324 = 25$

$0.0655x = 25 - 12.324$

$0.0655x = 12.676$

$x = \dfrac{12.676}{0.0655} = 193.527 \approx 194$

In approximately the year 2144 (1950 +194), life expectancy for a 65 year old male will be 90 years.

e.   A life expectancy of 81 years translates into $81 - 65 = 16$ years beyond age 65.

$g(x) \le 16$

$0.0655x + 12.324 \le 16$

$0.0655x \le 3.676$

$x \le \dfrac{3.676}{0.0655}$

$x \le 56.1$

Since 1950 + 56 = 2006, the average male could expect to live less than 81 years prior to 2007.

38. a. $m = \dfrac{1320 - 1000}{70 - 66} = \dfrac{320}{4} = 80$

$y - 1000 = 80(x - 66)$

$y - 1000 = 80x - 5280$

$y = 80x - 4280$

for $66 \le x \le 70$

b.

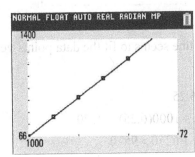

c.   This would be an exact model.

39. a.   Using a graphing calculator and the linear regression function, the equation of the model will be

$y = 72.25x - 3751.94$

for $62 \le x \le 70$

b.

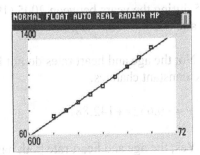

c.   This would be an approximate model.

40. a.

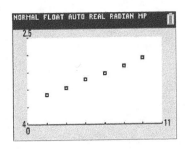

b. $y = 0.209x - 0.195$

c.

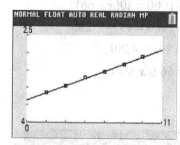

The line seems to fit the data points very well.

41. $y > 6.25$

$132x + 1000(6.25) > 9570$

$132x + 6250 > 9570$

$132x > 3320$

$x > 25.15 \approx 25$

$y < 3.63$

$132x + 1000(3.63) < 9570$

$132x + 3630 < 9570$

$132x < 5940$

$x < 45$

The marriage rate with be between 3.63 and 6.25 during the years between 2005 (1980 + 25) and 2025 (1980 + 45).

42. a. No; the age and heart rates do not both have constant changes.

b. $y = -0.652x + 142.881$

c. Yes; the age and heart rate both have constant increments. The function will be $y = -0.65x + 143$.

43. Let $3(12) \le x \le 5(12)$ or $36 \le x \le 60$. Years converted to months. Then

$0.554(36) - 2.886 \le y \le 0.554(60) - 2.886$.

Therefore, $17.058 \le y \le 30.354$. Or, rounding to the months that include this interval, $17 \le y \le 31$.

The criminal is expected to serve between 17 and 31 months inclusive.

44.

Let $x =$ the amount in the safer fund, and $y =$ the amount in the riskier fund.

$$\begin{cases} x + y = 240,000 & (Eq\,1) \\ 0.08x + 0.12y = 23,200 & (Eq\,2) \end{cases}$$

$$\begin{cases} -0.08x - 0.08y = -19,200 & -0.08 \times (Eq\,1) \\ 0.08x + 0.12y = 23,200 & (Eq\,2) \end{cases}$$

$0.04y = 4000$

$y = \dfrac{4000}{0.04} = 100,000$

Substituting to calculate x

$x + 100,000 = 240,000$

$x = 140,000$

They should invest $140,000 in the safer fund, and $100,000 in the riskier fund.

45.

Let $x =$ number of units.

$R = C$

$565x = 6000 + 325x$

$240x = 6000$

$x = 25$

25 units must be produced and sold to give a break-even point.

46. Let x = dosage of Medication A, and let y = dosage of Medication B.

$$\begin{cases} 6x + 2y = 25.2 & (Eq1) \\ \dfrac{x}{y} = \dfrac{2}{3} & (Eq2) \end{cases}$$

Solving $(Eq2)$ for x yields

$3x = 2y$

$x = \dfrac{2}{3}y$

Substituting

$6\left(\dfrac{2}{3}y\right) + 2y = 25.2$

$4y + 2y = 25.2$

$6y = 25.2$

$y = 4.2$

Substituting to calculate x

$x = \dfrac{2}{3}(4.2)$

$x = 2.8$

Medication A dosage is 2.8 mg while Medication B dosage is 4.2 mg.

47.

Let p = price and q = quantity.

$$\begin{cases} 3q + p = 340 & (Eq1) \\ -4q + p = -220 & (Eq2) \end{cases}$$

$$\begin{cases} -3q - 1p = -340 & -1\times(Eq1) \\ -4q + 1p = -220 & (Eq2) \end{cases}$$

$-7q = -560$

$q = \dfrac{-560}{-7} = 80$

Substituting to calculate p

$3(80) + p = 340$

$240 + p = 340$

$p = 100$

Equilibrium occurs when the price is $100, and the quantity is 80 pairs.

48. Let p = price and q = quantity.

$$\begin{cases} p = \dfrac{q}{10} + 8 & (Eq1) \\ 10p + q = 1500 & (Eq2) \end{cases}$$

Substituting

$10\left(\dfrac{q}{10} + 8\right) + q = 1500$

$q + 80 + q = 1500$

$2q = 1420$

$q = 710$

Substituting to calculate p

$p = \dfrac{710}{10} + 8$

$p = 79$

Equilibrium occurs when the price is $79, and the quantity is 710 units.

49. a. $x + y = 2600$

b. $40x$

c. $60y$

d. $40x + 60y = 120,000$

e.

$$\begin{cases} x + y = 2600 & (Eq1) \\ 40x + 60y = 120,000 & (Eq2) \end{cases}$$

$$\begin{cases} -40x - 40y = -104,000 & -40\times(Eq1) \\ 40x + 60y = 120,000 & (Eq2) \end{cases}$$

$20y = 16,000$

$y = \dfrac{16,000}{20} = 800$

Substituting to calculate x

$x + 800 = 2600$

$x = 1800$

The promoter must sell 1800 tickets at $40 per ticket and 800 tickets at $60 per ticket to yield $120,000.

50. a. $x + y = 500,000$

b. $0.12x$

c. $0.15y$

d. $0.12x + 0.15y = 64,500$

e.
$$\begin{cases} x + y = 500,000 & (Eq1) \\ 0.12x + 0.15y = 64,500 & (Eq2) \end{cases}$$

$$\begin{cases} -0.12x - 0.12y = -60,000 & -0.12 \times (Eq1) \\ 0.12x + 0.15y = 64,500 & (Eq2) \end{cases}$$

$$0.03y = 4500$$

$$y = \frac{4500}{0.03} = 150,000$$

Substituting to calculate x

$$x + 150,000 = 500,000$$

$$x = 350,000$$

Invest \$350,000 in the 12% property and \$150,000 in the 15% property.

Group Activity/Extended Application I: Taxes

1. Domain is Taxable Income: {63,700; 63,800; 63,900; 64,000; 64,100; 64,200; 64,300}
Range is Income Tax Due: {8779, 8804, 8829, 8854, 8879, 8904, 8929}

2.

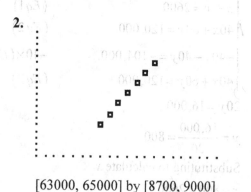

[63000, 65000] by [8700, 9000]

3. Yes, the points appear to lie on a line.

4. Yes, the inputs change by \$100, and the outputs change by \$25.

5. Yes, the rate of change is constant, and is \$.25 per \$1.00 of income.

6. Yes, a linear function will fit the data points exactly.

7. Using the first data point, $(63700, 8779)$

$$y - 8779 = .25(x - 63700)$$

$$y - 8779 = .25x - 15925$$

$$y = .25x - 7146$$

8. Using $x = 63900$

$$y = .25(63900) - 7146 = 8829,$$

Using $x = 64100$

$$y = .25(64100) - 7146 = 8879$$

Both results match the table values.

9. The table is a discrete function.

10. Yes, the model can be used for any taxable income between $63,700 and $64,300, and therefore is a continuous function.

11. For $x = 64150$

$y = .25(64150) - 7146 = 8891.50.$

Thus the tax due on $64,150 is $8891.50.

$x = 500,000 - y$

$0.12(500,000 - y) + 0.15y = 64,500.$

$60000 - 0.12y + 0.15y = 64,500$

$0.03 = 4500 \quad 0.27y = 4500.$

$\boxed{y = 150000}$

$y = \dfrac{4500}{0.27}.$

$\boxed{x = 350,000}$

Chapter 3
Quadratic, Piecewise-Defined, and Power
Functions

Toolbox Exercises

1. $\left(\dfrac{2}{3}\right)^{-2} = \left(\dfrac{3}{2}\right)^2 = \dfrac{3^2}{2^2} = \dfrac{9}{4}$

2. $\left(\dfrac{3}{2}\right)^{-3} = \left(\dfrac{2}{3}\right)^3 = \dfrac{2^3}{3^3} = \dfrac{8}{27}$

3. $10^{-2} \times 10^0 = \dfrac{1}{10^2} \times 1 = \dfrac{1}{100}$

4. $8^{-2} \times 8^0 = \dfrac{1}{8^2} \times 1 = \dfrac{1}{64}$

5. $\left(2^{-1}\right)^3 = 2^{-1\times 3} = 2^{-3} = \dfrac{1}{2^3} = \dfrac{1}{8}$

6. $\left(4^{-2}\right)^2 = 4^{-2\times 2} = 4^{-4} = \dfrac{1}{4^4} = \dfrac{1}{256}$

7. $\left(2x^{-2}y\right)^{-4} = 2^{1\times -4}x^{-2\times -4}y^{1\times -4}$
$= 2^{-4}x^8 y^{-4}$
$= \dfrac{x^8}{2^4 y^4}$
$= \dfrac{x^8}{16 y^4}$

8. $\left(\dfrac{x^2}{y^3}\right)^5 = \dfrac{x^{2\times 5}}{y^{3\times 5}} = \dfrac{x^{10}}{y^{15}}$

9. $\left(-32x^5\right)^{-2} = (-32)^{1\times -2}x^{5\times -2}$
$= (-32)^{-2}x^{-10}$
$= \dfrac{1}{(-32)^2 x^{10}}$
$= \dfrac{1}{1024 x^{10}}$

10. $\left(4x^3 y^{-1} z\right)^0 = 1$

11. $\left(\dfrac{x^{-2}y}{z}\right)^{-3} = \dfrac{x^{-2\times -3}y^{1\times -3}}{z^{1\times -3}}$
$= \dfrac{x^6 y^{-3}}{z^{-3}}$
$= \dfrac{x^6 z^3}{y^3}$

12. $\left(\dfrac{a^{-2}b^{-1}c^{-4}}{a^4 b^{-3}c^0}\right)^{-3} = \dfrac{a^{-2\times -3}b^{-1\times -3}c^{-4\times -3}}{a^{4\times -3}b^{-3\times -3}c^{0\times -3}}$
$= \dfrac{a^6 b^3 c^{12}}{a^{-12}b^9 c^0}$
$= a^{6-(-12)}b^{3-9}c^{12-0}$
$= a^{18}b^{-6}c^{12}$
$= \dfrac{a^{18}c^{12}}{b^6}$

13. $\left|-6\right| = 6$

14. $\left|7-11\right| = \left|-4\right| = 4$

15. a. $\sqrt{x^3} = \sqrt[2]{x^3} = x^{\frac{3}{2}}$

   b. $\sqrt[4]{x^3} = x^{\frac{3}{4}}$

   c. $\sqrt[5]{x^3} = x^{\frac{3}{5}}$

d. $\sqrt[6]{27y^9}$

$\quad = \left(27y^9\right)^{\frac{1}{6}}$

$\quad = \left(3^3 y^9\right)^{\frac{1}{6}}$

$\quad = 3^{\frac{3}{6}} y^{\frac{9}{6}}$

$\quad = 3^{\frac{1}{2}} y^{\frac{3}{2}}$

e. $27\sqrt[6]{y^9} = 27y^{\frac{9}{6}} = 27y^{\frac{3}{2}}$

16. a. $a^{\frac{3}{4}} = \sqrt[4]{a^3}$

b. $-15x^{\frac{5}{8}} = -15\sqrt[8]{x^5}$

c. $(-15x)^{\frac{5}{8}} = \sqrt[8]{(-15x)^5}$

17. $(6x-5)^2 = (6x-5)(6x-5)$

$\quad = 36x^2 - 30x - 30x + 25$

$\quad = 36x^2 - 60x + 25$

18. $(7s-2t)(7s+2t)$

$\quad = 49s^2 + 14st - 14st - 4t^2$

$\quad = 49s^2 - 4t^2$

19. $(8w+2)^2 = (8w+2)(8w+2)$

$\quad = 64w^2 + 16w + 16w + 4$

$\quad = 64w^2 + 32w + 4$

20. $(6x-y)^2 = (6x-y)(6x-y)$

$\quad = 36x^2 - 6xy - 6xy + y^2$

$\quad = 36x^2 - 12xy + y^2$

21. $(2y+5z)^2 = (2y+5z)(2y+5z)$

$\quad = 4y^2 + 10yz + 10yz + 25z^2$

$\quad = 4y^2 + 20yz + 25z^2$

22. $(3x^2+5y)(3x^2-5y)$

$\quad = 9x^4 - 15x^2 y + 15x^2 y - 25y^2$

$\quad = 9x^4 - 25y^2$

23. Remove common factor $3x$

$\quad 3x^2 - 12x = 3x(x-4)$

24. Remove common factor $12x^3$

$\quad 12x^5 - 24x^3 = 12x^3\left(x^2 - 2\right)$

25. Difference of two squares

$\quad 9x^2 - 25m^2 = (3x+5m)(3x-5m)$

26. Find two numbers whose product is 15 and whose sum is -8.

$\quad x^2 - 8x + 15 = (x-5)(x-3)$

27. Find two numbers whose product is -35 and whose sum is -2.

$\quad x^2 - 2x - 35 = (x-7)(x+5)$

28.

To factor by grouping, first multiply the 2nd degree term by the constant term:

$3x^2(-2) = -6x^2$.

Then, find two terms whose product is $-6x^2$ and whose sum is $-5x$, the middle term. $(-6x)$ and $(1x)$

$3x^2 - 5x - 2$

$= 3x^2 - 6x + 1x - 2$

$= (3x^2 - 6x) + (1x - 2)$

$= 3x(x-2) + 1(x-2)$

$= (3x+1)(x-2)$

29.

To factor by grouping, first multiply the 2nd degree term by the constant term:

$8x^2(5) = 40x^2$.

Then, find two terms whose product is $40x^2$ and whose sum is $-22x$, the middle term. $(-2x)$ and $(-20x)$

$8x^2 - 22x + 5$

$= 8x^2 - 2x - 20x + 5$

$= (8x^2 - 2x) + (-20x + 5)$

$= 2x(4x-1) + (-5)(4x-1)$

$= (2x-5)(4x-1)$

30.

$6n^2 + 18 + 39n$

$= 6n^2 + 39n + 18$

$= 3(2n^2 + 13n + 6)$

To factor by grouping, first remove the common factor 3, then multiply the 2nd degree term by the constant term:

$2n^2(6) = 12n^2$.

Then, find two terms whose product is $12n^2$ and whose sum is $13n$, the middle term. $(1n)$ and $(12n)$

$3(2n^2 + 13n + 6)$

$= 3(2n^2 + 1n + 12n + 6)$

$= 3[(2n^2 + 1n) + (12n + 6)]$

$= 3[n(2n+1) + 6(2n+1)]$

$= 3(n+6)(2n+1)$

31. Remove common factor $3b$

$9ab - 12ab^2 + 18b^2 = 3b(3a - 4ab + 6b)$

32. Remove common factor $8x$

$8x^2y - 160x + 48x^2 = 8x(xy - 20 + 6x)$

33. Remove common factor $4y^2$

$12y^3z + 4y^2x^2 - 8y^2z^3 = 4y^2(3yz + x^2 - 2z^3)$

34. Remove common factor $(x^2+4)^2$

$x^2(x^2+4)^2 + 2(x^2+4)^2 = (x^2+4)^2(x^2+2)$

35. Find two numbers whose product is 16 and whose sum is 8.

$x^2 + 8x + 16 = (x+4)(x+4)$

$\qquad\qquad\quad = (x+4)^2$

36. Find two numbers whose product is 4 and whose sum is −4.

$$x^2 - 4x + 4 = (x-2)(x-2)$$
$$= (x-2)^2$$

37.

To factor by grouping, first multiply the 2nd degree term by the constant term:

$$4x^2(25) = 100x^2.$$

Then, find two terms whose product is $100x^2$ and whose sum is $20x$, the middle term. $(10x)$ and $(10x)$

$$4x^2 + 20x + 25$$
$$= 4x^2 + 10x + 10x + 25$$
$$= (4x^2 + 10x) + (10x + 25)$$
$$= 2x(2x+5) + (5)(2x+5)$$
$$= (2x+5)(2x+5)$$
$$= (2x+5)^2$$

38. Find two numbers whose product is 9 and whose sum is −6.

$$x^4 - 6x^2 y + 9y^2 = (x^2 - 3y)(x^2 - 3y)$$
$$= (x^2 - 3y)^2$$

39. First, remove common factor $5x$

$$5x^5 - 80x = 5x(x^4 - 16)$$

Next, find two numbers whose product is −16 and whose sum is 0.

$$5x(x^4 - 16) = 5x(x^2 + 4)(x^2 - 4)$$

Finally, find two numbers whose product is −4 and whose sum is 0.

$$5x(x^2 + 4)(x^2 - 4) = 5x(x^2 + 4)(x+2)(x-2)$$

40. $3y^4 + 9y^2 - 12y^2 - 36$

$$= 3\left[y^4 + 3y^2 - 4y^2 - 12 \right]$$
$$= 3\left[(y^4 + 3y^2) + (-4y^2 - 12) \right]$$
$$= 3\left[y^2(y^2 + 3) + (-4)(y^2 + 3) \right]$$
$$= 3(y^2 - 4)(y^2 + 3)$$
$$= 3(y-2)(y+2)(y^2 + 3)$$

41. $18p^2 + 12p - 3p - 2$

$$= (18p^2 + 12p) + (-3p - 2)$$
$$= 6p(3p+2) + (-1)(3p+2)$$
$$= (6p-1)(3p+2)$$

42. $5x^2 - 10xy - 3x + 6y$

$$5x(x-2y) - 3(x-2y)$$
$$(x-2y)(5x-3)$$

43. a. Imaginary. The number has a non-zero real part and an imaginary part.

   b. Pure imaginary. The real part is zero.

   c. Real. The imaginary part is zero.

   d. Real. $2 - 5i^2 = 2 - 5(-1) = 7$

44. a. Imaginary. The number has a non-zero real part and an imaginary part.

   b. Real. The imaginary part is zero.

   c. Pure imaginary. The real part is zero.

   d. Imaginary. The number has a non-zero real part and an imaginary part. $2i^2 - i = 2(-1) - i = -2 - i$

45. $a + bi = 4 + 0i$. Therefore,
$a = 4, b = 0$.

46. $a + 3i = 15 - bi$
Therefore, $a = 15$ and
$-b = 3$ or $b = -3$.

47. $a + bi = 2 + 4i$
Therefore, $a = 2$ and $b = 4$.

Section 3.1 Skills Check

1. a. Yes. The equation fits the form
$f(x) = ax^2 + bx + c, a \neq 0$.

b. Since $a = 2 > 0$, the graph opens up and is therefore concave up.

c. Since the graph is concave up, the vertex point is a minimum.

3. Not quadratic. The equation does not fit the form $f(x) = ax^2 + bx + c, a \neq 0$. The highest exponent is 3.

5. a. Yes. The equation fits the form
$f(x) = ax^2 + bx + c, a \neq 0$.

b. Since $a = -5 < 0$, the graph opens down and is therefore concave down.

c. Since the graph is concave down, the vertex point is a maximum.

7. a.

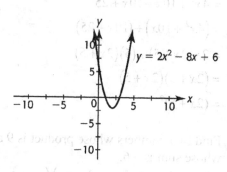

$y = 2x^2 - 8x + 6$

b. Yes.

9. a.

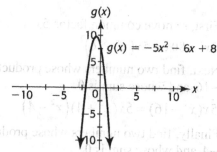

$g(x) = -5x^2 - 6x + 8$

b. Yes.

11. a.

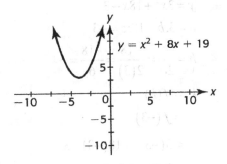

$y = x^2 + 8x + 19$

b. Yes.

13. a.

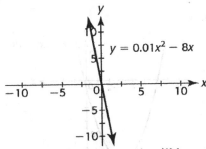

$y = 0.01x^2 - 8x$

b. No. The complete graph will be a parabola.

15.

Using the given vertex point $(2, -4)$ and the vertex form of the parabola equation, then $y = a(x-2)^2 - 4$. To solve for the value of a, plug in the other given point $(4, 0)$ to get $0 = a(4-2)^2 - 4$. Then $0 = a(2)^2 - 4 = 4a - 4$, so that $a = 1$. The final equation of the parabola in vertex form is $f(x) = 1(x-2)^2 - 4$.

17. The value of a is larger for y_1 since the graph is vertically stretched, causing it to appear more narrow than y_2.

19. Since $(-1, -7)$ and $(3, -7)$ are symmetric points on the graph, if it were drawn, the axis of symmetry of the parabola would be $x = 1$. The point $(1, 13)$ would then be the vertex of the parabola. Using the vertex form of the parabola equation, $y = a(x-1)^2 + 13$. To solve for the value of a, plug in the another given point from the table, $(5, -67)$, to get $-67 = a(5-1)^2 + 13$.

Then,
$-67 = a(4)^2 + 13$
$-67 = 16a + 13$ so that $a = -5$. The final equation of the parabola in vertex form is
$f(x) = -5(x-1)^2 + 13$
$f(x) = -5x^2 + 10x + 8$

21. a. $(1, 3)$

b.

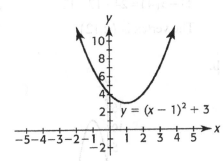

$y = (x - 1)^2 + 3$

23. a. $(-8, 8)$

b.

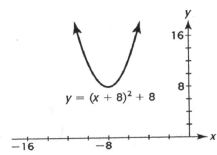

$y = (x + 8)^2 + 8$

25. a. $(4,-6)$

b.

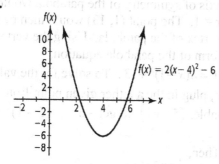

$f(x) = 2(x-4)^2 - 6$

27. a. $y = 12x - 3x^2$

$y = -3x^2 + 12x$

$a = -3, b = 12, c = 0$

$h = \dfrac{-b}{2a} = \dfrac{-12}{2(-3)} = \dfrac{-12}{-6} = 2$

$k = f(h) = f(2) = 12(2) - 3(2)^2 =$

$24 - 3(4) = 24 - 12 = 12$

The vertex is $(2,12)$.

b.

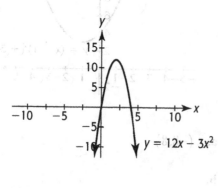

$y = 12x - 3x^2$

29. a. $y = 3x^2 + 18x - 3$

$a = 3, b = 18, c = -3$

$h = \dfrac{-b}{2a} = \dfrac{-18}{2(3)} = \dfrac{-18}{6} = -3$

$k = f(h)$

$= f(-3)$

$= 3(-3)^2 + 18(-3) - 3$

$= 3(9) - 54 - 3$

$= 27 - 54 - 3 = -30$

The vertex is $(-3,-30)$.

b.

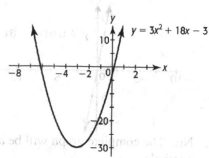

$y = 3x^2 + 18x - 3$

31. a. $y = 2x^2 - 40x + 10$

$a = 2, b = -40, c = 10$

$h = \dfrac{-b}{2a} = \dfrac{-(-40)}{2(2)} = \dfrac{40}{4} = 10$

b.

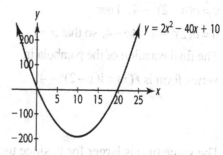

$y = 2x^2 - 40x + 10$

c. The vertex is located at $(10,-190)$

33. a. $y = -0.2x^2 - 32x + 2$

$a = -0.2, b = -32, c = 2$

$$h = \frac{-b}{2a} = \frac{-(-32)}{2(-0.2)} = \frac{32}{-0.4} = -80$$

b.

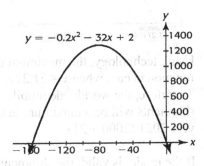

c.   The vertex is located at $(-80, 1282)$

35.

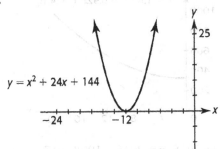

The vertex is located at $(-12, 0)$

37.

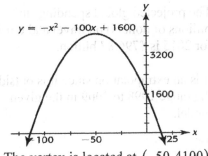

The vertex is located at $(-50, 4100)$

39.

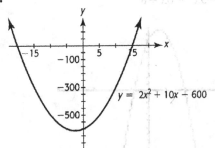

The vertex is located at $(-2.5, -612.5)$

41. $y = 2x^2 - 8x + 6$

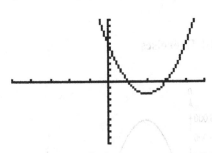

[–5, 5] by [–10, 10]

The x-intercepts are $(1, 0), (3, 0)$.

43. $y = x^2 - x - 110$

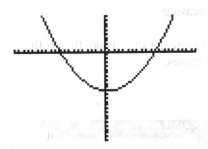

[–20, 20] by [–250, 100]

The x-intercepts are $(11, 0), (-10, 0)$.

45. $y = -5x^2 - 6x + 8$

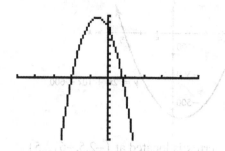

$[-5, 5]$ by $[-10, 10]$

The x-intercepts are $(-2, 0), (0.8, 0)$.

Section 3.1 Exercises

47. a.

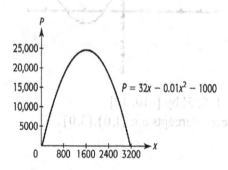

b. For x between 1 and 1600, the profit is increasing.

c. For x greater than 1600, the profit decreases.

49. a.

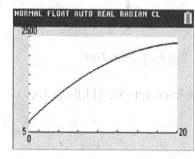

b.

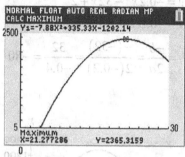

Using technology, the maximum of the function occurs when $x = 21.277$. Therefore, the worldwide mobile shipments will be a maximum in the year 2021 (2000 + 21).

c. If the model is valid, the shipments will decrease.

51. a.

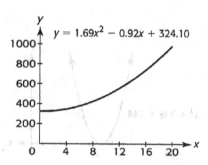

b. In 2015, $x = 2015 - 1998 = 17$.
$$y = 1.69x^2 - 0.92x + 324.10$$
$$y = 1.69(17)^2 - 0.92(17) + 324.10$$
$$y = 796.87$$
The projected global spending (in billions of dollars) on travel and tourism for 2015 is $796.87 billion.

c. It is an extrapolation since it is outside the range 1998 to 2009 in the given model.

53. a. $S(t) = 30 + 39.2t - 9.8t^2$

$$t = \frac{-b}{2a} = \frac{-39.2}{2(-9.8)} = \frac{-39.2}{-19.6} = 2$$

$$S = S(t) = S(2)$$

$$= 30 + 39.2(2) - 9.8(2)^2$$

$$= 30 + 78.4 - 39.2$$

$$= 69.2$$

The vertex is $(2, 69.2)$.

b. The ball reaches its maximum height of 69.2 meters in 2 seconds.

c. The function is increasing until $t = 2$ seconds. The ball rises for two seconds, at which time it reaches its maximum height. After two seconds, the ball falls.

55. a. $y = -0.001x^2 + 0.037x + 1.949$

$$x = \frac{-b}{2a} = \frac{-0.037}{2(-0.001)} = \frac{-0.037}{-0.002} = 18.5$$

$$y = -0.001x^2 + 0.037x + 1.949$$

$$= -0.001(18.5)^2 + 0.037(18.5) + 1.949$$

$$= -0.34225 + 0.6845 + 1.949$$

$$\approx 2.291$$

The vertex is $(18.5, 2.291)$; the maximum number of barrels of crude oil projected to be produced during this period is 2.291 billion barrels during 2029.

b. $x = 2033 - 2010 = 23$

$$y = -0.001x^2 + 0.037x + 1.949$$

$$= -0.001(23)^2 + 0.037(23) + 1.949$$

$$= -0.529 + 0.851 + 1.949$$

$$= 2.271$$

In the year 2033, there will be 2.271 billion barrels of crude oil produced.

c.

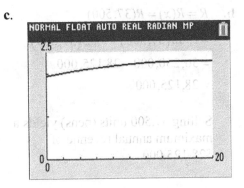

57. Note that the maximum profit occurs at the vertex of the quadratic function, since the function is concave down.

a. $P(x) = 40x - 3000 - 0.01x^2$

$$x = \frac{-b}{2a} = \frac{-40}{2(-0.01)} = \frac{-40}{-0.02} = 2000$$

b. $P = P(x) = P(2000)$

$$= 40(2000) - 3000 - 0.01(2000)^2$$

$$= 80,000 - 3000 - 0.01(4,000,000)$$

$$= 80,000 - 3000 - 40,000$$

$$= 37,000$$

Producing and selling 2000 units (MP3 players) yields a maximum profit of $37,000.

59. Note that the maximum revenue occurs at the vertex of the quadratic function, since the function is concave down.

a. $R(x) = 1500x - 0.02x^2$

$$x = \frac{-b}{2a} = \frac{-1500}{2(-0.02)} = \frac{-1500}{-0.04} = 37,500$$

b. $R = R(x) = R(37,500)$

$= 1500(37,500) - 0.02(37,500)^2$

$= 56,250,000 - 28,125,000$

$= 28,125,000$

Selling 37,500 units (pens) yields a maximum annual revenue of $28,125,000.

61. a. Yes. $A = x(100 - x) = 100x - x^2$. Note that A fits the form
$f(x) = ax^2 + bx + c, a \neq 0.$

b. The maximum area will occur at the vertex of graph of function A.

$x = \dfrac{-b}{2a} = \dfrac{-100}{2(-1)} = \dfrac{-100}{-2} = 50$

$A = A(x) = A(50)$

$= 100(50) - (50)^2$

$= 5000 - 2500$

$= 2500$

The maximum area of the pen is 2500 square feet.

63. a.

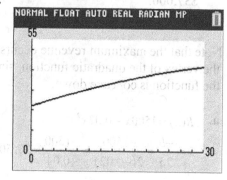

b.

$y = -0.010x^2 + 0.971x + 22.1$

$x = \dfrac{-b}{2a} = \dfrac{-0.971}{2(-0.010)} = \dfrac{-0.971}{-0.02} = 48.55$

$y = -0.010(48.55)^2 + 0.971(48.55) + 22.1$

$y \approx 45.671$

Thus the vertex is $(48.55, 45.671)$; the percent of Americans who are obese will reach a maximum of 45.671% in 2049.

c. If the model remains valid, the percent of Americans who are obese will continue to increase until 2049 and then it will begin to decrease.

65. a. Since $a = 0.114 > 0$ in the equation
$y = 0.114x^2 - 2.322x + 45.445$, the graph is concave up, and the vertex is a minimum.

b.

$y = 0.114x^2 - 2.322x + 45.445$

$x = \dfrac{-b}{2a} = \dfrac{-(-2.322)}{2(0.114)} = \dfrac{2.322}{0.228} = 10.184$

$y = 0.114(10.184)^2 - 2.322(10.184) + 45.445$

$y = 33.621.$

Thus the vertex is $(10.184, 33.621)$.

When x is 10.184, it would be in the 11th year after 1990. Therefore, during the year 2001, 33.621 million people in the U.S. lived below the poverty level.

c.

67. a.

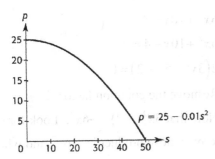

b. Decreasing. The graph is falling as s increases.

c. $(0, 25)$

d. When the wind speed is zero, the amount of particulate pollution is 25 ounces per cubic yard.

69.

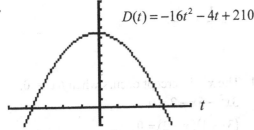

$$D(t) = -16t^2 - 4t + 210$$

$[-5, 5]$ by $[-50, 300]$

The t-intercepts are approximately $(-3.75, 0)$ and $(3.5, 0)$. The $(3.5, 0)$ makes sense because time is understood to be positive. In the context of the question, the tennis ball will hit the pool in 3.5 seconds.

71. a. $y = -16t^2 + 32t + 3$

b. $y = -16t^2 + 32t + 3$

$$t = \frac{-b}{2a} = \frac{-(32)}{2(-16)} = \frac{-32}{-32} = 1$$

$$y = -16(1)^2 + 32(1) + 3 = 19$$

The maximum height of the ball is 19 feet when $t = 1$ second.

73. a.

Rent	Number of Apartments Rented	Total Revenue
$1200	100	$120,000
$1240	98	$121,520
$1280	96	$122,880
$1320	94	$124,080

b. Yes. Revenue = Rent multiplied by Number of Apartments Rented. $(1200 + 40x)$ represents the rent amount, while $(100 - 2x)$ represents the number of apartments rented.
$$R = 120,000 + 1600x - 80x^2$$

c. $x = \dfrac{-b}{2a} = \dfrac{-(1600)}{2(-80)} = \dfrac{-1600}{-160} = 10$

The maximum occurs when $x = 10$. Therefore the most profitable rent to charge is $1200 + 40(10) = \$1600$.

75. a.

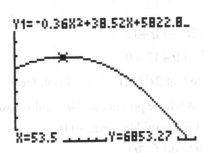

$[0, 200]$ by $[0, 10,000]$

The vertex is $(53.5, 6853.3)$.

b. The model predicts that in 2044 (1990 + 54), the population of the world will by 6853.3 million, or 6,853,300,000.

c. The model predicts a population increase from 1990 until 2044. After 2044, based on the model, the population of the world will decrease.

Section 3.2 Skills Check

1. $x^2 - 3x - 10 = 0$
    $(x-5)(x+2) = 0$
    $x - 5 = 0, x + 2 = 0$
    $x = 5, x = -2$

3. $x^2 - 11x + 24 = 0$
    $(x-8)(x-3) = 0$
    $x - 8 = 0, x - 3 = 0$
    $x = 8, x = 3$

5. $2x^2 + 2x - 12 = 0$
    $2(x^2 + x - 6) = 0$
    $2(x+3)(x-2)$
    $x + 3 = 0, x - 2 = 0$
    $x = -3, x = 2$

7.
    $0 = 2t^2 - 11t + 12$
    $2t^2 - 11t + 12 = 0$
    Note that $2t^2(12) = 24t^2$. Look for two
    terms whose product is $24t^2$ and whose
    sum is the middle term, $-11t$.
    $(-8t)$ and $(-3t)$
    $2t^2 - 8t - 3t + 12 = 0$
    $(2t^2 - 8t) + (-3t + 12) = 0$
    $2t(t-4) - 3(t-4) = 0$
    $(2t-3)(t-4) = 0$
    $2t - 3 = 0, t - 4 = 0$
    $t = \dfrac{3}{2}, t = 4$

9.
    $6x^2 + 10x = 4$
    $6x^2 + 10x - 4 = 0$
    $2(3x^2 + 5x - 2) = 0$
    Remove the common factor, 2.
    Note that $3x^2(-2) = -6x^2$. Look for two
    terms whose product is $-6x^2$ and whose
    sum is the middle term, $5x$.
    $(6x)$ and $(-1x)$
    $2(3x^2 + 6x - 1x - 2) = 0$
    $2[(3x^2 + 6x) + (-1x - 2)] = 0$
    $2[3x(x+2) + (-1)(x+2)] = 0$
    $2(3x-1)(x+2) = 0$
    $3x - 1 = 0, x + 2 = 0$
    $x = \dfrac{1}{3}, x = -2$

11. The x – intercept occurs when $f(x) = 0$.
    $3x^2 - 5x - 2 = 0$
    $(3x+1)(x-2) = 0$
    $3x + 1 = 0, x - 2 = 0$
    $x = -\frac{1}{3}, x = 2$

13. The x – intercept occurs when $f(x) = 0$.
    $4x^2 - 9 = 0$
    $(2x+3)(2x-3) = 0$
    $2x + 3 = 0, 2x - 3 = 0$
    $x = -\frac{3}{2}, x = \frac{3}{2}$

15. The x – intercept occurs when $f(x) = 0$.
    $3x^2 + 4x - 4 = 0$
    $(3x-2)(x+2) = 0$
    $3x - 2 = 0, x + 2 = 0$
    $x = \frac{2}{3}, x = -2$

17. $y = x^2 + 7x + 10$

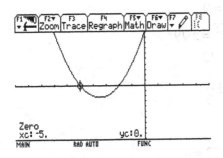

[−10, 5] by [−10, 10]

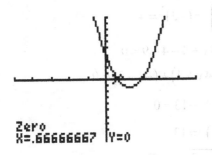

[−10, 5] by [−10, 10]

The x-intercepts are $(-2,0)$ and $(-5,0)$.

19. $y = 3x^2 - 8x + 4$

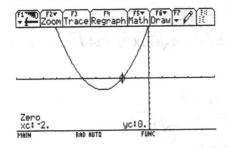

[−5, 5] by [−10, 10]

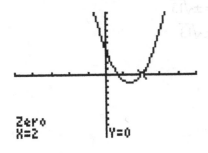

[−5, 5] by [−10, 10]

The x-intercepts are $\left(\dfrac{2}{3},0\right)$ and $(2,0)$.

21. $y = 2x^2 + 7x - 4$

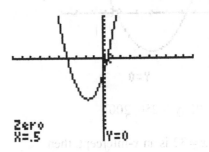

[−10, 10] by [−20, 10]

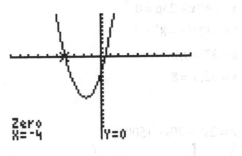

[−10, 10] by [−20, 10]

The x-intercepts are $(-4,0)$ and $(0.5,0)$.

23. $y = 2w^2 - 5w - 3$

[−10, 10] by [−10, 10]

Since $w = 3$ is an x-intercept, then
$w - 3$ is a factor.
$$2w^2 - 5w - 3 = 0$$
$$(w-3)(2w+1) = 0$$
$$w - 3 = 0, 2w + 1 = 0$$
$$w = 3, w = -\frac{1}{2}$$

25. $y = x^2 - 40x + 256$

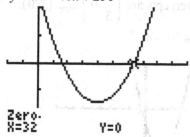

Zero.
X=32   Y=0

$[-10, 50]$ by $[-250, 200]$

Since $x = 32$ is an x-intercept, then
$x - 32$ is a factor.
$x^2 - 40x + 256 = 0$
$(x - 32)(x - 8) = 0$
$x - 32 = 0, x - 8 = 0$
$x = 32, x = 8$

27. $y = 2s^2 - 70s - 1500$

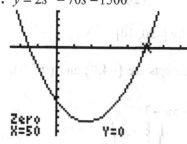

Zero
X=50   Y=0

$[-25, 75]$ by $[-2500, 1000]$

Since $x = 50$ is an x-intercept for
$2s^2 - 70s - 1500 = 0$, then $s - 50$
is a factor.
$(s - 50)(2s + 30) = 0$
$2(s - 50)(s + 15) = 0$
$s - 50 = 0, s + 15 = 0$
$s = 50, s = -15$

29. $4x^2 - 9 = 0$
$4x^2 = 9$
$\sqrt{4x^2} = \pm\sqrt{9}$
$2x = \pm 3$
$x = \pm\dfrac{3}{2}$

31. $x^2 - 32 = 0$
$x^2 = 32$
$\sqrt{x^2} = \pm\sqrt{32}$
$x = \pm\sqrt{32} = \pm\sqrt{(16)(2)} = \pm 4\sqrt{2}$

33. $(x - 5)^2 = 9$
$\sqrt{(x - 5)^2} = \pm\sqrt{9}$
$x - 5 = \pm 3$
$x - 5 = 3, x - 5 = -3$
$x = 8, x = 2$

35. $x^2 - 4x - 9 = 0$
$\left(\dfrac{-4}{2}\right)^2 = (-2)^2 = 4$
$x^2 - 4x + 4 - 4 - 9 = 0$
$\left(x^2 - 4x + 4\right) + (-4 - 9) = 0$
$(x - 2)^2 - 13 = 0$
$(x - 2)^2 = 13$
$\sqrt{(x - 2)^2} = \pm\sqrt{13}$
$x - 2 = \pm\sqrt{13}$
$x = 2 \pm\sqrt{13}$

37. $x^2 - 3x + 2 = 0$

$$\left(\frac{-3}{2}\right)^2 = \frac{9}{4}$$

$$x^2 - 3x + \frac{9}{4} - \frac{9}{4} + 2 = 0$$

$$\left(x^2 - 3x + \frac{9}{4}\right) + \left(-\frac{9}{4} + 2\right) = 0$$

$$\left(x - \frac{3}{2}\right)^2 + \left(-\frac{9}{4} + \frac{8}{4}\right) = 0$$

$$\left(x - \frac{3}{2}\right)^2 - \frac{1}{4} = 0$$

$$\left(x - \frac{3}{2}\right)^2 = \frac{1}{4}$$

$$\sqrt{\left(x - \frac{3}{2}\right)^2} = \pm\sqrt{\frac{1}{4}}$$

$$x - \frac{3}{2} = \pm\frac{1}{2}$$

$$x = \frac{3}{2} \pm \frac{1}{2}$$

$$x = \frac{3}{2} + \frac{1}{2}, x = \frac{3}{2} - \frac{1}{2}$$

$$x = 2, x = 1$$

39. $x^2 - 5x + 2 = 0$

$$a = 1, b = -5, c = 2$$

$$x = \frac{-b \pm \sqrt{b^2 - 4ac}}{2a}$$

$$x = \frac{-(-5) \pm \sqrt{(-5)^2 - 4(1)(2)}}{2(1)}$$

$$x = \frac{5 \pm \sqrt{25 - 8}}{2}$$

$$x = \frac{5 \pm \sqrt{17}}{2}$$

41. $5x + 3x^2 = 8$

$$3x^2 + 5x - 8 = 0$$

$$a = 3, b = 5, c = -8$$

$$x = \frac{-b \pm \sqrt{b^2 - 4ac}}{2a}$$

$$x = \frac{-(5) \pm \sqrt{(5)^2 - 4(3)(-8)}}{2(3)}$$

$$x = \frac{-5 \pm \sqrt{25 + 96}}{6}$$

$$x = \frac{-5 \pm \sqrt{121}}{6}$$

$$x = \frac{-5 \pm 11}{6}$$

$$x = \frac{-5 + 11}{6}, x = \frac{-5 - 11}{6}$$

$$x = 1, x = -\frac{16}{6} = -\frac{8}{3}$$

43. $y = 2x^2 + 2x - 12$

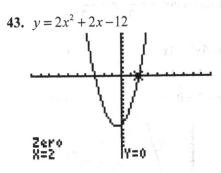

[–10, 10] by [–20, 10]

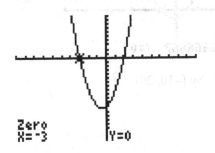

[–10, 10] by [–20, 10]

The solutions are $x = 2$, $x = -3$.

45. $y = 6x^2 + 5x - 6$

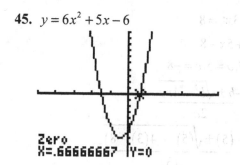

[−5, 5] by [−10, 10]

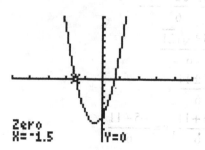

[−5, 5] by [−10, 10]

The solutions are $x = \dfrac{2}{3}, x = -\dfrac{3}{2}$.

47. $4x + 2 = 6x^2 + 3x$

$0 = 6x^2 + 3x - 4x - 2$

$6x^2 - x - 2 = 0$

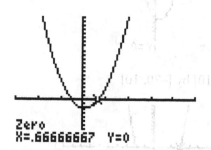

[−3, 5] by [−10, 20]

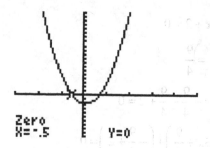

[−3, 5] by [−10, 20]

The solutions are $x = \dfrac{2}{3}, x = -\dfrac{1}{2}$.

49. $x^2 + 25 = 0$

$x^2 = -25$

$\sqrt{x^2} = \pm\sqrt{-25}$

$x = \pm 5i$

Graphical check

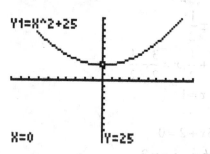

[−10, 10] by [−100, 100]

Note that the graph has no x-intercepts.

51. $(x-1)^2 = -4$

$\sqrt{(x-1)^2} = \pm\sqrt{-4}$

$x - 1 = \pm 2i$

$x = 1 \pm 2i$

Graphical check

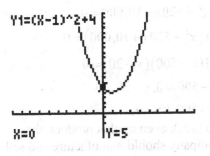

Y1=(X-1)^2+4

X=0   Y=5

[–10, 10] by [–5, 20]

Note that the graph has no x-intercepts.

53. $x^2 + 4x + 8 = 0$

$$x = \frac{-b \pm \sqrt{b^2 - 4ac}}{2a}$$

$$x = \frac{-4 \pm \sqrt{4^2 - 4(1)(8)}}{2(1)}$$

$$x = \frac{-4 \pm \sqrt{-16}}{2}$$

$$x = \frac{-4 \pm 4i}{2} = -2 \pm 2i$$

Graphical check

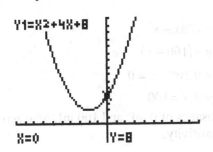

Y1=X2+4X+8

X=0   Y=8

[–10, 10] by [–5, 30]

Note that the graph has no x-intercepts.

55. a. Since the graph shows two x-intercepts at $(-2,0)$ and $(3,0)$, by definition the value of the discriminant is positive.

b. And, there are two real solutions.

c. And, the solutions for $f(x) = 0$ are $x = -2$ and $x = 3$.

57. a. Since the graph shows one x-intercepts at $(3,0)$, by definition the value of the discriminant is zero.

b. And, there is one real solution, with multiplicity 2.

c. And, the solutions for $f(x) = 0$ is $x = 3$.

59. a. $b^2 - 4ac = (-5)^2 - 4(3)(-2)$

$$= 25 + 24$$

$$= 49$$

b. Since the discriminant is positive, by definition, there are two real solutions.

61. a. $b^2 - 4ac = (4)^2 - 4(4)(1)$

$$= 16 - 16$$

$$= 0$$

b. Since the discriminant is zero, by definition, there is one real solution with multiplicity 2.

Section 3.2 Exercises

63. Let $S(t) = 228$ and solve for t.

$$228 = 100 + 96t - 16t^2$$

$$16t^2 - 96t + 128 = 0$$

$$16(t^2 - 6t + 8) = 0$$

$$16(t - 4)(t - 2) = 0$$

$$t - 4 = 0, t - 2 = 0$$

$$t = 4, t = 2$$

The ball is 228 feet high after 2 seconds and 4 seconds.

65. Let $P(x) = 0$ and solve for x.

$-12x^2 + 1320x - 21,600 = 0$

$-12(x^2 - 110x + 1800) = 0$

$-12(x - 90)(x - 20) = 0$

$x - 90 = 0, x - 20 = 0$

$x = 90, x = 20$

Producing and selling either 20 or 90 of the readers produces a profit of zero dollars. Therefore 20 readers or 90 readers represent the break-even point for manufacturing and selling this product.

67. a. $P(x) = R(x) - C(x)$

$P(x) = 550x - (10,000 + 30x + x^2)$

$P(x) = 550x - 10,000 - 30x - x^2$

$P(x) = -x^2 + 520x - 10,000$

b. Let $x = 18$.

$P(18) = -(18)^2 + 520(18) - 10,000$

$P(18) = -324 + 9360 - 10,000$

$p(18) = -964$

When 18 refrigerators are produced and sold, there is a loss of $964.

c. Let $x = 32$.

$P(32) = -(32)^2 + 520(32) - 10,000$

$P(32) = -1024 + 16,640 - 10,000$

$p(32) = 5616$

When 32 refrigerators are produced and sold, there is a profit of $5616.

d. Let $P(x) = 0$ and solve for x.

$-x^2 + 520x - 10,000 = 0$

$-1(x^2 - 520x + 10,000) = 0$

$-1(x - 500)(x - 20) = 0$

$x - 500 = 0, x - 20 = 0$

$x = 500, x = 20$

To break even on this product, the company should manufacture and sell either 20 or 500 refrigerators.

69. a. Let $p = 0$ and solve for s.

$0 = 25 - 0.01s^2$

$-25 = -0.01s^2$

$\dfrac{-25}{-0.01} = \dfrac{-0.01}{-0.01}s^2$

$2500 = s^2$

$s = \pm\sqrt{2500} = \pm 50$

When s is 50 or -50, $p = 0$.

b. When $p = 0$, the pollution in the air above the power plant is zero.

c. Only the positive solution, $s = 50$, makes sense because wind speed must be positive.

71. a. $0 = 100x - x^2$

$0 = x(100 - x)$

$x = 0, 100 - x = 0$

$x = 0, x = 100$

Dosages of 0 mL and 100 mL give zero sensitivity.

b. When x is zero, there is no amount of drug in a person's system, and therefore no sensitivity to the drug. When x is 100 mL the amount of drug in a person's system is so high that the person may be overdosed on the drug and therefore has no sensitivity to the drug.

73. Equilibrium occurs when demand is equal to supply. Solve

$$109.70 - 0.10q = 0.01q^2 + 5.91$$

$$0 = 0.01q^2 + 0.10q - 103.79$$

$$a = 0.01, b = 0.10, c = -103.79$$

$$q = \frac{-b \pm \sqrt{b^2 - 4ac}}{2a}$$

$$q = \frac{-0.10 \pm \sqrt{(0.10)^2 - 4(0.01)(-103.79)}}{2(0.01)}$$

$$q = \frac{-0.10 \pm \sqrt{0.01 + 4.1516}}{0.02}$$

$$q = \frac{-0.10 \pm \sqrt{4.1616}}{0.02}$$

$$q = \frac{-0.10 \pm 2.04}{0.02}$$

$$q = \frac{-0.10 + 2.04}{0.02}, q = \frac{-0.10 - 2.04}{0.02}$$

$$q = 97, q = -107$$

Since q represents the quantity of trees in hundreds at equilibrium, q must be positive. A q-value of -107 does not make sense in the physical context of the question. Selling 9700 trees, ($q = 97$), creates an equilibrium price. The equilibrium price is given by:

$$p = 109.70 - 0.10q$$

$$p = 109.70 - 0.10(97)$$

$$p = 109.70 - 9.70$$

$$p = 100.00 \text{ or } \$100.00 \text{ per tree.}$$

75.

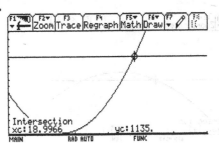

[0, 30] by [0, 1500]

$$y = 9.3x^2 - 148.8x + 605.595$$

$$1135 = 9.3x^2 - 148.8x + 605.595$$

$$0 = 9.3x^2 - 148.8x - 529.405$$

$$x = 18.9996$$

$$x \approx 19$$

Thus bill payment using smart phones is projected to reach \$1135 billion in the year 2019, (2000 +1 9).

77. a. The solution $x = 20$, where x is the number of years after 1970, means that in 1990, energy consumption in the U.S. was 87.567 quadrillion BTU's.

b. $y = -0.013x^2 + 1.281x + 67.147$

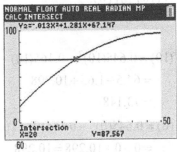

c. Yes, if a second solution is found, it would show that after the year 2020, when the maximum BTU's are consumed, the consumption would again reach 87.567 about 28.5 years later, during the year 2049.

79.

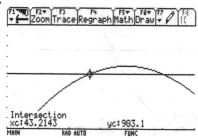

[30, 50] by [850, 950]

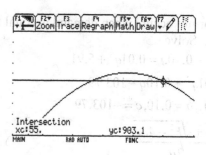

[30, 50] by [850, 950]

$$y = -0.224x^2 + 22x + 370.7$$
$$903.1 = -0.224x^2 + 22x + 370.7$$
$$0 = -0.224x^2 + 22x - 532.4$$
$$x = 43.2144 \text{ and } x = 55$$
$$x \approx 44 \text{ and } x = 55$$

Thus this population is projected to reach 903.1 million in the years 2014 (1970 + 44) and 2025 (1970 + 55).

81. a. To determine the increase in the amount of federal funds, in billions, spent on child nutrition programs between 2000 and 2010, calculate:

$$N(2010 - 2000) - N(2000 - 2000)$$
$$= N(10) - N(0).$$

$$N(10) = 0.645(10)^2 - 0.165(10) + 10.298$$
$$= 64.5 - 1.65 + 10.298$$
$$= 73.148$$
$$N(0) = 0.645(0)^2 - 0.165(0) + 10.298$$
$$= 0 - 0 + 10.298 = 10.298$$

$$N(10) - N(0) = 73.148 - 10.298$$
$$= 62.85$$

There is a increase of approximately 62.85 billion dollars.

b.

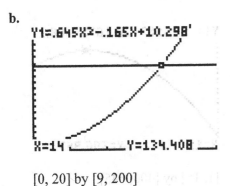

[0, 20] by [9, 200]

In 2014 (2000 + 14), the amount of funds spent will be 134.408 billion dollars.

c. $N(20) = 0.645(20)^2 - 0.165(20)$
$$+10.298$$
$$= 258 - 3.3 + 10.298$$
$$= 264.998$$

Assuming the model continues to be valid until 2020, the amount spent in that year would be 264.998 billion dollars, an extrapolation, beyond the current year, of the function.

83. a.

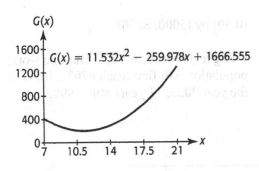

$G(x) = 11.532x^2 - 259.978x + 1666.555$

b.

The price of gold in 2020 corresponds to

$G(30) = 11.532(30)^2 - 259.978(30)$
$$+1666.555$$
$$= 10,378.8 - 7799.34 + 1666.555$$
$$= 4246.015 \text{ or } \$4246.02 \text{ U.S. dollars}$$

c.

X	Y₁
23	1787.5
24	2069.5
25	2374.6
26	2702.8
27	3054
28	3428.3
29	3825.6

X=26

Thus, 26 years after 1990, in the year 2016, the price of gold will reach $2702.80.

85. a.

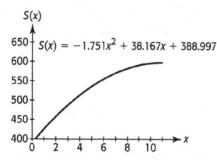

b. According to the complete model below in the window [1, 20] by [400, 645], the sales will be 550 billion dollars in the years following $x = 5.7$ (2006) and $x = 16.078$ (2017).

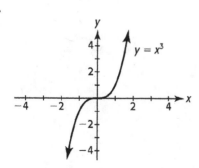

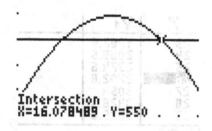

c.

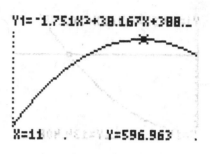

[1, 15] by [400, 645]

No, the model indicates the drop in sales to be nearer the year 2011, than to 2008.

87. $y = -0.36x^2 + 38.52x + 5822.86$

$-0.36x^2 + 38.52x + 5822.86 = 6702$

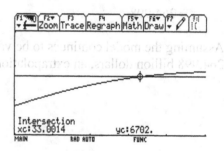

[0,50] by [5000, 8000]

Using technology, $x = 33$. Thus the world population will first reach 6702 million in the year 2023, 33 years after 1990.

Section 3.3 Skills Check

1.

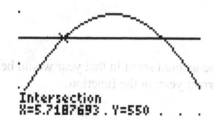

3.

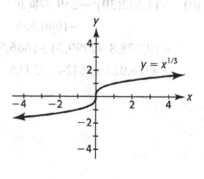

5.

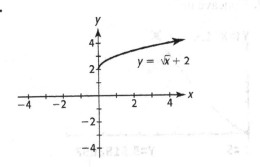

$y = \sqrt{x} + 2$

7.

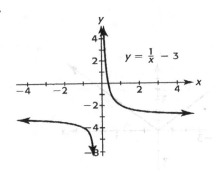

$y = \frac{1}{x} - 3$

9.

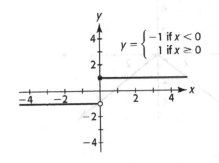

$y = \begin{cases} -1 \text{ if } x < 0 \\ 1 \text{ if } x \geq 0 \end{cases}$

11. a.

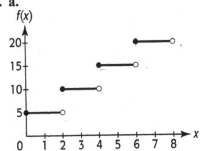

b. It is a piecewise, or step, function.

13. a.

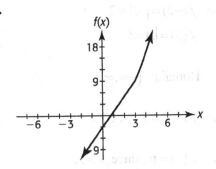

b. $f(2) = 4(2) - 3 = 8 - 3 = 5$

$f(4) = (4)^2 = 16$

c. Domain: $(-\infty, \infty)$

15. a. $f(0.2) = [\![0.2]\!] = 0$

$f(3.8) = [\![3.8]\!] = 3$

$f(-2.6) = [\![-2.6]\!] = -3$

$f(5) = [\![5]\!] = 5$

b.

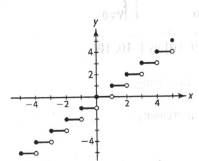

Each step should have a solid dot on its left end and an open dot on its right end.

17. a.

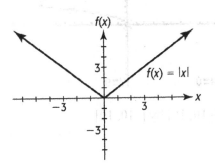

$f(x) = |x|$

b. $f(-2) = |-2| = 2$

$f(5) = |5| = 5$

c. Domain: $(-\infty, \infty)$

19. a. $f(-1) = 5$, since $x \le 1$.

b. $f(3) = 6$, since $x > 1$.

21. a. $f(-1) = (-1)^2 - 1 = 0$, since $x \le 0$.

b. $f(3) = (3)^3 + 2 = 29$, since $x > 0$.

23. The function is increasing for all values of x.

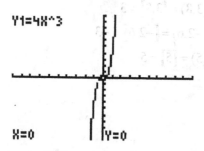

$[-10, 10]$ by $[-10, 10]$

a. increasing

b. increasing

25. Concave down.

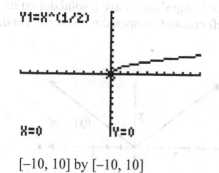

$[-10, 10]$ by $[-10, 10]$

27. Concave up.

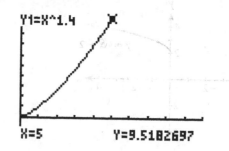

$[0, 10]$ by $[-2, 10]$

29.

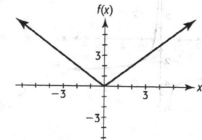

31.

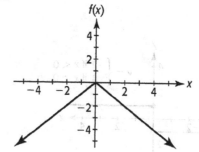

33. They are the same.

35. $\left| x - \dfrac{1}{2} \right| = 3$

$x - \dfrac{1}{2} = 3 \quad \Bigg| \quad x - \dfrac{1}{2} = -3$

$x = 3 + \dfrac{1}{2} \quad \Bigg| \quad x = -3 + \dfrac{1}{2}$

$x = \dfrac{7}{2} \quad \Bigg| \quad x = -\dfrac{5}{2}$

Using the intersections of graphs method to check the solution graphically yields,

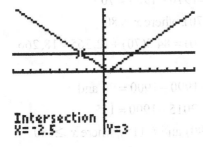

[–10, 10] by [–10, 10]

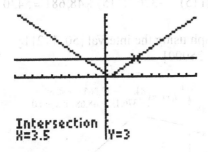

[–10, 10] by [–10, 10]

37. $\left| 3x - 1 \right| = 4x$

$3x - 1 = 4x \quad \Big| \quad 3x - 1 = -4x$

$-x = 1 \quad \Big| \quad 7x = 1$

$x = -1 \quad \Big| \quad x = \dfrac{1}{7}$

Note that –1 does not check.

$x = -1$

$\left| 3(-1) - 1 \right| = 4(-1)$

$\left| -3 - 1 \right| = -4$

$\left| -4 \right| = -4$

$4 \neq -4$

The only solution is $x = \dfrac{1}{7}$.

Using the intersections of graphs method to check the solution graphically yields,

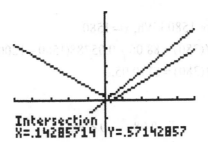

[–5, 5] by [–10, 25]

39. $S = kT^{\frac{2}{3}}$

$64 = k(64)^{\frac{2}{3}} = k(16)$

$\dfrac{64}{16} = k = 4$

$S = 4T^{\frac{2}{3}}$

$S = 4(8)^{\frac{2}{3}} = 4(4) = 16$

41. $y = \dfrac{k}{x^4}$

$5 = \dfrac{k}{(-1)^4}$

$k = 5$

$y = \dfrac{5}{(.5)^4} = 80$

Section 3.3 Exercises

43. a.

$$M(x) = \begin{cases} 7.10 + 0.06747x & x \le 1200 \\ 88.06 + 0.05788(x - 1200) & x > 1200 \end{cases}$$

b. For 960 kWh, $x = 960$

$M(960) = 7.10 + 0.06747(960)$

$M(960) = \$71.87.$

c. For 1580 kWh, $x = 1580$

$M(380) = 88.06 + 0.05788(1580 - 1200)$

$M(380) = \$110.05.$

45. a.

$$P(w) = \begin{cases} 0.98 & 0 < x \le 1 \\ 1.19 & 1 < x \le 2 \\ 1.40 & 2 < x \le 3 \\ 1.61 & 3 < x \le 4 \end{cases}$$

b. $p(1.2) = 1.19.$ The cost of mailing a 1.2-ounce letter in a large envelope is $1.19.

c. Domain: $(0, 4]$ or $0 < x \le 4.$

d. $P(2) = 1.19$

$p(2.01) = 1.40$

e. The cost of mailing a 2-ounce letter in a large envelope is $1.19, and the cost for a 2.01-ounce letter is $1.40.

47. a. Using the given function, where x is the number of years from 1900,

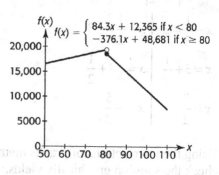

$x = 1970 - 1900 = 70$

$f(70),$ where $x < 80$

$f(70) = 84.3(70) + 12,365 = 18,266$

$x = 1990 - 1900 = 90$ and

$x = 2015 - 1900 = 115$

$f(90)$ and $f(115),$ where $x \ge 80$

$f(90) = -376.1(90) + 48,681 = 14,832$

$f(115) = -376.1(115) + 48,681 = 5430$

b. graph using the interval [50, 112] by [0, 25000]

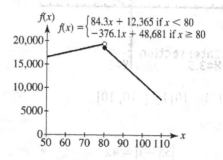

49. a. Since x is raised to a power, this is classified as a power function.

b. Since x is the number of years after 1960, $F(35)$ represents the number of female physicians 35 years after 1960, or in the year, 1995. Thus,

$F(x) = 0.623x^{1.552}$

$F(35) = 0.623(35)^{1.552} = 155.196$

Therefore, in 1995, there were 155.196 thousand (155,196) female physicians.

c. For the year 2020,

$$x = 2020 - 1960 = 60$$

$$F(60) = 0.623(60)^{1.552} = 358.244$$

According to the model, in the year 2020, there will be 358.244 thousand (358,244) female physicians.

51. a. For the year 2000,

$$t = 2000 - 1999 = 1$$

$$f(t) = 61.925t^{0.041}$$

$$f(1) = 61.925(1)^{0.041} = 61.925$$

For the year 2008,

$$t = 2008 - 1999 = 9$$

$$f(9) = 61.925(9)^{0.041} = 67.763$$

Thus in 2000, the total percent of individuals aged 16 to 24 enrolled in college as of October, was 61.925%, and in the year 2008, it was 67.763%.

b.

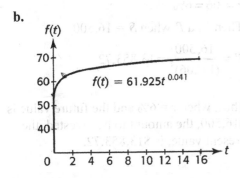

c. Yes, but very far in the future.

53. a.

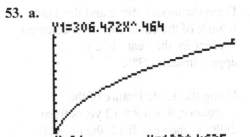

[0, 20] by [0, 1500]

According to the model, the function is increasing for the years 2000 to 2020.

b. According to the model, the function is concave down. This means that the graph is increasing at a decreasing rate.

c.

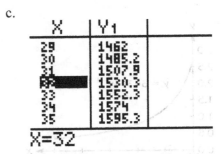

Approximately 32 years after 2000, in the year 2032, the personal consumption expenditures will reach $1532.35.

55. a.

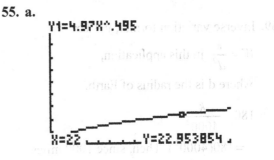

[0, 30] by [0, 100]

Although concave down, the model indicates that the percent of the United States adult population with diabetes is projected to increase.

b. From the model above and the Trace feature of the calculator, the projected percent for the year 2022 is approximately 23%.

c. Using the Table feature of the calculator, when $x = 12$ years from 2000, in the year 2012, the projection is to be 17%.

X	Y1
9	14.747
10	15.537
11	16.287
12	17.004
13	17.691
14	18.352
15	18.99

X=12

57. a.

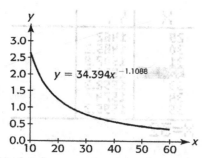

$y = 34.394x^{-1.1088}$

b. As shown in the model using the Trace feature of the calculator, the purchasing power of a 1983 dollar in 2020 is 0.367 dollars, or 37 cents.

59. Inverse variation format, $y = k/x$:

$W = \dfrac{k}{d^2}$ in this application, $d = r^2$

where d is the radius of Earth.

$180 = \dfrac{k}{4000^2}$

$k = 180(4000^2)$ Then, since 1000 miles above Earth makes $d = 5000$,

$W = \dfrac{180(4000^2)}{(5000)^2} = 115.2$

Thus, a 180-pound man would weigh 115.2 pounds 1000 miles above the surface of Earth.

61. Direct variation format, $y = kx$:

$S = k(1+r)^3$

$6298.56 = k(1+.08)^3$

$6298.56 = k(1.259712)$

$k = 5000$

Then find r, given

$5955.08 = 5000(1+r)^3$

$(1+r)^3 = \dfrac{5955.08}{5000} = 1.191016$

$1+r = \sqrt[3]{1.191016} = 1.06$

$r = 1.06 - 1 = .06 = 6\%$

63. Inverse variation format, $y = k/x$:

$P = \dfrac{S}{(1+r)^3}$

$8396.19 = \dfrac{10000}{(1+r)^3}$

$(1+r)^3 = \dfrac{10000}{8396.19} = 1.191016401$

$(1+r) = \sqrt[3]{1.191016401} = 1.06$

$r = .06 = 6\%$

Then find P when $S = 16,500$

$P = \dfrac{16,500}{(1+.06)^3} = 13,853.72$

Thus, when r = 6% and the future value is $16,500, the amount to be invested, the present value, is $13,853.72.

Section 3.4 Skills Check

1. Given the quadratic model: $y = ax^2 + bx + c$, the following system of equations can be determined from the given points and then solved for a, b, and c:

$1 = a(0)^2 + b(0) + c = 0a + 0b + c$

$10 = a(3)^2 + b(3) + c = 9a + 3b + c$

$15 = a(-2)^2 + b(-2) + c = 4a - 2b + c$

so that $a = 2, b = -3, c = 1$ and the quadratic equation is $y = 2x^2 - 3x + 1$.

3. Given the quadratic model: $y = ax^2 + bx + c$, the following system of equations can be determined from the given points and then solved for a, b, and c:

$30 = a(6)^2 + b(6) + c = 36a + 6b + c$

$-3 = a(0)^2 + b(0) + c = 0a + 0b + c$

$7.5 = a(-3)^2 + b(-3) + c = 9a - 3b + c$

so that $a = 1, b = -0.5, c = -3$ and the quadratic equation is $y = x^2 - 0.5x - 3$.

5. Given the quadratic model: $y = ax^2 + bx + c$, the following system of equations can be determined from the given points and then solved for a, b, and c:

$6 = a(0)^2 + b(0) + c = 0a + 0b + c$

$\dfrac{22}{3} = a(2)^2 + b(2) + c = 4a + 2b + c$

$\dfrac{99}{2} = a(-9)^2 + b(-9) + c = 81a - 9b + c$

so that $a = 1/2, b = -1/3, c = +6$ and the quadratic equation is $y = \dfrac{x^2}{2} - \dfrac{x}{3} + 6$.

7. Three points of this function, a parabola, can be located at (0, 48), its starting point with t = 0, at (1, 64), the height of 64 feet after 1 second, and at (3, 0), the height at ground level after 3 seconds. Solving the system of three equations as in problems 1 through 6 produces the quadratic equation:

$h = -16t^2 + 32t + 48$.

9. Given the quadratic model: $y = ax^2 + bx + c$, the following system of equations can be determined from the given points and then solved for a, b, and c:

$6 = a(-1)^2 + b(-1) + c = 1a - 1b + c$

$3 = a(2)^2 + b(2) + c = 4a + 2b + c$

$10 = a(3)^2 + b(3) + c = 9a + 3b + c$

so that $a = 2, b = -3, c = 1$ and the quadratic equation is $y = 2x^2 - 3x + 1$.

11. Choose any three points from the table and solve using the technique in the first 10 exercises, producing the quadratic equation:

$y = 3x^2 - 2x$.

13. The x-values are not equally spaced.

15.

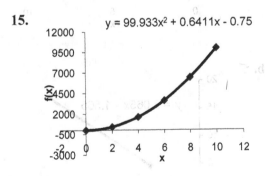

$y = 99.933x^2 + 0.6411x - 0.75$

17. a.

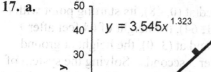

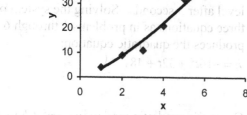

$y = 3.545x^{1.323}$

b.

$y = 8.114x - 8.067$

c.   The power function is a better fit.

19. a.

$y = 1.292x^{1.178}$

b.

$y = 2.065x - 1.565$

c.   Both models appear to be good fits.

21.

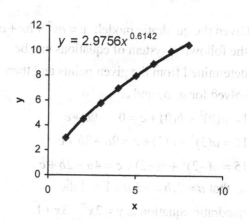

$y = 2.9756x^{0.6142}$

Section 3.4 Exercises

23. a. $y = 0.221x^2 - 0.381x + 4.180$

b.

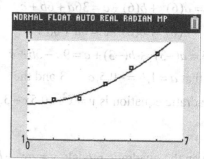

c.   In 2018, $x = 2018 - 2010 = 8$

$y = 0.221(8)^2 - 0.381(8) + 4.180$

$y = 15.276$

$y \approx \$15.3$ billion

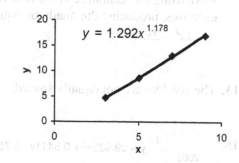

25. a. $y = 9.286x^2 - 147.619x + 605.595$

b.

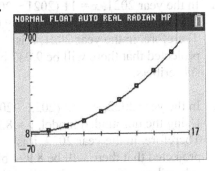

The model appears to fit the data points very well.

c. In 2018, $x = 2018 - 2000 = 18$

$y = 9.286(18)^2 - 147.619(18) + 605.595$

$y = 957.117$

$y \approx \$957$ billion

27. a. $y = 0.0052x^2 - 0.62x + 15.0$

b.

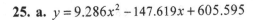

Using the Table feature of the calculator, at 50 mph, the wind chill will be minus 3 degrees.

c.

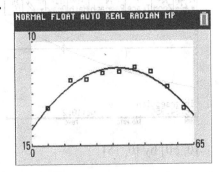

No, it appears the model predicts the wind chill temperature will be a minimum near 59.6 mph, and it is illogical that it would be warmer for winds greater than 60 mph.

29. a. $y = -0.00834x^2 + 0.690x - 6.81$

b.

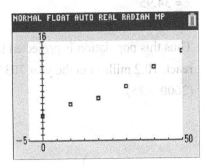

c. Using the maximum function on a calculator, the maximum occurs at age 42 (when $x = 41.38$).

31. a.

b. $y = 0.002107x^2 + 0.09779x + 4.207$

c.

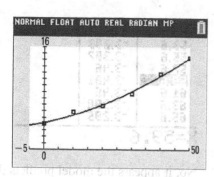

d. In the year 2056, $x = 56$ (2056 − 2000). Using the unrounded model, $y = 16.3$. Therefore, in the year 2056, it is projected that 16.3 million Americans will have an Alzheimer's or other dementia disease.

e.

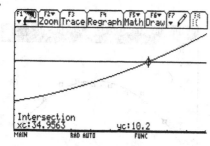

[0, 50] by [0, 15]

$y = 0.002107x^2 + 0.09779x + 4.207$

$10.2 = 0.002107x^2 + 0.09779x + 4.207$

$0 = 0.002107x^2 + 0.09779x - 5.993$

$x = 34.95$

$x \approx 35$

Thus this population is projected to reach 10.2 million in the year 2035 (2000 + 35).

33. a. $Q(t) = -0.032t^2 + 0.682t + 5.41$

b. In the year 2021, $x = 11$ (2021 − 2010). Using the unrounded model, $y = 9.047$. Therefore, in the year 2021, it is projected that there will be 9.047 billion subscribers.

c. In the year 2022, $x = 12$ (2022 − 2010). Using the unrounded model, $y = 8.994$. Therefore, in the year 2022, it is projected that there will be 8.994 billion subscribers. These results tell us that the number of subscribers is decreasing.

d. Using the maximum function on a calculator, the maximum occurs in the year 2021 (when $x = 10.7$).

e. No; the model is not a good predictor because the number of subscribers should continue to increase.

35. a. $y = -0.224x^2 + 22.005x + 370.705$

b. Using the maximum function on a calculator, the maximum occurs at the point (49.11, 911). Therefore, the maximum population is projected to be 911 million in the year 2020 (1970 + 50)

37. a. $y = 0.068x^2 + 1.746x + 96.183$

b.

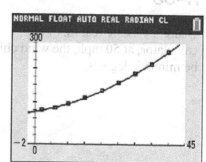

c. In the year 2022, $x = 12$ (2022 − 2010). Using the unrounded model, $y = 126.98$. Therefore, in the year 2022, it is projected that the CPI will be 126.98.

d.

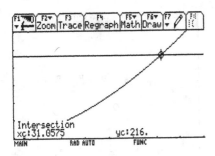

[0, 40] by [100, 250]

$y = 0.068x^2 + 1.746x + 96.183$

$216 = 0.068x^2 + 1.746x + 96.183$

$0 = 0.068x^2 + 1.746x - 119.817$

$x = 31.05$

$x \approx 31$

Thus the CPI is projected to reach 216 in the year 2041 (2010 + 31).

39. a.

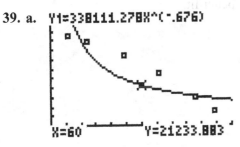

[0, 120] by [−5000, 60,000]

b.

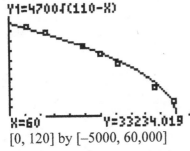

[0, 120] by [−5000, 60,000]

Equation b) fits the data much better.

41. a. $y = 0.00785x^2 - 1.22x + 66.792$

b.

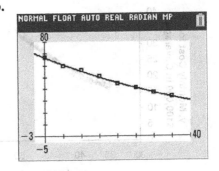

c. Using the minimum function on a calculator, the minimum occurs when $x = 77.6$ or in the year 2078 (2000 + 78).

43. a.

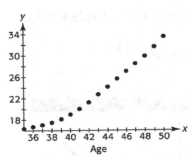

b. Yes. It appears that a quadratic function will fit the data.

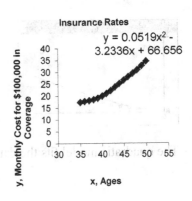

c.

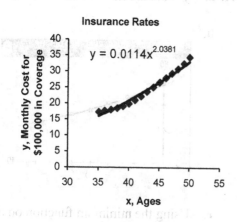

Insurance Rates

$y = 0.0114x^{2.0381}$

x, Ages

d. It appears that the quadratic model, based on the scatter plots, is the best fit.

45. a. $y = 0.116x^2 - 3.792x + 45.330$

b. $y = 0.711x^{1.342}$

c.

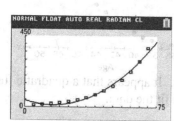

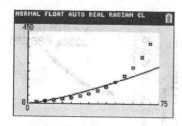

The quadratic model fits the data better.

47. a.

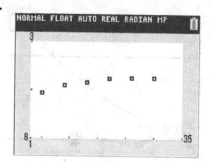

b. $y = 1.428x^{1.408}$

c. $y = -0.001x^2 + 0.058x + 1.473$

d. In 2040, $x = 40$, and the unrounded power model estimate of the number of barrels of crude oil is 2.4 billion.

e. In 2040, $x = 40$, and the unrounded quadratic model estimate of the number of barrels of crude oil is 2.1 billion.

f. Based on the data and the answers to parts d and e, he quadratic model is the better fit.

Chapter 3 Skills Check

1. a. The vertex is at $(5, 3)$.

b.

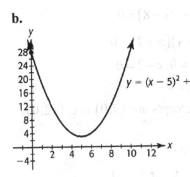

3. a. The vertex is at $(1, -27)$.

b.

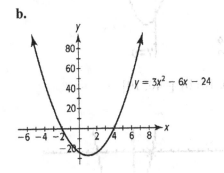

$$f(x) = 3x^2 - 6x - 24$$

$$x = \frac{-b}{2a} = \frac{-(-6)}{2(3)} = \frac{6}{6} = 1$$

$$y = f(x) = f(1)$$

$$f(1) = 3(1)^2 - 6(1) - 24$$

$$f(1) = 3 - 6 - 24 = -27$$

The vertex is $(1, -27)$.

5. a. The vertex is at $(15, 80)$.

b.

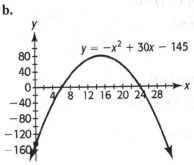

$$f(x) = -x^2 + 30x - 145$$

$$x = \frac{-b}{2a} = \frac{-(30)}{2(-1)} = \frac{-30}{-2} = 15$$

$$y = f(x) = f(15)$$

$$f(15) = -(15)^2 + 30(15) - 145$$

$$f(15) = -225 + 450 - 145$$

The vertex is $(15, 80)$.

7. a. The vertex is at $(0.05, -60.0005)$. Use the procedure in #3 to find the vertex point.

b.

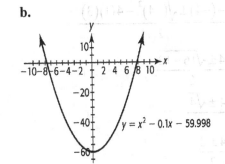

9. $x^2 - 5x + 4 = 0$

$(x-4)(x-1) = 0$

$x = 4, x = 1$

11. $5x^2 - x - 4 = 0$

$(x-1)(5x+4) = 0$

$x = 1, x = -\dfrac{4}{5}$

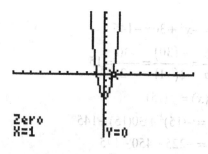

Zero
X=1   Y=0

$[-10, 10]$ by $[-10, 10]$

13. $x^2 - 4x + 3 = 0$

$a = 1, b = -4, c = 3$

$x = \dfrac{-b \pm \sqrt{b^2 - 4ac}}{2a}$

$x = \dfrac{-(-4) \pm \sqrt{(-4)^2 - 4(1)(3)}}{2(1)}$

$x = \dfrac{4 \pm \sqrt{16 - 12}}{2}$

$x = \dfrac{4 \pm \sqrt{4}}{2}$

$x = \dfrac{4 \pm 2}{2}$

$x = 3, x = 1$

15. a. Algebraically:

Let $f(x) = 0$

$3x^2 - 6x - 24 = 0$

$3(x^2 - 2x - 8) = 0$

$3(x-4)(x+2) = 0$

$x - 4 = 0, x + 2 = 0$

$x = 4, x = -2$

The x-intercepts are $(4, 0)$ and $(-2, 0)$.

Graphically:

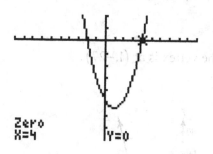

Zero
X=4   Y=0

$[-10, 10]$ by $[-40, 10]$

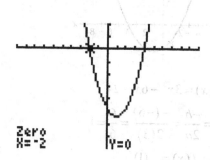

Zero
X=-2   Y=0

$[-10, 10]$ by $[-40, 10]$

Again, the x-intercepts are $(4, 0)$ and $(-2, 0)$.

b. Solving $f(x) = 0$ produces the x-intercepts, as shown in part a). The x-intercepts are $(4, 0)$ and $(-2, 0)$.

17. $5x^2 - 20 = 0$

$5x^2 = 20$

$x^2 = 4$

$x = \pm 2$

19. $z^2 - 4z + 6 = 0$

$a = 1, b = -4, c = 6$

$z = \dfrac{-b \pm \sqrt{b^2 - 4ac}}{2a}$

$z = \dfrac{-(-4) \pm \sqrt{(-4)^2 - 4(1)(6)}}{2(1)}$

$z = \dfrac{4 \pm \sqrt{16 - 24}}{2}$

$z = \dfrac{4 \pm \sqrt{-8}}{2}$

$z = \dfrac{4 \pm 2i\sqrt{2}}{2}$

$z = 2 \pm i\sqrt{2}$

21. $4x^2 - 5x + 3 = 0$

$a = 4, b = -5, c = 3$

$x = \dfrac{-b \pm \sqrt{b^2 - 4ac}}{2a}$

$x = \dfrac{-(-5) \pm \sqrt{(-5)^2 - 4(4)(3)}}{2(4)}$

$x = \dfrac{5 \pm \sqrt{25 - 48}}{8}$

$x = \dfrac{5 \pm \sqrt{-23}}{8}$

$x = \dfrac{5 \pm i\sqrt{23}}{8}$

23.

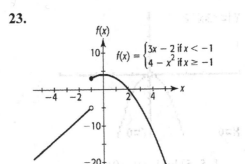

25.

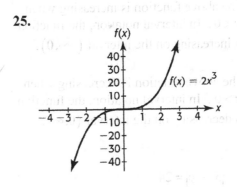

27.

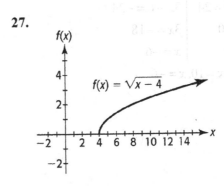

29.

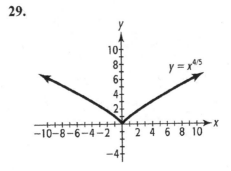

31.   Y1=-3X^2

[-5, 5] by [-10, 10]

X=0   Y=0

a. The above function is increasing when $x < 0$. In interval notation, the function is increasing on the interval $(-\infty, 0)$.

b. The above function is decreasing when $x > 0$. In interval notation, the function is decreasing on the interval $(0, \infty)$.

33. $|3x - 6| = 24$

$3x - 6 = 24$ | $3x - 6 = -24$
$3x = 30$ | $3x = -18$
$x = 10$ | $x = -6$

$x = 10, x = -6$

35.

Given the quadratic model: $y = ax^2 + bx + c$, the following system of equations can be determined from the given points and then solved for a, b, and c:

$-2 = a(0)^2 + b(0) + c = 0a + 0b + c$

$12 = a(-2)^2 + b(-2) + c = 4a - 2b + c$

$7 = a(3)^2 + b(3) + c = 9a + 3b + c$

so that $a = 2, b = -3, c = -2$ and the quadratic equation is $y = 2x^2 - 3x - 2$.

37.

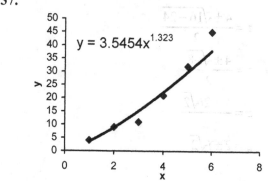

$y = 3.5454x^{1.323}$

39. $q = kp^{\frac{3}{2}}$

$16 = k(4)^{\frac{3}{2}} = k(8)$

$\dfrac{16}{8} = k = 2$

$q = 2p^{\frac{3}{2}}$

$q = 2(16)^{\frac{3}{2}} = 2(64) = 128$

Chapter 3 Review Exercises

41. a. The maximum profit occurs at the vertex.

$$P(x) = -0.01x^2 + 62x - 12,000$$

$$x = \frac{-b}{2a} = \frac{-62}{2(-0.01)} = \frac{-62}{-0.02} = 3100$$

$$P(3100) = -0.01(3100)^2 + 62(3100) - 12,000$$

$$P(3100) = -96,100 + 192,200 - 12,000$$

$$P(3100) = 84,100$$

The vertex is $(3100,\ 84,100)$.

Producing and selling 3100 units produces the maximum profit of $84,100.

b. The maximum possible profit is $84,100.

42. a.

$$P(x) = R(x) - C(x)$$

$$= (200x - 0.01x^2) - (38x + 0.01x^2 + 16,000)$$

$$= 200x - 0.01x^2 - 38x - 0.01x^2 - 16,000$$

$$= -0.02x^2 + 162x - 16,000$$

$$x = \frac{-b}{2a} = \frac{-162}{2(-0.02)} = \frac{-162}{-0.04} = 4050$$

$$P(4050) = -0.02(4050)^2 + 162(4050)$$
$$-16,000$$

$$P(4050) = -328,050 + 656,100 - 16,000$$

$$P(4050) = 312,050$$

The vertex is $(4050,\ 312,050)$.

Producing and selling 4050 units gives the maximum profit of $312,050.

b. The maximum possible profit is $312,050.

43. a. $h = \dfrac{-b}{2a} = \dfrac{-64}{2(-16)} = 2$. The ball reaches its maximum height in 2 seconds.

b. $S(2) = 192 + 64(2) - 16(2)^2$
$$= 256$$
The ball reaches its maximum height at 256 feet.

44. a. $h = \dfrac{-b}{2a} = \dfrac{-29.4}{2(-9.8)} = 1.5$.
The ball reaches its maximum height in 1.5 seconds.

b. $S(1.5) = 60 + 29.4(1.5) - 9.8(1.5)^2$
$$= 82.05$$
The ball reaches its maximum height at 82.05 meters.

45. a. $y = -1.48x^2 + 38.901x - 118.429$

$$x = \frac{-b}{2a} = \frac{-38.901}{2(-1.48)} = \frac{-38.901}{-2.96} = 13.1$$

The year is 13.1 after 1990 = 2004.

b. $y = -1.48x^2 + 38.901x - 118.429$

$$y = -1.48(13.1)^2 + 38.901(13.1)$$
$$-118.429$$

$$y = -253.9828 + 509.6031 - 118.429$$

$$y = 137.1913 = 137 \text{ thousand visas.}$$

c.   In 2009 (1990 + 19), the number of visas will be 100 thousand.

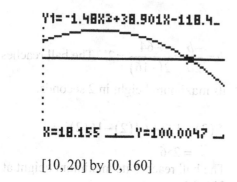

Y1=-1.48X²+38.901X-118.4_

X=18.155   Y=100.0047

[10, 20] by [0, 160]

46. $3600 - 150x + x^2 = 0$

$$x^2 - 150x + 3600 = 0$$

$$(x - 30)(x - 120) = 0$$

$$x = 30, x = 120$$

Break-even occurs when 30 or 120 units are produced and sold.

47.   The ball is on the ground when $S = 0$.

$$400 - 16t^2 = 0$$

$$-16t^2 = -400$$

$$t^2 = 25$$

$$t = \pm 5$$

In the physical context of the question, $t \geq 0$. The ball reaches the ground in 5 seconds.

48.

$$-0.3x^2 + 1230x - 120,000 = 324,000$$

$$-0.3x^2 + 1230x - 444,000 = 0$$

$$x = \frac{-b \pm \sqrt{b^2 - 4ac}}{2a}$$

$$= \frac{-(1230) \pm \sqrt{(1230)^2 - 4(-0.3)(-444,000)}}{2(-0.3)}$$

$$= \frac{-1230 \pm \sqrt{1,512,900 - 532,800}}{-0.6}$$

$$= \frac{-1230 \pm \sqrt{980,100}}{-0.6}$$

$$= \frac{-1230 \pm 990}{-0.6}$$

$$x = 400, x = 3700$$

A profit of $324,000 occurs when 400 or 3700 units are produced and sold.

49. a.

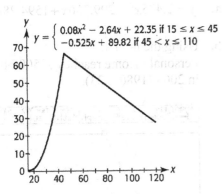

$$y = \begin{cases} 0.08x^2 - 2.64x + 22.35 & \text{if } 15 \le x \le 45 \\ -0.525x + 89.82 & \text{if } 45 < x \le 110 \end{cases}$$

b.

Since 1990 is 90 years after 1900, $x = 90$. Since 90 fits the second piece of the function, substitute 90 into:

$T(x) = -0.525x + 89.82$

$T(90) = -0.525(90) + 89.82 = 42.57$

Thus, the tax rate for a millionaire head of household in 1990 was 42.57%.

c.

Since 2010 is 110 years after 1900, $x = 110$. Since 110 fits the second piece of the function, substitute 110 into:

$T(x) = -0.525x + 89.82$

$T(110) = -0.525(110) + 89.82 = 32.07$

Thus, the tax rate for a millionaire head of household in 2010 was 32.07%, so the model is fairly good.

50. a. $y = -0.00482x^2 + 0.754x + 8.512$

b.   In 2022, $x = 2022 - 2000 = 22$

$y = -0.00482(22)^2 + 0.754(22)$
$+8.512 = 22.767 \approx 22.8\%$

c.

X	Y1
35	28.998
36	29.409
37	29.811
38	30.204
39	30.587
40	30.96
41	31.324

X=38

Using the table feature of the calculator, the model predicts the percent to 30.2% in the year 2038.

51. a.

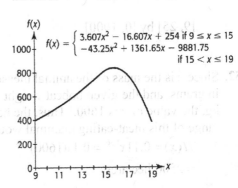

$$f(x) = \begin{cases} 3.607x^2 - 16.607x + 254 & \text{if } 9 \le x \le 15 \\ -43.25x^2 + 1361.65x - 9881.75 \\ & \text{if } 15 < x \le 19 \end{cases}$$

b.   Using $x = 13$ for the year 2003, the number of deaths in that year were approximately 648.

Y1=3.607X²-16.607X+254/(_

X=13 _____ Y=647.692 ___

c.   Using $x = 19$ for the year 2009, the number of deaths in that year were approximately 376.

Y2=-43.25X²+1361.65X-98_

X=19 _____ Y=376.35 ___

d. Extending the graphing window to $x = 25$, and using the Intersect feature of the calculator, the model predicts the number of deaths to be 200, when $x = 20$, in the year 2010.

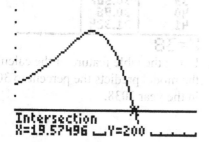

Intersection
X=19.57496 Y=200

[9, 25] by [0, 1000]

52. Since x is the mass of the animal measured in grams, and the given bobcat weight is 1.6 kg, the value of x is 1600. Thus, the home range of this meat-eating mammal would be:
$$H(x) = 0.11x^{1.36} = 0.11(1600)^{1.36}$$
$$= 2506 \text{ hectares}$$

53. a. $y = 0.065x^2 + 1.561x + 6.689$

b.

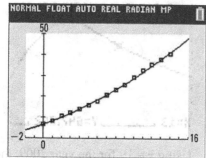

c. Yes. The function seems to fit the data very well.

d. Since the percentage is part of the range, $y = 55$.
$$y = 0.065x^2 + 1.561x + 6.689$$
$$55 = 0.065x^2 + 1.561x + 6.689$$
$$x = 17.78, \text{ and the year will be 2018.}$$

54. a. $y = 2.475x^2 + 299.256x + 1594.282$

b. Using the model:
Personal income reaches $9500 billion in 2003 (1980 + 23).

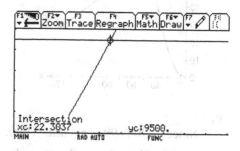

Intersection
xc: 22.3037   yc: 9500.

[0, 50] by [5000, 10000]

c. The personal income level in 2009 was $12,026 billion. Twice that amount is $24,052 billion. Using the model, the personal income level doubles during 2033 (1980 + 53).

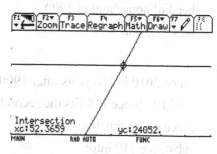

Intersection
xc: 52.3659   yc: 24052.

[30, 70] by [20000, 26000]

55. a. $y = 15.8425x^{0.5176}$

b.

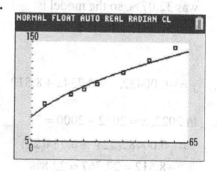

Yes. The function seems to fit the data very well.

c. Since 2033 − 1990 = 43, x = 43.

$$y = 15.8425(43)^{0.5176}$$

$$y = 15.8425(7.0062)$$

$$y \approx 111$$

Therefore, in 2033, the number of individuals in the U.S. civilian non-institutional population 16 years and older who are non-White or Hispanic is projected to be 111 million.

56. a.

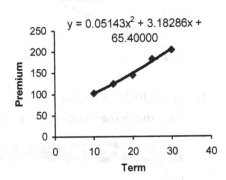

Insurance Premiums

y = 0.05143x² + 3.18286x + 65.40000

b. Applying the intersection of graphs method using unrounded model and y = 130

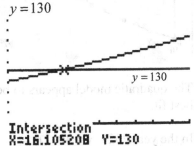

[10, 30] by [−50, 250]

A premium of $130 would purchase a term of 16 years.

57. a.

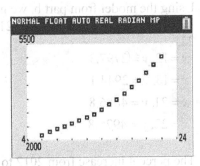

b. The quadratic equation would be the most appropriate:

$$y = 7.897x^2 - 50.650x + 2267.916$$

c.

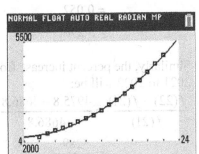

d. Using the model from part b, we get the following values for x = 6, 12, and 22.

x = 6, y = 2248.3

x = 12, y = 2797.3

x = 22, y = 4975.8

The average rate of change from 2006 to 2012 will be:

$$\frac{f(12) - f(6)}{12 - 6} = \frac{2797.3 - 2248.3}{12 - 6}$$

$$= \frac{549}{6}$$

$$= 91.5$$

Similarly, the average rate of change from 2012 to 2022 will be:

$$\frac{f(22) - f(12)}{22 - 12} = \frac{4975.76 - 2797.28}{22 - 12}$$

$$= \frac{2178.46}{10}$$

$$= 217.8$$

e. Using the model from part b, we get the following values for $x = 12, 13, 21$, and 22.

$x = 12, y = 2797.3$

$x = 13, y = 2944.1$

$x = 21, y = 4686.8$

$x = 22, y = 4975.8$

The percent increase from 2012 to 2013 will be:

$$\frac{f(13) - f(12)}{f(12)} = \frac{2944.1 - 2797.3}{2797.3}$$

$$= \frac{146.8}{2797.3}$$

$$= 0.052$$

$$= 5.2\%$$

Similarly, the percent increase from 2021 to 2022 will be:

$$\frac{f(22) - f(21)}{f(21)} = \frac{4975.8 - 4686.8}{4686.8}$$

$$= \frac{289}{4686.8}$$

$$= 0.062$$

$$= 6.2\%$$

Yes, the statement is reasonable because the percent starts below 5.8% and finishes above the 5.8%.

58. a. $y = 749.95x^{0.167}$

b. The graph is concave down.

c.

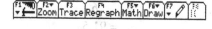

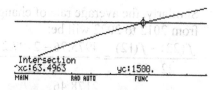

[30, 80] by [1300, 1700]

In 2043, 63 years after 1980, the Hawaiian population will rise to 1.5 million people.

59. a. $y = 2.095x + 6.265$, good fit to the data.

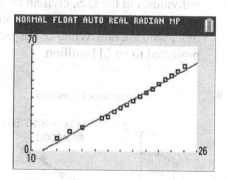

b. $y = 0.037x^2 + 1.026x + 12.644$, better fit than the linear function to the data.

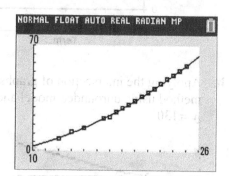

c. The quadratic model appears to be the best fit.

d. In the year 2020, $x = 30$. Thus,

$y = 2.095(30) + 6.265 = 69.115 = 69.1$

$y = 0.037(30)^2 + 1.026(30) + 12.644$

$= 76.724 = 76.7$

Each model predicts the amount of money Americans will spend on their pets in 2020 to be, 69.1 billion dollars and 76.7 billion dollars, respectively.

60. a.

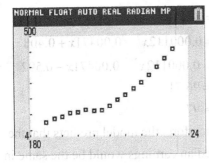

b. $y = -0.99x^2 + 24.27x + 114.11$

c. $y = 0.93x^2 - 11.55x + 256.47$

d.

$$y = \begin{cases} -0.99x^2 + 24.27x + 114.11 & 6 \le x \le 13 \\ 0.93x^2 - 11.55x + 256.47 & 13 < x \le 22 \end{cases}$$

e.   For 2010, $x = 10$ (2010 – 2000) and for 2022, $x = 22$. Therefore,

$y = -0.99x^2 + 24.27x + 114.11$

$y = -0.99(10)^2 + 24.27(10) + 114.11$

$y = -99 + 242.7 + 114.11$

$y = 257.81$

$y = 0.93x^2 - 11.55x + 256.47$

$y = 0.93(22)^2 - 11.55(22) + 256.47$

$y = 450.12 - 254.1 + 256.47$

$y = 452.49$

Therefore, the prescription drug expenditures for 2010 and 2022 will be $257.8 billion and $452.5 billion, respectively.

Group Activity/Extended Application II

1. $f(x) = 0.977x + 37.898$

2. $g(x) = -0.0102x^2 + 2.240x + 1.642$

3.

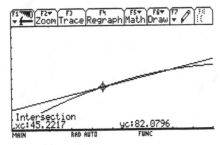

[0, 100] by [0, 100]

From the intersection of these graphs, in the year 1986 (1940 + 46) the number of female employees reached the number of male employees.

4.

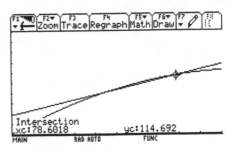

[0, 100] by [0, 100]

From the intersection of these graphs, in the year 2019 (1940 + 79) the number of male employees will again exceed the number of female employees.

5.

Year	Female-to-Male Earnings Ratio
1948	0.421
1950	0.371
1960	0.309
1970	0.335
1980	0.393
1990	0.496
2000	0.567
2010	0.645
2012	0.635
2013	0.626

6. $y = 0.000112x^2 - 0.00471x + 0.408$

7.

$y = 0.000112x^2 - 0.00471x + 0.408$

$1 = 0.000112x^2 - 0.00471x + 0.408$

$0 = 0.000112x^2 - 0.00471x - 0.592$

$x = 96.71$

$x \approx 97$

Therefore, the model predicts that the median earnings would be the same in the year 2037 (1940 + 97).

8. Answers may vary.

Chapter 4
Additional Topics with Functions

Toolbox Exercises

1. Since the reciprocal function has the form, $f(x) = \dfrac{1}{x}$, and since x cannot $= 0$, the domain of the reciprocal function is $(-\infty, 0) \cup (0, \infty)$. In this function, $f(x)$ also cannot $= 0$, so the range of the reciprocal function is also $(-\infty, 0) \cup (0, \infty)$.

2. Since the constant function has the form, $g(x) = k$, there is no restriction on the value of x while the value of $g(x)$ can only equal whatever the value of k is. Thus the domain of the constant function is $(-\infty, \infty)$ and the range is $\{k\}$.

3. As shown in the model, there is no interval where the reciprocal function increases, so the function decreases on its entire domain, $(-\infty, 0) \cup (0, \infty)$.

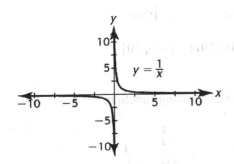

4. As shown in the model, the absolute value function increases on the interval $(0, \infty)$, and it decreases on the interval $(-\infty, 0)$.

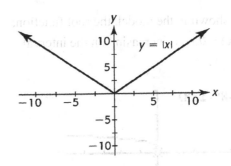

5. As shown in the model, the range of the squaring function is $[0, \infty)$.

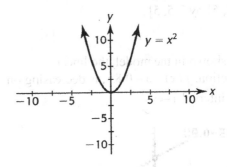

6. As shown in the model above, the domain of the squaring function is $(-\infty, \infty)$.

7. As shown in the model, the root function, $g(x) = \sqrt[5]{x}$, is increasing on the interval $(-\infty, \infty)$.

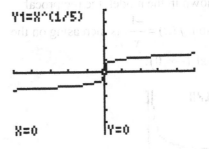

$[-5, 5]$ by $[-5, 5]$

8. As shown in the model, the root function, $h(x) = \sqrt[4]{x}$, is increasing on the interval $[0, \infty)$.

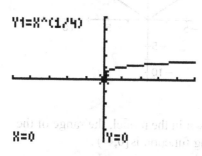

[–5, 5] by [–5, 5]

9. As shown in the model, the linear function, $f(x) = 5 - 0.8x$, is decreasing on the interval $(-\infty, \infty)$.

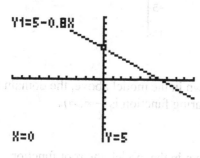

[–10, 10] by [–10, 10]

10. As shown in the model, the reciprocal function, $f(x) = \dfrac{-1}{x}$, is increasing on the interval $(-\infty, 0)$.

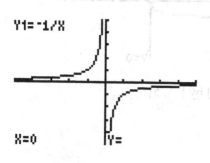

[–5, 5] by [–5, 5]

11. As shown in the model, the power function, $g(x) = -2x^{\frac{2}{3}}$, is decreasing on the interval $(0, \infty)$.

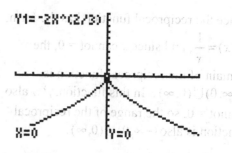

[–10, 10] by [–10, 10]

12. As shown in the model, the power function, $h(x) = 3x^{\frac{1}{6}}$, is increasing on the interval $[0, \infty)$.

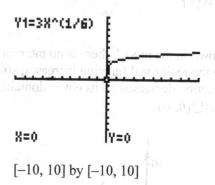

[–10, 10] by [–10, 10]

13. This is a cubing function.

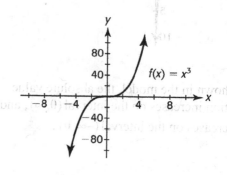

14. This is a square root function.

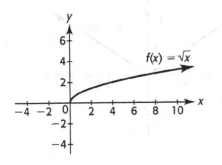

15. This is a power function, with power 1/6.

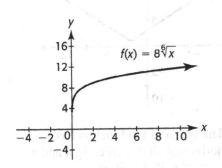

16. This is a power function, with power -3.

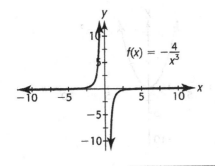

Section 4.1 Skills Check

1. a.

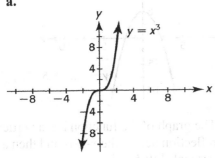

17. No, the graph would fail the vertical line test.

18. No, the graph would fail the vertical line test.

19. Yes, this is a function since it passes the vertical line test.

20. Yes, this is a function since it passes the vertical line test.

21. Yes, this is a function since it passes the vertical line test.

22. Yes, this is a function since it passes the vertical line test.

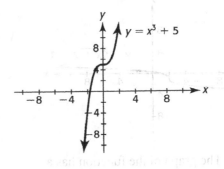

b. The graph of the function has a vertical shift 5 units up.

3. a.

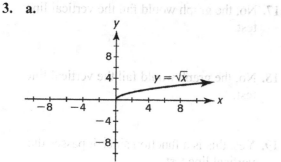

7. a.

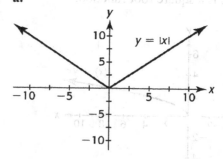

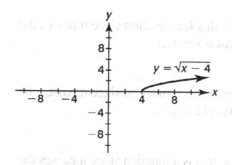

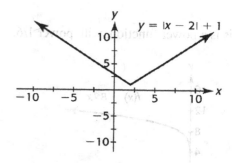

b. The graph of the function has a horizontal shift 4 units right.

b. The graph of the function has a horizontal shift 2 units right and a vertical shift 1 unit up.

5. a.

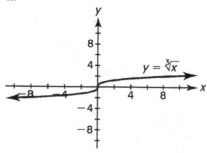

9. a.

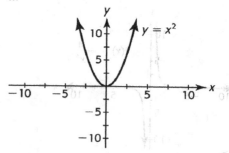

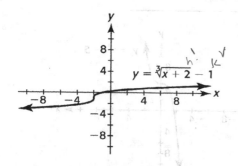

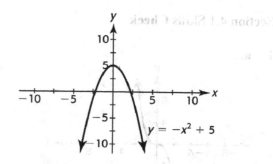

b. The graph of the function has a horizontal shift 2 units left and a vertical shift 1 unit down.

b. The graph of the function has a vertical reflection across the x-axis and then a vertical shift 5 units up.

11. a.

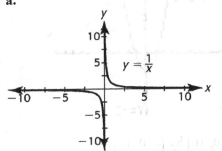

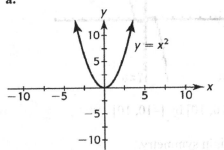

b. The graph of the function has a vertical shift 3 units down.

13. a.

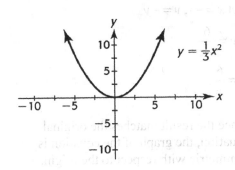

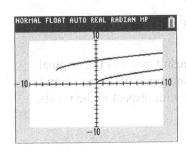

b. The graph of the function has a vertical compression by a factor of 1/3.

15. a.

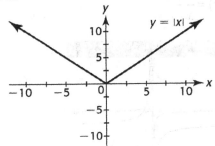

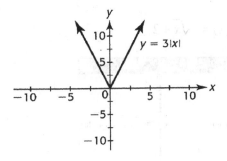

b. The graph of the function has a vertical stretch by a factor of 3.

17. The graph of the function is shifted 2 units right and 3 units up.

19. $y = (x+4)^{\frac{3}{2}}$

21. $y = 3x^{\frac{3}{2}} + 5$

23. $f(x) = \sqrt{x+6} + 3$

25. $f(x) = \dfrac{4}{x} - 3$

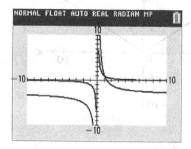

27. $f(x) = -\sqrt{x} + 2$

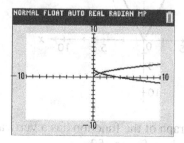

29. $g(x) = -x^2 + 2$

31. $g(x) = |x+3| - 2$

33. The graph has y-axis symmetry and is an even function.

35. The graph has x-axis symmetry it is not a function.

37. Yes, the graph would have y-axis symmetry.

39. y-axis symmetry.

Let $x = -x$

$$y = 2(-x)^2 - 3 = 2x^2 - 3$$

Since the result matches the original equation, the graph of the equation is symmetric with respect to the y-axis.

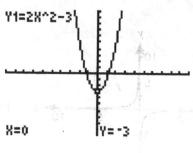

$[-10, 10]$ by $[-10, 10]$

41. Origin symmetry.

Let $x = -x, y = -y$

$$-y = (-x)^3 - (-x)$$
$$-y = -x^3 + x$$
$$y = x^3 - x$$

Since the result matches the original equation, the graph of the equation is symmetric with respect to the origin.

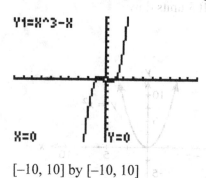

$[-10, 10]$ by $[-10, 10]$

43. Origin symmetry.

Let $x = -x, y = -y$

$$-y = \dfrac{6}{-x}$$
$$y = \dfrac{6}{x}$$

Since the result matches the original equation, the graph of the equation is symmetric with respect to the origin.

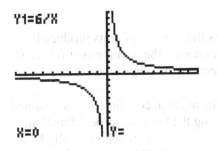

Y1=6/X

X=0   Y=

[−10, 10] by [−10, 10]

Y1=√(25−X^2)

X=0   Y=5

[−10, 10] by [−10, 10]

45. x-axis symmetry.

Let $y = -y$

$$x^2 + (-y)^2 = 25$$

$$x^2 + y^2 = 25$$

Since the result matches the original equation, the graph of the equation is symmetric with respect to the x-axis.

y-axis symmetry.

Let $x = -x$

$$(-x)^2 + y^2 = 25$$

$$x^2 + y^2 = 25$$

Since the result matches the original equation, the graph of the equation is symmetric with respect to the y-axis.

origin symmetry.

Let $x = -x, y = -y$

$$(-x)^2 + (-y)^2 = 25$$

$$x^2 + y^2 = 25$$

Since the result matches the original equation, the graph of the equation is symmetric with respect to the origin.

Since the given equation is not a function, it can not be easily graphed using the graphing calculator. The equation must be rewritten

as $y = \pm\sqrt{25 - x^2}$

47. $f(-x) = |-x| - 5$

$$= |-1(x)| - 5$$

$$= |-1||x| - 5$$

$$= |x| - 5$$

$$= f(x)$$

Since $f(-x) = f(x)$, the function has y-axis symmetry and is even.

49. $g(-x) = \sqrt{(-x)^2 + 3}$

$$= \sqrt{x^2 + 3}$$

Since $g(-x) = g(x)$, the function has y-axis symmetry and is even.

51. $g(-x) = \dfrac{5}{-x}$

$$= -\dfrac{5}{x}$$

$$= -\left(\dfrac{5}{x}\right)$$

Since $g(-x) = -g(x)$, the function has origin symmetry and is odd.

53. The graph has y-axis symmetry and is therefore even.

55. Even though this equation's graph would be a parabola opening to the side and it would not be a function, it is symmetric about the line $y = -1$.

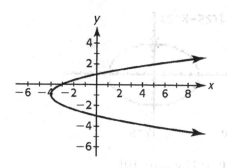

Section 4.1 Exercises

57. a. This function is a shifted graph of
$M(x) = x^2$ $\left(y = x^2\right)$

b. $M(3) = -0.062(3-4.8)^2 + 25.4$

$= -0.062(-1.8)^2 + 25.4$

$= 25.199$

In 2003, 25.199 million (25,199,000) people 12 and older used marijuana.

c.

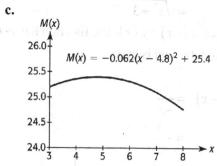

59. a. Since q is in the numerator with an exponent of one, $p = \dfrac{180+q}{6}$ is a linear function.

b. Since there is a variable in the denominator, $p = \dfrac{30,000}{q} - 20$ is a shift of the reciprocal function. It can be obtained from the basic reciprocal function by a vertical stretch by a factor of 30,000 and a shift 20 units down.

61. a. As the number of units produced increases, the average cost function will decrease in value

b. The average cost function is obtained using the basic reciprocal function, followed by a stretch vertically by a factor of 30,000, and a vertical shift 150 units up

63. a. Shift 10 units left and one unit down. Reflect about the t-axis and a vertical stretch by a factor of 1000.

b.

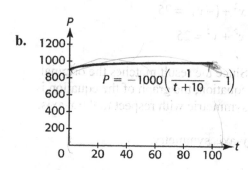

65. $C(x) = 306.472(x+5)^{0.464}$

67. Since in the given function x represents the years after 1980, $x+10$ represents the year 1990 (since it is 10 years after 1980). Therefore, the new function would be
$T(x) = -13.898(x+10)^2 + 255.467(x+10)$
$+ 5425.618$

69. a. $t = 2016 - 1980 = 36$

$S(36) = 0.00061(36)^{3.9}$

≈ 715.994

The number of subscribers in 2016 will be approximately 715.994 million.

b. Since in the given function t represents the years after 1980, $t+5$ represents the year 1985 (since it is 5 years after 1980). Therefore, the new function would be

$$C(t) = 0.00061(t+5)^{3.9}$$

c. $t = 2016 - 1985 = 31$

$$C(31) = 0.00061(31+5)^{3.9}$$

$$= 0.00061(36)^{3.9} = 715.994$$

Using the shifted model yields the same result as part a) above.

71. a. $x = 2017 - 1980 = 37$

$$P(37) = 2029\sqrt{37}$$

$$\approx 12{,}342$$

The poverty threshold in 2017 will be approximately \$12,342.

b. Since in the given function x represents the years since 1980, $x+5$ represents the year 1985(since it is 5 years after 1980). Therefore, the new function would be $S(x) = 2029\sqrt{x+5}, x \geq -5$.

c. $x = 2017 - 1985 = 32$

$$S(32) = 2029\sqrt{32+5}$$

$$= 2029\sqrt{37} \approx 12{,}342$$

Using the shifted model yields the same result.

Section 4.2 Skills Check

1. a. $(f+g)(x)$

$= f(x) + g(x)$

$= (3x - 5) + (4 - x)$

$= 2x - 1$

b. $(f-g)(x)$

$= f(x) - g(x)$

$= (3x - 5) - (4 - x)$

$= 3x - 5 - 4 + x$

$= 4x - 9$

c. $(f \cdot g)(x)$

$= f(x) \cdot g(x)$

$= (3x - 5)(4 - x)$

$= -3x^2 + 17x - 20$

d. $\left(\dfrac{f}{g}\right)(x)$

$= \dfrac{f(x)}{g(x)}$

$= \dfrac{3x - 5}{4 - x}$

e. $g(x) \neq 0$

$4 - x = 0$

$x = 4$

Domain: $(-\infty, 4) \cup (4, \infty)$

3. a. $(f+g)(x)$

$= f(x) + g(x)$

$= (x^2 - 2x) + (1 + x)$

$= x^2 - x + 1$

b. $(f-g)(x)$

$= f(x) - g(x)$

$= (x^2 - 2x) - (1 + x)$

$= x^2 - 2x - 1 - x$

$= x^2 - 3x - 1$

c. $(f \cdot g)(x)$

$= f(x) \cdot g(x)$

$= (x^2 - 2x)(1 + x)$

$= x^3 - x^2 - 2x$

d. $\left(\dfrac{f}{g}\right)(x)$

$= \dfrac{f(x)}{g(x)}$

$= \dfrac{x^2 - 2x}{1 + x}$

e. $g(x) \neq 0$

$1 + x = 0$

$x = -1$

Domain: $(-\infty, -1) \cup (-1, \infty)$

5. a.

$(f + g)(x)$

$= f(x) + g(x)$

$= \left(\dfrac{1}{x}\right) + \left(\dfrac{x+1}{5}\right)$

$LCD: 5x$

$= \dfrac{5}{5}\left(\dfrac{1}{x}\right) + \dfrac{x}{x}\left(\dfrac{x+1}{5}\right)$

$= \left(\dfrac{5}{5x}\right) + \left(\dfrac{x^2 + x}{5x}\right)$

$= \dfrac{x^2 + x + 5}{5x}$

b.

$(f - g)(x)$

$= f(x) - g(x)$

$= \left(\dfrac{1}{x}\right) - \left(\dfrac{x+1}{5}\right)$

$LCD: 5x$

$= \dfrac{5}{5}\left(\dfrac{1}{x}\right) - \dfrac{x}{x}\left(\dfrac{x+1}{5}\right)$

$= \left(\dfrac{5}{5x}\right) - \left(\dfrac{x^2 + x}{5x}\right)$

$= \dfrac{-x^2 - x + 5}{5x}$

c. $(f \cdot g)(x)$

$= f(x) \cdot g(x)$

$= \left(\dfrac{1}{x}\right)\left(\dfrac{x+1}{5}\right)$

$= \dfrac{x+1}{5x}$

$\left(\dfrac{f}{g}\right)(x)$

$= \dfrac{f(x)}{g(x)}$

$= \dfrac{\dfrac{1}{x}}{\dfrac{x+1}{5}}$

d. $= \dfrac{1}{x} \cdot \dfrac{5}{x+1}$

$= \dfrac{5}{x(x+1)}$

e. $g(x) \neq 0$

$x(x+1) = 0$

$x = 0, -1$

Domain: $(-\infty, -1) \cup (-1, 0) \cup (0, \infty)$

7. **a.** $(f+g)(x)$

$= f(x)+g(x)$

$= (\sqrt{x})+(1-x^2)$

$= \sqrt{x}+1-x^2$

b. $(f-g)(x)$

$= f(x)-g(x)$

$= (\sqrt{x})-(1-x^2)$

$= \sqrt{x}-1+x^2$

c. $(f \cdot g)(x)$

$= f(x) \cdot g(x)$

$= (\sqrt{x})(1-x^2)$

d. $\left(\dfrac{f}{g}\right)(x)$

$= \dfrac{f(x)}{g(x)}$

$= \dfrac{\sqrt{x}}{1-x^2}$

e. $g(x) \neq 0$

$1-x^2 \neq 0$

$x \neq -1,1$

And in $f(x)=\sqrt{x},\ x \geq 0$

Domain: $[0,1) \cup (1,\infty)$

9. **a.** $(f+g)(2)$

$= f(2)+g(2)$

$= (2^2 - 5(2))+(6-(2)^3)$

$= -6-2$

$= -8$

b. $(g-f)(-1)$

$= g(-1)-f(-1)$

$= (6-(-1)^3)-((-1)^2 - 5(-1))$

$= 7-6$

$= 1$

c. $(f \cdot g)(-2)$

$= f(-2) \cdot g(-2)$

$= ((-2)^2 - 5(-2)) \cdot (6-(-2)^3)$

$= (14)(14)$

$= 196$

d. $\left(\dfrac{g}{f}\right)(3)$

$= \dfrac{g(3)}{f(3)}$

$= \dfrac{(6-(3)^3)}{(3^2 - 5(3))}$

$= \dfrac{-21}{-6}$

$= 3.5$

11. **a.** $(f \circ g)(x)$

$= f(g(x))$

$= 2(3x-1)-6$

$= 6x-8$

b. $(g \circ f)(x)$

$= g(f(x))$

$= 3(2x-6)-1$

$= 6x-19$

13. a. $(f \circ g)(x)$

$= f(g(x))$

$= \left(\dfrac{1}{x}\right)^2$

$= \dfrac{1}{x^2}$

b. $(g \circ f)(x)$

$= g(f(x))$

$= \dfrac{1}{x^2}$

15. a. $(f \circ g)(x)$

$= f(g(x))$

$= \sqrt{(2x-7)-1}$

$= \sqrt{2x-8}$

b. $(g \circ f)(x)$

$= g(f(x))$

$= 2(\sqrt{x-1})-7$

$= 2\sqrt{x-1}-7$

17. a. $(f \circ g)(x)$

$= f(g(x))$

$= |(4x)-3|$

$= |4x-3|$

b. $(g \circ f)(x)$

$= g(f(x))$

$= 4(|x-3|)$

$= 4|x-3|$

19. a. $(f \circ g)(x)$

$= f(g(x))$

$= \dfrac{3\left(\dfrac{2x-1}{3}\right)+1}{2}$

$= \dfrac{2x-1+1}{2}$

$= \dfrac{2x}{2}$

$= x$

b. $(g \circ f)(x)$

$= g(f(x))$

$= \dfrac{2\left(\dfrac{3x+1}{2}\right)-1}{3}$

$= \dfrac{3x+1-1}{3}$

$= \dfrac{3x}{3}$

$= x$

21. a. $f(g(2)) = 2\left(\dfrac{2-5}{3}\right)^2 = 2(-1)^2 = 2$

b. $g(f(-2)) = \dfrac{\left[2(-2)^2\right]-5}{3}$

$= \dfrac{8-5}{3}$

$= \dfrac{3}{3}$

$= 1$

23. a. $(f+g)(2) = f(2)+g(2) = 1+(-3) = -2$

b. $(f \circ g)(-1) = f(g(-1)) = f(0) = -1$

c. $\left(\dfrac{f}{g}\right)(4) = \dfrac{f(4)}{g(4)} = \dfrac{3}{-1} = -3$

d. $(f \circ g)(1) = f(g(1)) = f(-2) = -3$

e. $(g \circ f)(-2) = g(f(-2)) = g(-3) = 2$

Section 4.2 Exercises

25. a. $P(x) = R(x) - C(x)$
$$= (89x) - (23x + 3420)$$
$$= 89x - 23x - 3420$$
$$= 66x - 3420$$

b. $P(150) = 66(150) - 3420 = 6480$

The profit on the production and sale of 150 bicycles is $6480.

c. The average profit per bicycle is the total profit divided by the number sold:
$$\overline{P}(x) = \frac{P(x)}{x}$$
$$= \frac{66x - 3420}{x}$$

d. Let $x = 150$ using the function from part c:
$$\overline{P}(x) = \frac{66x - 3420}{x}$$
$$\overline{P}(150) = \frac{66(150) - 3420}{150}$$
$$= 43.20$$
The average profit per bicycle if 150 are produced and sold would be $43.20.

27. a. The revenue function is linear, while the cost function is quadratic. Note that $C(x)$ fits the form
$$f(x) = ax^2 + bx + c, a \neq 0.$$

b. $P(x) = R(x) - C(x)$
$$P(x) = 1050x - (10,000 + 30x + x^2)$$
$$P(x) = 1050x - 10,000 - 30x - x^2$$
$$P(x) = -x^2 + 1020x - 10,000$$

c. Quadratic. Note that $P(x)$ fits the form $f(x) = ax^2 + bx + c, a \neq 0.$

29. a. $P(x) = R(x) - C(x)$
$$P(x) = 550x - (10,000 + 30x + x^2)$$
$$P(x) = 550x - 10,000 - 30x - x^2$$
$$P(x) = -x^2 + 520x - 10,000$$

b. Note that the maximum profit occurs at the vertex of the quadratic function, since the function is concave down.
$$x = \frac{-b}{2a} = \frac{-520}{2(-1)} = \frac{-520}{-2} = 260$$

c. $y = P(x)$
$$= P(260)$$
$$= -(260)^2 + 520(260) - 10,000$$
$$= -67,600 + 135,200 - 10,000$$
$$= 57,600$$

Producing and selling 260 cameras yields a maximum profit of $57,600.

31. a. $\overline{C}(x)$ fits the form $\left(\frac{f}{g}\right)(x)$ where $f(x) = C(x)$ and $g(x) = x$. Note that $\overline{C}(x) = \frac{C(x)}{x}.$

b. Let $x = 3000$ and calculate $\overline{C}(x)$.
$$\overline{C}(3000) = \frac{50,000 + 105(3000)}{3000}$$
$$= \frac{365,000}{3000}$$
$$= 121.\overline{6} \approx \$121.67 \text{ per set}$$

33. a. $\overline{C}(x) = \frac{C(x)}{x} = \frac{3000 + 72x}{x}$

b. $\overline{C}(100) = \dfrac{C(100)}{100}$

$= \dfrac{3000 + 72(100)}{100}$

$= \dfrac{3000 + 7200}{100}$

$= \dfrac{10,200}{100}$

$= 102$ or \$102 per printer

35. a. Let $T(p)$ represent the total number of tickets for a home football game.

$T(p) = S(p) + N(p)$

$= (62p + 8500) + (0.5p^2 + 16p + 4400)$

$= 0.5p^2 + 78p + 12,900$

b. Since p represents the winning percentage for the football team, $0 \le p \le 100$. Therefore the domain of the function is $[0,100]$.

c. $T(90) = 0.5(90)^2 + 78(90) + 12,900$

$= 4050 + 7020 + 12,900$

$= 23,970$

The stadium holds 23,970 people.

37. a. $B(8) = 6(8+1)^{\frac{3}{2}} = 6(27) = 162$

On May 8th the number of bushels of tomatoes harvested was 162.

b. $P(8) = 8.5 - 0.12(8) = 7.54$

On May 8th the price per bushel of tomatoes was \$7.54.

c. $(B \cdot P)(x)$ represents the worth of the tomatoes on the xth day of May.

$(B \cdot P)(8) = B(8) \cdot P(8)$

$= 162 \cdot 7.54$

$= 1221.48$

On May 8th the worth was \$1221.48.

d. $W(x) = (B \cdot P)(x)$

$= B(x) \cdot P(x)$

$= \left[6(x+1)^{\frac{3}{2}} \right] \cdot (8.5 - 0.12x)$

$= 6(x+1)^{\frac{3}{2}} (8.5 - 0.12x)$

39. a. $P(x) = R(x) - C(x)$

$= (592x) - (32,000 + 432x)$

$= 592x - 32,000 - 432x$

$= 160x - 32,000$

b. $P(600) = 160(600) - 32,000 = 64,000$

The profit for producing and selling 600 satellite systems in one month is \$64,000.

c. Since the function is linear, the rate of change is constant. For every one unit increase in production, the profit increases by \$160 per unit.

41. a. No. Adding the percentages is not valid since the percentages are based on different populations of people. Adding the number of males and the number of females completing college and then dividing by the total number of people is a legitimate approach.

b. Based on results using the Table feature of a graphing calculator, the percentage in 2002 is 64.78%, and the percentage in 2012 is 68.79 %.

x	y1				
3.	64.778				
5.	66.149				
7.	67.068				
9.	67.763				
11.	68.322				
13.	68.792				
15.	69.197				
17.	69.553				

x=3.

MAIN   RAD AUTO   FUNC

c. $2002 \Rightarrow \dfrac{62.1 + 68.4}{2} = 65.25$

$2012 \Rightarrow \dfrac{61.3 + 71.3}{2} = 66.3$

The percentages are relatively close but not the same.

43. For a)–d), consider the output from the function.

   a. Meat put in a container.

   b. Meat ground.

   c. Meat ground and then ground again.

   d. Meat ground and put in a container.

   e. Meat put in a container, and then both ground.

   f. Only part d), unless the reader enjoys ground styrofoam!

45. Let $B(x)$ convert a Japanese shoe size, x, into a British shoe size, $B(x)$.

Japanese $\to$ U.S. $\to$ British

$B(x) = (p \circ s)(x)$

$\qquad = p(s(x))$

$\qquad = (x - 17) - 1.5$

$\qquad = x - 18.5$

47.

Chilean pesos $\to$ Japanese yen $\to$ Russian rubles

Let $R(x) = 0.34954x$, where x is yen and $R(x)$ is rubles.
Let $S(x) = 0.171718x$, where x is pesos and $S(x)$ is yen.
Let $V(x) = (R \circ S)(x)$, where x is pesos and $V(x)$ is rubles.

$V(x) = R(S(x))$

$\qquad = 0.34954(0.171718x)$

$\qquad = 0.060022x$

$V(100) = 0.060022(100)$

$\qquad\quad = 6 \text{ rubles}$

49. The function is $\left(\dfrac{g}{f}\right)(x)$. Note that

$\left(\dfrac{g}{f}\right)(x) = \dfrac{g(x)}{f(x)}$.

51. The function is $(f + g)(x) = f(x) + g(x)$.

53. The normal price is $0.50x$ where x represents retail price. Since the sale price is 20% off the normal price, the sale price is $0.50x - (0.20)(0.50x) = 0.50x - 0.10x = 0.40x$.
Therefore the books are on sale for 40% of retail price.

Section 4.3 Skills Check

1. Yes, the function is one-to-one and has an inverse.

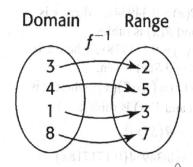

Domain   Range

f^{-1}

3. No, the function is not one-to-one, and therefore has no inverse. Inputs of 5 and 6 correspond with an output of 2.

5. a. $f(g(x)) = (f \circ g)(x) = 3\left(\dfrac{x}{3}\right) = x$

$g(f(x)) = (g \circ f)(x) = \dfrac{(3x)}{3} = x$

b. Yes, since $(f \circ g)(x) = (g \circ f)(x) = x$.

7. Is $(f \circ g)(x) = (g \circ f)(x) = x$?

$(f \circ g)(x) = f(g(x))$

$\qquad = \left(\sqrt[3]{x-1}\right)^3 + 1$

$\qquad = x - 1 + 1$

$\qquad = x$

$(g \circ f)(x) = g(f(x))$

$\qquad = \sqrt[3]{x^3 + 1 - 1}$

$\qquad = \sqrt[3]{x^3}$

$\qquad = x$

Yes, f and g are inverse functions.

9. See the completed tables below.

x	$f(x)$	x	$f^{-1}(x)$
−1	−7	−7	−1
0	−4	−4	0
1	−1	−1	1
2	2	2	2
3	5	5	3

Note that values for x in the second table are the values for $f(x)$ in the first table.

11. Not one-to-one since (1, 5) and (4, 5) make the output 5 have more than one input.

13.

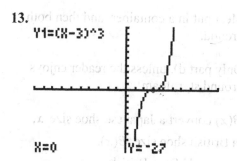

Y1=(X−3)^3

X=0   Y=−27

[−10, 10] by [−10, 10]

Since the graph of $f(x) = (x - 3)^3$ passes the horizontal line test, this is a one-to-one function.

15. Since the given graph passes both the vertical line test and the horizontal line test, this is a one-to-one function.

17.

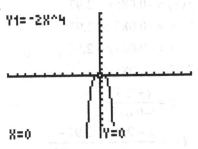

[–10, 10] by [–10, 10]

Since the graph of $f(x) = -2x^4$ fails the horizontal line test, this is not a one-to-one function.

19. a.
$$f(x) = 3x - 4$$
$$y = 3x - 4$$
$$x = 3y - 4$$
$$3y = x + 4$$
$$y = \frac{x+4}{3}$$
$$f^{-1}(x) = \frac{x+4}{3}$$

b. Yes. Substituting the x-values from the table into $f^{-1}(x)$ generates the $f^{-1}(x)$ outputs found in the table.

21. $h^{-1}(-2) = 3 \Leftrightarrow h(3) = -2$, if h and h^{-1} are inverse functions.

23.
$$g(x) = 4x + 1$$
$$y = 4x + 1$$
$$x = 4y + 1$$
$$4y = x - 1$$
$$y = \frac{x-1}{4}$$
$$g^{-1}(x) = \frac{x-1}{4}$$

25.
$$g(x) = x^2 - 3$$
$$y = x^2 - 3$$
$$x = y^2 - 3$$
$$y^2 = x + 3$$
$$y = \pm\sqrt{x+3}$$

Since $x \geq 0$ is given for $g(x)$, then $y \geq 0$ for $g^{-1}(x)$.

Thus only the positive solution represents the inverse function.

$$g^{-1}(x) = \sqrt{x+3}$$

27.

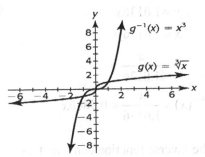

29. Y1=2X^3+1

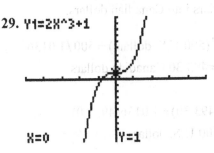

X=0   Y=1

[–5, 5] by [–10, 10]

The function passes the horizontal line test, is one-to-one, and has an inverse function.

31.

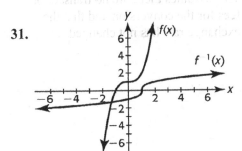

Section 4.3 Exercises

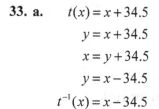

33. a. $t(x) = x + 34.5$

$y = x + 34.5$

$x = y + 34.5$

$y = x - 34.5$

$t^{-1}(x) = x - 34.5$

b. $t^{-1}(43) = 43 - 34.5 = 8.5$

The U.S. shoe size is 8.5 or $8\frac{1}{2}$.

35. a. $f(x) = 1.0136x$

$y = 1.0136x$

$x = 1.0136y$

$y = \dfrac{x}{1.0136}$

$f^{-1}(x) = \dfrac{x}{1.0136} = 0.9866x$

The inverse function converts U.S. dollars into Canadian dollars.

b. $f^{-1}(500 \text{ U.S. dollars}) = 500 / 1.0136$

$y = 493.30$ Canadian dollars

$f(493.30) = 1.0136(493.30)$

$= 500$ U.S. dollars

If you convert $500 from U.S. to Canadian dollars and then convert the money back to U.S. dollars, you will still have $500 U.S currency. (Note: This assumes there are no transaction fees for the conversion and that the exchange rate has not changed

37. a. $f(x) = -0.085x + 2.97$

$y = -0.085x + 2.97$

$x = -0.085y + 2.97$

$x - 2.97 = -0.085y$

$y = \dfrac{x - 2.97}{-0.085}$

$f^{-1}(x) = \dfrac{x - 2.97}{-0.085} = \dfrac{2.97 - x}{0.085}$

The inverse function will calculate the number of years after 2000 in which the percent of children ages 0 to 19 taking antidepressants from 2004 to 2009 reaches a given level.

b. $f^{-1}(2.3) = \dfrac{2.97 - 2.3}{0.085} = 7.88 \approx 8$

The percent will reach 2.3% in 2008 $(2000 + 8)$.

39. a. $W(x) = 0.002x^3$

$y = 0.002x^3$

$x = 0.002y^3$

$y^3 = \dfrac{x}{0.002}$

$y = \sqrt[3]{\dfrac{x}{0.002}} = \sqrt[3]{500x}$

$W^{-1}(x) = \sqrt[3]{500x}$

b.   Given the weight, the inverse function calculates the length.

c. $W^{-1}(2) = \sqrt[3]{500(2)} = \sqrt[3]{1000} = 10$

The length of the fish is 10 inches.

d.   Both the domain and the range are $(0, \infty)$. The weight and the length must be greater than zero or there is no fish to measure.

41. $C(x) = 3x + 2$

$y = 3x + 2$

$x = 3y + 2$

$3y = x - 2$

$y = \dfrac{x-2}{3}$

$C^{-1}(x) = \dfrac{x-2}{3}$

The decoded numerical sequence is
{13 1 11 5 27 13 25 27 4 1 25},
which translates into "MAKE MY DAY".

43. No. The given function is not one-to-one.
More than one check could correspond to
the same dollar amount.

45. a.   Yes. The graph of the equation would
pass the horizontal line test.

b. $f(x) = \dfrac{4}{3}\pi x^3$

$y = \dfrac{4}{3}\pi x^3$

$x = \dfrac{4}{3}\pi y^3$

$3(x) = 3\left(\dfrac{4}{3}\pi y^3\right)$

$3x = 4\pi y^3$

$y^3 = \dfrac{3x}{4\pi}$

$y = \sqrt[3]{\dfrac{3x}{4\pi}}$

$f^{-1}(x) = \sqrt[3]{\dfrac{3x}{4\pi}}$

c.   Both the domain and range are $(0, \infty)$.
If the domain is less than or equal to
zero, then there is no sphere.

d.   Given its volume, the inverse function
can be used to calculate the radius of a
sphere.

e. $f^{-1}(65,450) = \sqrt[3]{\dfrac{3(65,450)}{4\pi}}$

≈ 25

If the volume is 65,450 cubic inches, the
radius is approximately 25 inches.

47. a. $y = 53.265x^{0.170}$

$x = 53.265y^{0.170}$

$\dfrac{x}{53.265} = y^{0.170}$

$\left(\dfrac{x}{53.265}\right)^{\frac{1}{0.170}} = \left(y^{0.170}\right)^{\frac{1}{0.170}}$

$f^{-1}(x) = (0.01877x)^{\frac{1}{0.170}}$

b. $f(x) = 53.265x^{0.170}$

$f(18) = 53.265(18)^{0.170}$

$= 87.064$

$f^{-1}(x) = (0.01877x)^{\frac{1}{0.170}}$

$f^{-1}(87.064) = (0.01877(87.064))^{\frac{1}{0.170}}$

$= (1.63419128)^{\frac{1}{0.170}}$

$= 18$

49. a.   No. The function is not one-to-one.
Note that $I(-1) = I(1) = 300,000$.

b.   In the given physical context, since x
represents distance, the domain is
$(0, \infty)$.

$x \neq 0$, since $\dfrac{300,000}{x^2}$ is undefined.

c.   Yes. Based on the restricted domain
$(0, \infty)$, the function is one-to-one.

d. $I(x) = \dfrac{300{,}000}{x^2}, x > 0$

$$y = \dfrac{300{,}000}{x^2}$$

$$x = \dfrac{300{,}000}{y^2}$$

$$xy^2 = 300{,}000$$

$$y^2 = \dfrac{300{,}000}{x}$$

$$y = \pm\sqrt{\dfrac{300{,}000}{x}}$$

Based on the physical context,

$$I^{-1}(x) = \sqrt{\dfrac{300{,}000}{x}}$$

$$I^{-1}(75{,}000) = \sqrt{\dfrac{300{,}000}{75{,}000}}$$

$$= \sqrt{4}$$

$$= 2$$

When the distance is 2 feet, the intensity of light is 75,000 candlepower.

51. a. $f(x) = 1.6249x$

$$y = 1.6249x$$

$$x = 1.6249y$$

$$y = \dfrac{x}{1.6249} = \dfrac{1}{1.6249}x$$

$$f^{-1}(x) = 0.6154x$$

The inverse function converts U.S. dollars into British pounds.

b. $f^{-1}(1000 \text{ U.S. dollars}) = 0.6154(1000)$ If $y = 615.40$ U.K. pounds

$$f(615.40) = 1.6249(615.40)$$

$$= 1000 \text{ U.S. dollars}$$

If you convert $1000 from U.S. to British currency and then convert the money back to U.S. dollars, you will still have $1000 U.S currency. (Note: This assumes there are no transaction fees for the conversion and that the exchange rate has not changed.)

Section 4.4 Skills Check

1. $\sqrt{2x^2 - 1} - x = 0$

$$\sqrt{2x^2 - 1} = x$$

$$\left(\sqrt{2x^2 - 1}\right)^2 = (x)^2$$

$$2x^2 - 1 = x^2$$

$$x^2 - 1 = 0$$

$$(x+1)(x-1) = 0$$

$$x = 1, x = -1$$

-1 does not check

$$\sqrt{2(-1)^2 - 1} - (-1) =$$

$$\sqrt{2(1) - 1} + 1 = \sqrt{1} + 1 = 2 \neq 0$$

Applying the intersection of graphs method to check graphically:

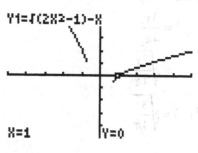

Y1=√(2X²-1)-X

X=1   Y=0

$[-5, 5]$ by $[-5, 5]$

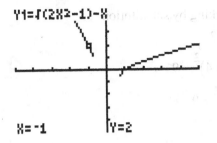

$X=-1$ $Y=2$

$[-5, 5]$ by $[-5, 5]$

The only solution that checks is $x = 1$.

3. $\sqrt[3]{x-1} = -2$

$\left(\sqrt[3]{x-1}\right)^3 = (-2)^3$

$x - 1 = -8$

$x = -7$

Applying the intersection of graphs method to check graphically:

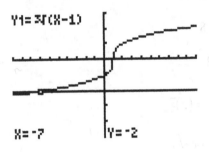

$X=-7$ $Y=-2$

$[-10, 10]$ by $[-5, 3]$

The solution is $x = -7$.

5. $\sqrt{3x-2} + 2 = x$

$\sqrt{3x-2} = x - 2$

$\left(\sqrt{3x-2}\right)^2 = (x-2)^2$

$3x - 2 = x^2 - 4x + 4$

$x^2 - 7x + 6 = 0$

$(x-6)(x-1) = 0$

$x = 6, x = 1$

Applying the intersection of graphs method to check graphically:

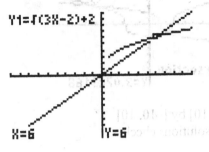

$X=6$ $Y=6$

$[-10, 10]$ by $[-10, 10]$

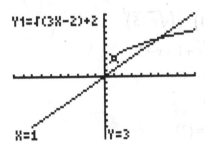

$X=1$ $Y=3$

$[-10, 10]$ by $[-10, 10]$

There is only one solution. The x-value of 1 does not check. The solution is $x = 6$.

7. $\sqrt[3]{4x+5} = \sqrt[3]{x^2-7}$

$\left(\sqrt[3]{4x+5}\right)^3 = \left(\sqrt[3]{x^2-7}\right)^3$

$4x + 5 = x^2 - 7$

$x^2 - 4x - 12 = 0$

$(x-6)(x+2) = 0$

$x = 6, x = -2$

Applying the intersection of graphs method to check graphically:

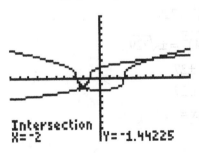

Intersection
$X=-2$ $Y=-1.44225$

$[-10, 10]$ by $[-10, 10]$

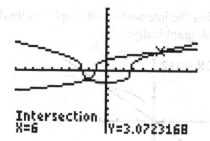

Intersection
X=6   Y=3.0723168

[–10, 10] by [–10, 10]
Both solutions check.

9. $\sqrt{x} - 1 = \sqrt{x-5}$

$\left(\sqrt{x} - 1\right)^2 = \left(\sqrt{x-5}\right)^2$

$x - 2\sqrt{x} + 1 = x - 5$

$-2\sqrt{x} = -6$

$\sqrt{x} = 3$

$\left(\sqrt{x}\right)^2 = (3)^2$

$x = 9$

Checking by substitution
$x = 9$

$\sqrt{x} - 1 \stackrel{?}{=} \sqrt{x-5}$

$\sqrt{9} - 1 \stackrel{?}{=} \sqrt{9-5}$

$2 = 2$
The solution is $x = 9$.

11. $(x+4)^{\frac{2}{3}} = 9$

$\sqrt[3]{(x+4)^2} = 9$

$\left[\sqrt[3]{(x+4)^2}\right]^3 = [9]^3$

$(x+4)^2 = 729$

$\sqrt{(x+4)^2} = \pm\sqrt{729}$

$x + 4 = \pm 27$

$x = -4 \pm 27$

$x = 23, x = -31$

Checking by substitution
$x = 23$

$(23+4)^{\frac{2}{3}} = 9$

$(27)^{\frac{2}{3}} = 9$

$9 = 9$

$x = -31$

$(-31+4)^{\frac{2}{3}} = 9$

$(-27)^{\frac{2}{3}} = 9$

$(-3)^2 = 9$

$9 = 9$

The solutions are $x = 23, x = -31$.

13. $4x^{\frac{5}{2}} - 8 = 0$

$4x^{\frac{5}{2}} = 8$

$x^{\frac{5}{2}} = 2$

$\left(x^{\frac{5}{2}}\right)^{\frac{2}{5}} = (2)^{\frac{2}{5}}$

$x = 2^{\frac{2}{5}}$ or $\sqrt[5]{4}$

Checking by substitution
$x = 2^{\frac{2}{5}}$

$4\left(2^{\frac{2}{5}}\right)^{\frac{5}{2}} - 8 = 0$

$4(2) - 8 = 0$

$8 - 8 = 0$

$0 = 0$

The solution is $x = 2^{\frac{2}{5}}$ or $\sqrt[5]{4}$.

15. $x^4 - 5x^2 + 4 = 0$

   $(x^2 - 4)(x^2 - 1) = 0$

   $(x - 2)(x + 2)(x - 1)(x + 1) = 0$

   $x = 2, -2, 1, -1$

Checking by substitution

$x = 2$

$(2)^4 - 5(2)^2 + 4 = 0$

$16 - 20 + 4 = 0$

$0 = 0$

$x = -2$

$(-2)^4 - 5(-2)^2 + 4 = 0$

$16 - 20 + 4 = 0$

$0 = 0$

$x = 1$

$(1)^4 - 5(1)^2 + 4 = 0$

$1 - 5 + 4 = 0$

$0 = 0$

$x = -1$

$(-1)^4 - 5(-1)^2 + 4 = 0$

$1 - 5 + 4 = 0$

$0 = 0$

The solutions are $x = 2, -2, 1, -1$.

17. Let $u = x^{-1}$ then

   $x^{-2} - x^{-1} - 30 = 0$

   $u^2 - u - 30 = 0$

   $(u - 6)(u + 5) = 0$

   $u = 6, -5$

   $x^{-1} = 6, x^{-1} = -5$ $\Rightarrow$ $\dfrac{1}{x} = 6$ or $x = \dfrac{1}{6}$

   $x = \frac{1}{6}, -\frac{1}{5}$ flip

Checking by substitution

$x = \frac{1}{6}$

$\left(\frac{1}{6}\right)^{-2} - \left(\frac{1}{6}\right)^{-1} - 30 = 0$

$36 - 6 - 30 = 0$

$0 = 0$

$x = -\frac{1}{5}$

$\left(-\frac{1}{5}\right)^{-2} - \left(-\frac{1}{5}\right)^{-1} - 30 = 0$

$25 + 5 - 30 = 0$

$0 = 0$

The solutions are $x = \frac{1}{6}, -\frac{1}{5}$.

19. $2x^{\frac{2}{3}} + 5x^{\frac{1}{3}} - 12 = 0$

   $(2x^{\frac{1}{3}} - 3)(x^{\frac{1}{3}} + 4) = 0$

   $(x^{\frac{1}{3}} - \frac{3}{2})(x^{\frac{1}{3}} + 4) = 0$

   $(x^{\frac{1}{3}} - \frac{3}{2}) = 0, \quad (x^{\frac{1}{3}} + 4) = 0$

   $x^{\frac{1}{3}} = \frac{3}{2}, \quad x^{\frac{1}{3}} = -4$

   $x = \frac{27}{8}, -64$

Checking by substitution

$x = \frac{27}{8}$

$2\left(\frac{27}{8}\right)^{\frac{2}{3}} + 5\left(\frac{27}{8}\right)^{\frac{1}{3}} - 12 = 0$

$\frac{9}{2} + \frac{15}{2} - 12 = 0$

$0 = 0$

$x = -64$

$2(-64)^{\frac{2}{3}} + 5(-64)^{\frac{1}{3}} - 12 = 0$

$32 - 20 - 12 = 0$

$0 = 0$

The solutions are $x = \frac{27}{8}, -64$.

21. $x^2 + 4x < 0$

$x(x+4) < 0$

$x(x+4) = 0$

$x = 0, x = -4$

sign of x $\leftarrow ---_{-4} ---_0 +++\rightarrow$

sign of $(x+4)$ $\leftarrow ---_{-4} +++_0 +++\rightarrow$

sign of $x(x+4)$ $\leftarrow +++_{-4} ---_0 +++\rightarrow$

Considering the inequality symbol in the original question, the solution is $(-4,0)$.

23. $9 - x^2 \geq 0$

$-1(x^2 - 9) \geq 0$

$\dfrac{-1(x^2-9)}{-1} \leq \dfrac{0}{-1}$

$\boxed{x^2 - 9 \leq 0}$ $\leftarrow$ Solve the simplified question

$(x+3)(x-3) \leq 0$

$(x+3)(x-3) = 0$

$x = -3, x = 3$

sign of $(x+3)$ $\leftarrow ---_{-3} +++_3 +++\rightarrow$

sign of $(x-3)$ $\leftarrow ---_{-3} ---_3 +++\rightarrow$

sign of $(x+3)(x-3)$ $\leftarrow +++_{-3} ---_3 +++\rightarrow$

Considering the inequality symbol in the simplified question, the solution is $[-3,3]$.

25. $-x^2 + 9x - 20 > 0$

$-1(x^2 - 9x + 20) > 0$

$\dfrac{-1(x^2 - 9x + 20)}{-1} < \dfrac{0}{-1}$

$\boxed{x^2 - 9x + 20 < 0}$ $\leftarrow$ Solve the simplified question

$(x-5)(x-4) < 0$

$(x-5)(x-4) = 0$

$x = 5, x = 4$

sign of $(x-5)$ $\leftarrow ---_4 ---_5 +++\rightarrow$

sign of $(x-4)$ $\leftarrow ---_4 +++_5 +++\rightarrow$

sign of $(x-5)(x-4)$ $\leftarrow +++_4 ---_5 +++\rightarrow$

Considering the inequality symbol in the simplified question, the solution is $(4,5)$.

27. $2x^2 - 8x \geq 24$

$\boxed{2x^2 - 8x - 24 \geq 0}$ $\leftarrow$ Solve the simplified question

$2(x^2 - 4x - 12) \geq 0$

$2(x-6)(x+2) \geq 0$

$2(x-6)(x+2) = 0$

$2 \neq 0, x = 6, x = -2$

sign of $(x-6)$ $\leftarrow ---_{-2} ---_6 +++\rightarrow$

sign of $(x+2)$ $\leftarrow ---_{-2} +++_6 +++\rightarrow$

sign of $(x-6)(x+2)$ $\leftarrow +++_{-2} ---_6 +++\rightarrow$

Considering the inequality symbol in the simplified question, the solution is $(-\infty, -2] \cup [6, \infty)$.

29. $x^2 - 6x < 7$

$\boxed{x^2 - 6x - 7 < 0}$ $\leftarrow$ Solve the simplified question

$(x-7)(x+1) < 0$

$(x-7)(x+1) = 0$

$x = 7, x = -1$

sign of $(x-7)$ $\leftarrow ---_{-1} ---_7 +++\rightarrow$

sign of $(x+1)$ $\leftarrow ---_{-1} +++_7 +++\rightarrow$

sign of $(x-7)(x+1)$ $\leftarrow +++_{-1} ---_7 +++\rightarrow$

Considering the inequality symbol in the simplified question, the solution is $(-1,7)$.

31. $2x^2 - 7x + 2 \geq 0$

Solving graphically means to find the x-intercepts, in this case, 0.314 and 3.186. Then the solution to $2x^2 - 7x + 2 \geq 0$ is the interval where the graph is above the x-axis, in this case to the left of 0.314 and to the right of 3.186, including both values. Thus the solution to the inequality is $(-\infty, 0.314] \cup [3.186, \infty)$

Y1=2X^2-7X+2

$(0.314, 0)$ $(3.186, 0)$

X=0   Y=2

[-5, 5] by [-10, 10]

33. $5x^2 \geq 2x + 6$

$\boxed{5x^2 - 2x - 6 \geq 0}$ ← Solve the simplified question

Solving graphically means to find the x-intercepts, in this case, $x = -0.914$ and $x = 1.314$. Then the solution to $5x^2 - 2x - 6 \geq 0$ is the interval where the graph is above the x-axis, in this case to the left of -0.914 and to the right of 1.314, including both values. Thus the solution to the inequality is $(-\infty, -0.914] \cup [1.314, \infty)$

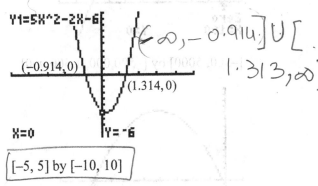

$(-\infty, -0.914] \cup [1.313, \infty)$

$[-5, 5]$ by $[-10, 10]$

35. $(x+1)^3 < 4$

$\sqrt[3]{(x+1)^3} < \sqrt[3]{4}$

$x + 1 < \sqrt[3]{4}$

$x < \sqrt[3]{4} - 1$

Applying the intersections of graphs method

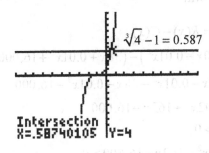

$[-10, 10]$ by $[-10, 10]$

Considering the graph, the solution is $\left(-\infty, \sqrt[3]{4} - 1\right)$ or approximately $(-\infty, 0.587)$.

37. $(x-3)^5 < 32$

$\sqrt[5]{(x-3)^5} < \sqrt[5]{32}$

$x - 3 < 2$

$x < 5$

Applying the intersections of graphs method

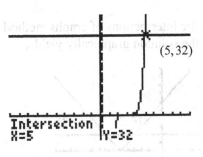

$[-10, 10]$ by $[-10, 40]$

Considering the graph, the solution is $(-\infty, 5)$.

39. $|2x - 1| < 3$

AND

$2x - 1 < 3 \quad | \quad 2x - 1 > -3$

$2x < 4 \quad | \quad 2x > -2$

$x < 2 \quad | \quad x > -1$

$x < 2 \text{ and } x > -1$

$(-1, 2)$

Using the intersections of graphs method to check the solution graphically yields,

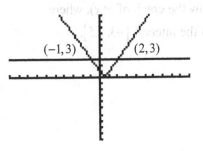

$[-10, 10]$ by $[-10, 10]$

41. $|x-6| \geq 2$

OR

$$x-6 \geq 2 \quad | \quad x-6 \leq -2$$
$$x \geq 8 \quad | \quad x \leq 4$$

$$x \leq 4 \ \text{or} \ x \geq 8$$

$$(-\infty, 4] \cup [8, \infty)$$

Using the intersections of graphs method to check the solution graphically yields,

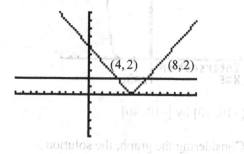

(4, 2)   (8, 2)

$[-10, 15]$ by $[-5, 10]$

43. a. $x \leq -2$ or $x \geq 3$, or

$(-\infty, -2] \cup [3, \infty)$

b. $-2 < x < 3$, or $(-2, 3)$

45. a.   no solution, since no part of the graph is above the x- axis

b.   all real numbers $(-\infty, \infty)$, since the entire graph is below the x- axis

47. $f(x) \leq g(x)$, that is, the graph of $f(x)$ is below the graph of $g(x)$, where x is in the interval $[-3, 2.5]$.

Section 4.4 Exercises

49. $P(x) > 0$

$-0.3x^2 + 1230x - 120,000 > 0$

Applying the x-intercept method

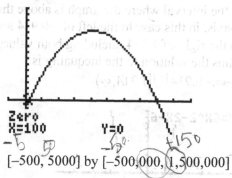

Zero
X=100   Y=0

$[-500, 5000]$ by $[-500,000, 1,500,000]$

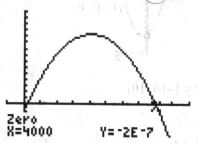

Zero
X=4000   Y= -2E -7

$[-500, 5000]$ by $[-500,000, 1,500,000]$

Considering the graphs, the function is greater than zero over the interval $(100, 4000)$. Producing and selling between 100 and 4000 units, not inclusive, will result in a profit.

51. $P(x) = R(x) - C(x)$

$= (200x - 0.01x^2) - (38x + 0.01x^2 + 16,000)$

$= 200x - 0.01x^2 - 38x - 0.01x^2 - 16,000$

$= -0.02x^2 + 162x - 16,000$

$P(x) > 0$

$-0.02x^2 + 162x - 16,000 > 0$

negative

Applying the *x*-intercept method

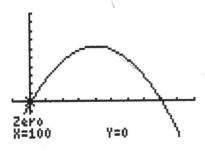

[−1000, 10,000] by [−200,000, 500,000]

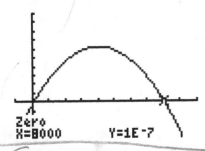

[−1000, 10,000] by [−200,000, 500,000]

Considering the graphs, the function is greater than zero over the interval $(100, 8000)$. Producing and selling between 100 and 8000 units, not inclusive, will result in a profit.

53. Applying the intersection of graphs method

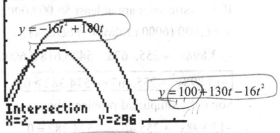

[0, 15] by [−10, 600]

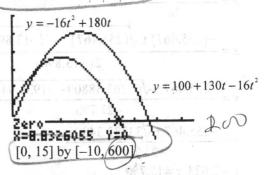

[0, 15] by [−10, 600]

Considering the graphs, the second projectile is above the first projectile over the interval $(2, 8.83)$. Between 2 seconds and 8.83 seconds the height of the second projectile exceeds the height of the first projectile.

$$\frac{-b}{2a}$$

$$x = \frac{-b}{2a}$$

$$\frac{-162}{2(-0.02)}$$ double

$$h = 4050$$

$$x = 3|2050$$

55.

If domestic sales are at least $6,000,000,000,
$y \geq 6,000$ (6000 million). Therefore,

$-13.898x^2 + 255.467x + 5425.618 \geq 6000$

$\boxed{-13.898x^2 + 255.467x - 574.382 \geq 0}$

Solve the simplified problem above.

$-13.898x^2 + 255.467x - 574.382 = 0$

$x = \dfrac{-b \pm \sqrt{b^2 - 4ac}}{2a}$

$x = \dfrac{-(255.467) \pm \sqrt{(255.467)^2 - 4(-13.898)(-574.382)}}{2(-13.898)}$

$x = \dfrac{-255.467 \pm \sqrt{65263.38809 - 31931.04414}}{-27.796}$

$x = \dfrac{-255.467 \pm \sqrt{33332.34395}}{-27.796}$

$x = 2.623, x = 15.759$

Applying the x-intercept method

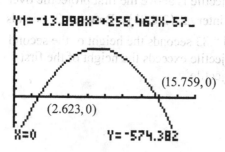

[0, 20] by [−500, 1000]

Considering the graph, the equation is greater than or equal to zero over the interval $[2.623, 15.759]$. Therefore, the sales (in millions) of fine-cut cigarettes in Canada after the year 1980 total at least $6 billion between 1983 and 1995, inclusive.

57.

If the percent is at most 13.49, then $y \geq 13.49$. Therefore,

$0.0027x^2 - 0.3692x + 18.914 \geq 13.49$

$\boxed{0.0027x^2 - 0.3692x + 5.424 \geq 0}$

Solve the simplified problem above.

$0.0027x^2 - 0.3692x + 5.424 = 0$

$x = \dfrac{-b \pm \sqrt{b^2 - 4ac}}{2a}$

$x = \dfrac{-(-0.3692) \pm \sqrt{(-0.3692)^2 - 4(0.0027)(5.424)}}{2(0.0027)}$

$x = \dfrac{0.3692 \pm \sqrt{0.13630864 - 0.0585792}}{0.0054}$

$x = \dfrac{0.3692 \pm \sqrt{0.07772944}}{0.0054}$

$x = \dfrac{0.3692 \pm 0.2788}{0.0054}$

$x = 120, x \approx 16.74$

Thus after the year 2020, the percentage of the U.S. population that is foreign born is at most 13.49%.

59. $y = 0.0052x^2 - 0.62x + 15.0$

Applying the intersection of graphs method

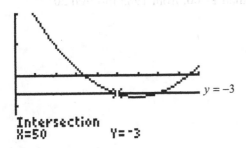

[0, 90] by [−10, 10]

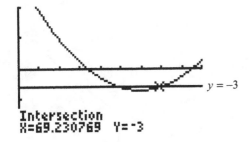

Considering the graphs, the wind chill temperature is −3°F or below when the wind speed is between 50 mph and 69.2 mph.

61. $y = -0.1967x^2 + 4.063x + 27.7455$

Applying the intersection of graphs method

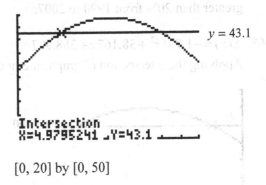

[0, 20] by [0, 50]

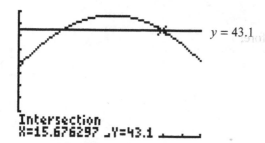

Considering the graphs, the % of marijuana use is above 43.1% over the interval $(4.98, 15.68)$. Thus between 1995 and 2005, the % of marijuana use is above 43.1%.

63. $y = -0.061x^2 + 0.275x + 33.698$

Applying the intersection of graphs method

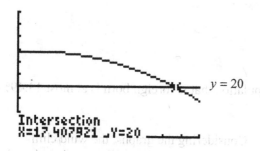

[0, 20] by [0, 50]

Therefore, the percent of high school students who smoked cigarettes on 1 or more of the 30 days preceding the survey, is greater than 20% from 1990 to 2007.

65. $S(x) = -1.751x^2 + 38.167x + 388.997$

Applying the intersection of graphs method

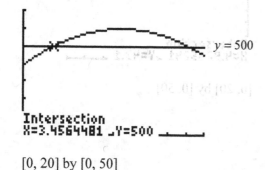

[0, 20] by [0, 50]

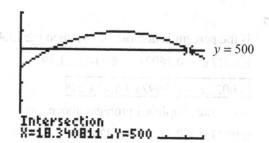

Therefore, retail sales are above 500 billion dollars between the years 2004 and 2018.

67. $y = 34.394x^{-1.109}$

Applying the intersection of graphs method

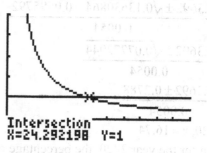

[0, 55] by [0, 5]

Therefore, based on the current model, the purchasing power of a 1983 dollar, is less than $1.00, from 1985 through 2012.

Chapter 4 Skills Check

1. The graph of the function is shifted 8 units right and 7 units up.

3. **a.**

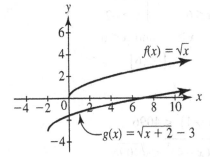

b. The graph of g(x) is shifted two units left and three units down in comparison to *f(x)*.

5. $y = (x - 6)^{1/3} + 4$

7. The graph matches equation f.

9. The graph matches equation e.

11. To test for *y*-axis symmetry:

Let $x = -x$ and check if $f(-x) = f(x)$
since $f(x) = -x^2 + 5$,
$f(-x) = -(-x)^2 + 5$
$f(-x) = -x^2 + 5$
Thus, $f(-x) = f(x)$

Since $f(-x) = f(x)$, the graph of the equation is symmetric with respect to the *y*-axis.

13. $(f + g)(x) =$
$(3x^2 - 5x) + (6x - 4) =$
$3x^2 + x - 4$

15. $(g \cdot f)(x) =$
$(6x - 4) \cdot (3x^2 - 5x) =$
$18x^3 - 42x^2 + 20x$

17. $(f - g)(x) =$
$(3x^2 - 5x) - (6x - 4) =$
$3x^2 - 11x + 4$, then
$(f - g)(-2) =$
$3(-2)^2 - 11(-2) + 4 = 38$

19. $(g \circ f)(x) = g(f(x))$
$6(3x^2 - 5x) - 4 =$
$18x^2 - 30x - 4$

21. **a.** $f(g(x)) = 2\left(\dfrac{x + 5}{2}\right) - 5 = x$

$g(f(x)) = \dfrac{(2x - 5) + 5}{2} = x$

b. Since $f(g(x)) = g(f(x))$, *f(x)* and *g(x)* are inverse functions.

23. $g(x) = \sqrt[3]{x - 1}$
$y = \sqrt[3]{x - 1}$
$x = \sqrt[3]{y - 1}$
$x^3 = \left(\sqrt[3]{y - 1}\right)^3$
$x^3 = y - 1$
$y = x^3 + 1$
$g^{-1}(x) = x^3 + 1$

25.

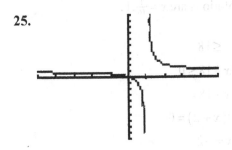

[–5, 5] by [–10, 10]

Since the function passes the horizontal line test. It is one-to-one.

27. $\sqrt{4x^2+1}=2x+2$

$$\left(\sqrt{4x^2+1}\right)^2=(2x+2)^2$$

$$4x^2+1=4x^2+8x+4$$

$$8x=-3$$

$$x=-\frac{3}{8}$$

29. $4x-5x^{\frac{1}{2}}+1=0$

$$(4x^{\frac{1}{2}}-1)(x^{\frac{1}{2}}-1)=0$$

$$(x^{\frac{1}{2}}-\tfrac{1}{4})(x^{\frac{1}{2}}-1)=0$$

$$(x^{\frac{1}{2}}-\tfrac{1}{4})=0,\ \ (x^{\frac{1}{2}}-1)=0$$

$$x^{\frac{1}{2}}=\tfrac{1}{4},\ \ x^{\frac{1}{2}}=1$$

$$x=\tfrac{1}{16},1$$

Checking by substitution

$$x=\tfrac{1}{16}$$

$$4(\tfrac{1}{16})-5(\tfrac{1}{16})^{\frac{1}{2}}+1=0$$

$$\tfrac{1}{4}-\tfrac{5}{4}+1=0$$

$$0=0$$

$$x=1$$

$$4(1)-5(1)^{\frac{1}{2}}+1=0$$

$$4-5+1=0$$

$$0=0$$

The solutions are $x=\tfrac{1}{16},1$.

31. $x^2-7x\le 18$

$$x^2-7x-18\le 0$$

$$x^2-7x-18=0$$

$$(x-9)(x+2)=0$$

$$x=9, x=-2$$

sign of $(x-9)$ $\leftarrow---_{-2}---_{9}+++\rightarrow$

sign of $(x+2)$ $\leftarrow---_{-2}+++_{9}+++\rightarrow$

sign of $(x-9)(x+2)$ $\leftarrow+++_{-2}---_{9}+++\rightarrow$

Considering the inequality symbol in the simplified question, the solution is $[-2,9]$.

33. $|2x-4|\le 8$

AND

$2x-4\le 8$	$2x-4\ge -8$
$2x\le 12$	$2x\ge -4$
$x\le 6$	$x\ge -2$

$$x\ge -2 \text{ and } x\le 6$$

$$[-2,6]$$

35. $(x-4)^3<4096$

$$\sqrt[3]{(x-4)^3}<\sqrt[3]{4096}$$

$$x-4<16$$

$$x<20$$

The solution is $(-\infty,20)$.

Chapter 4 Review Exercises

37. a.

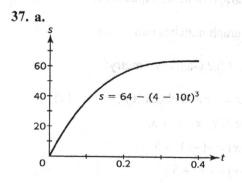

b. The bullet travels 64 inches in 0.4 seconds.

38. a. Note that the function is quadratic written in vertex form, $y=a(x-h)^2+k$. The vertex is $(h,k)=(4,380)$. Therefore the maximum height occurs 4 seconds into the flight of the rocket.

b. Referring to part *a*, the maximum height of the rocket is 380 feet.

c. In comparison to $y = t^2$ the graph is shifted 4 units right, stretched by a factor of 16, reflected across the x-axis, and 380 units up.

39. a. $f(x) = -7.232x^2 + 95.117x + 441.138$

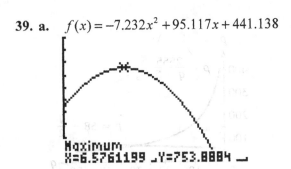

Maximum
X=6.5761199 Y=753.8884

[0, 20] by [0, 1000]

Thus, the year in which the number of passengers was at a maximum was 2007.

b. Applying the intersection of graphs method

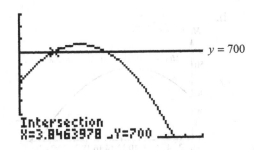

$y = 700$

Intersection
X=3.8463978 Y=700

[0, 20] by [0, 1000]

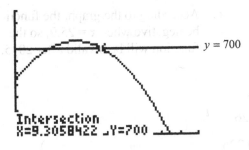

$y = 700$

Intersection
X=9.3058422 Y=700

From the graphs, the number of passengers was above 700 million between the years 2004 and 2009.

40. $y = 0.037x^2 + 1.018x + 12.721$
Applying the intersection of graphs method

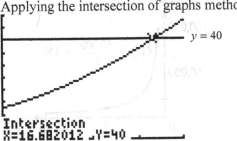

$y = 40$

Intersection
X=16.682012 Y=40

[0, 20] by [0, 50]

From 2007 through 2010, Americans spent more than 40 million dollars on their pets.

41. $y = 3.980x^2 - 17.597x + 2180.899$
Applying the intersection of graphs method

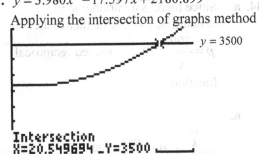

$y = 3500$

Intersection
X=20.549694 Y=3500

[0, 25] by [0, 4000]

From 1990 to 2010, the number of postsecondary degrees earned was less than 3500.

42. a. $\overline{C}(x) = \dfrac{30x + 3150}{x}$

b.

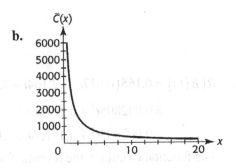

43. a.

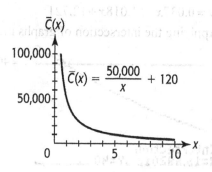

$$\bar{C}(x) = \frac{50,000}{x} + 120$$

b. This is a decreasing function.

c. The graph of this function has a vertical stretch by a factor of 50,000 and a vertical shift 120 units up.

44. a. Since there is a variable in the denominator of the demand function, $p = \dfrac{300}{q} - 20$ is a shifted reciprocal function.

b.

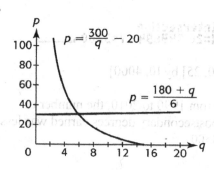

$$p = \frac{300}{q} - 20$$

$$p = \frac{180 + q}{6}$$

45. a. Since there is a variable in the denominator of the demand function, $p = \dfrac{2555}{q+5}$ is a shifted reciprocal function.

b.

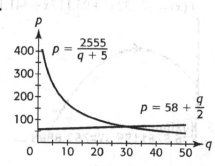

$$p = \frac{2555}{q + 5}$$

$$p = 58 + \frac{q}{2}$$

46. a. The transformation of $f(x) = x^2$ includes a horizontal shift to the right of 10.3 units, a vertical reflection across the x-axis, a vertical compression by a factor of 0.2, and a vertical shift up of 48.968 units.

b.

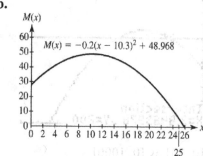

$$M(x) = -0.2(x - 10.3)^2 + 48.968$$

c. According to the graph, the function will be negative when $x = 25.9$, so the function will be invalid after 2015.

47. a.
$$R(E(t)) = 0.165(0.017t^2 + 2.164t + 8.061) - 0.226$$
$$= 0.002805t^2 + 0.35706t + 1.330065 - 0.226$$
$$= 0.002805t^2 + 0.35706t + 1.104065$$

The function calculates the revenue for Southwest Airlines given the number of years past 1990.

b. $R(E(t)) = 0.002805t^2 + 0.35706t + 1.104065$

$R(E(3)) = 0.002805(3)^2 + 0.35706(3) + 1.104065$

$R(E(3)) = 2.20049 \approx 2.2$

In 1993 Southwest Airlines had revenue of $2.2 billion.

c. $E(7) = 0.017(7)^2 + 2.164(7) + 8.061$

$= 24.042$

In 1997 Southwest Airlines had 24,042 employees.

d. $R(E(t)) = 0.002805t^2 + 0.35706t + 1.104065$

$R(E(7)) = 0.002805(7)^2 + 0.35706(7) + 1.104065$

$R(E(7)) = 3.74093 \approx 3.7$

In 1997 Southwest Airlines had revenue of $3.7 billion.

48. a.
$$f(x) = 0.554x - 2.886$$
$$y = 0.554x - 2.886$$
$$x = 0.554y - 2.886$$
$$0.554y = x + 2.886$$
$$y = \frac{x + 2.886}{0.554}$$
$$f^{-1}(x) = \frac{x + 2.886}{0.554}$$

b. The inverse function calculates the mean length of the original prison sentence given the mean time spent in prison.

49. a.
$$y = 53.265x^{0.170}$$
$$x = 53.265y^{0.170}$$
$$\frac{x}{53.265} = y^{0.170}$$
$$\left(\frac{x}{53.265}\right)^{\frac{1}{0.170}} = \left(y^{0.170}\right)^{\frac{1}{0.170}}$$
$$f^{-1}(x) = (0.01877x)^{\frac{1}{0.170}}$$

b.
$$f(x) = 53.265x^{0.170}$$
$$f(15) = 53.265(15)^{0.170}$$
$$= 84.407$$
$$f^{-1}(x) = (0.01877x)^{\frac{1}{0.170}}$$
$$f^{-1}(84.407) = (0.01877(84.407))^{\frac{1}{0.170}}$$
$$= (1.584315671)^{\frac{1}{0.170}}$$
$$= 15$$

50. $P(x) = 34.394x^{-1.109}$

a. $f(x) = \dfrac{50}{1.6x} - 0.2$

$f(x)$ is a transformed reciprocal function.

b. For $x = 50$, $(2010 - 1960)$

$P(50) = 34.394(50)^{-1.109} = \$0.449$

$f(50) = \dfrac{50}{1.6(50)} - 0.2 = \$0.425$

The function $P(x)$ gives the better estimate of the purchasing power of a 1983 dollar in 2010.

51. $P(x) = -0.01x^2 + 62x - 12{,}000$

$P(x) > 0$

$-0.01x^2 + 62x - 12{,}000 > 0$

Applying the *x*-intercept method

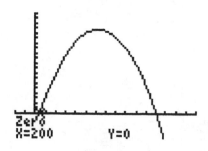

[–1000, 8000] by [–25,000, 100,000]

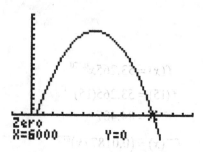

[–1000, 8000] by [–25,000, 100,000]

Manufacturing and producing between 200 and 6000 units will result in a profit.

52.

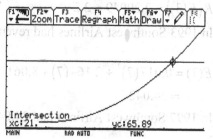

[0, 25] by [0, 100]

The model indicates that the revenue will be at least $65.89 billion from 2021 through 2025.

Group Activity/Extended Application I

1.

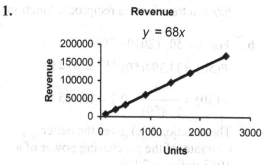

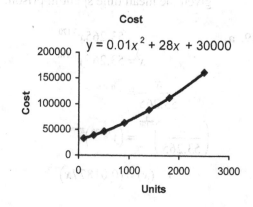

2. a. $P(x) = R(x) - C(x)$

$$= 68x - \left(0.01x^2 + 28x + 30,000\right)$$

$$= -0.01x^2 + 40x - 30,000$$

b.

0	−30,000
100	−26,100
600	−9600
1600	8400
2000	10,000
2500	7500

3. $R(x) = C(x)$

$$68x = 0.01x^2 + 28x + 30,000$$

$$0.01x^2 - 40x + 30,000 = 0$$

$$x = \frac{-b \pm \sqrt{b^2 - 4ac}}{2a}$$

$$x = \frac{-(-40) \pm \sqrt{(-40)^2 - 4(0.01)(30,000)}}{2(0.01)}$$

$$x = \frac{40 \pm \sqrt{1600 - 1200}}{0.02}$$

$$x = \frac{40 \pm \sqrt{400}}{0.02}$$

$$x = \frac{40 \pm 20}{0.02}$$

$$x = 3000, x = 1000$$

Producing and selling either 3000 or 1000 units forces the profit to equal the cost.

4.

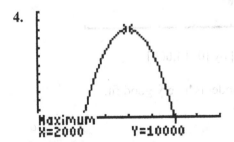

Maximum
X=2000   Y=10000

[0, 4000] by [−2000, 12,000]

Producing and selling 2000 units results in a maximum profit of $10,000.

5. a. $\overline{C}(x) = \dfrac{C(x)}{x} = \dfrac{0.01x^2 + 28x + 30,000}{x}$

b.

1	30,028.01
100	329.00
300	131.00
1400	63.43
2000	63.00
2500	65.00

c. Y1=(.01X^2+28X+30000)/X

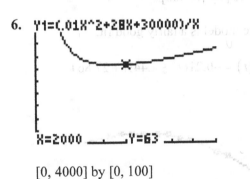

X=1250 _____ Y=64.5 _____

[0, 2500] by [0, 400]

6. Y1=(.01X^2+28X+30000)/X

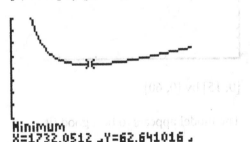

X=2000 _____ Y=63 _____

[0, 4000] by [0, 100]

Minimum
X=1732.0512 _Y=62.641016 _

[0, 4000] by [0, 100]

The minimum average cost is $62.64 which occurs when 1732 units are produced and sold.

7. The values are different. However, they are relatively close together. The number of units that maximizes profit is most important. While keeping costs low is important, the key to a successful business is generating profit.

Group Activity/Extended Application II

1. $f(t) = 0.336t^2 + 15.116t + 21.608$

2.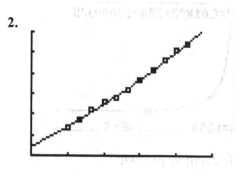

[0, 15] by [0, 300]

The model is a fairly good fit.

3. $g(t) = -0.216t^2 + 4.449t + 27.864$

4.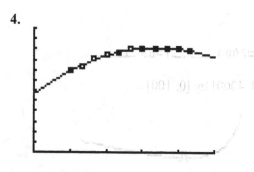

[0, 15] by [0, 60]

The model appears to be a good fit.

5. $(f \cdot g)(t) = -0.06t^4 - 1.7t^3 + 70.49t^2 + 516.53t + 602.64$

6.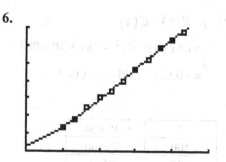

[0, 15] by [0, 14,000]

The model is a fairly good fit.

7.

Year	Average Revenue
1998	2728.91
1999	3548.58
2000	4956.07
2001	6081.12
2002	6813.07
2003	7921.82
2004	9223.57
2005	10390.64
2006	11782.55
2007	12754.48
2008	13459.93

8. $r(t) = 1.788t^2 + 1080.376t - 624.321$

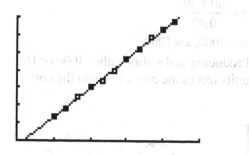

[0, 15] by [0, 14,000]

The model is a very good fit.

9. $s(t) = -1.722t^4 + 51.941t^3 - 540.319x^2$
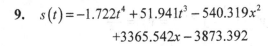
$\qquad\quad + 3365.542x - 3873.392$

[0, 15] by [0, 14,000]

Both models fit about the same. Neither is
an exact match.

Chapter 5
Exponential and Logarithmic Functions

Toolbox Exercises

1. a. $x^4 \cdot x^3 = x^{4+3} = x^7$

b. $\dfrac{x^{12}}{x^7} = x^{12-7} = x^5$

c. $(4ay)^4 = (4)^4 a^4 y^4 = 256 a^4 y^4$

d. $\left(\dfrac{3}{z}\right)^4 = \dfrac{3^4}{z^4} = \dfrac{81}{z^4}$

e. $2^3 \cdot 2^2 = 2^{3+2} = 2^5 = 32$

f. $\left(x^4\right)^2 = x^{4 \cdot 2} = x^8$

2. a. $y^5 \cdot y = y^5 \cdot y^1 = y^{5+1} = y^6$

b. $\dfrac{w^{10}}{w^4} = w^{10-4} = w^6$

c. $(6bx)^3 = (6)^3 b^3 x^3 = 216 b^3 x^3$

d. $\left(\dfrac{5z}{2}\right)^3 = \dfrac{5^3 z^3}{2^3} = \dfrac{125 z^3}{8}$

e. $3^2 \cdot 3^3 = 3^{2+3} = 3^5 = 243$

f. $\left(2y^3\right)^4 = 2^4 y^{3 \cdot 4} = 16 y^{12}$

3. $10^{5^0} = 10^{(1)} = 10$

4. $4^{2^2} = 4^4 = 256$

5. $x^{-4} \cdot x^{-3} = x^{-4 + -3} = x^{-7} = \dfrac{1}{x^7}$

6. $y^{-5} \cdot y^{-3} = y^{-5 + -3} = y^{-8} = \dfrac{1}{y^8}$

7. $\left(c^{-6}\right)^3 = c^{-6 \cdot 3} = c^{-18} = \dfrac{1}{c^{18}}$

8. $\left(x^{-2}\right)^4 = x^{-2 \cdot 4} = x^{-8} = \dfrac{1}{x^8}$

9. $\dfrac{a^{-4}}{a^{-5}} = a^{-4-(-5)} = a^1 = a$

10. $\dfrac{b^{-6}}{b^{-8}} = b^{-6-(-8)} = b^2$

11. $\left(x^{-\frac{1}{2}}\right)\left(x^{\frac{2}{3}}\right) = x^{-\frac{1}{2}+\frac{2}{3}} = x^{-\frac{3}{6}+\frac{4}{6}} = x^{\frac{1}{6}}$

12. $\left(y^{-\frac{1}{3}}\right)\left(y^{\frac{2}{5}}\right) = y^{-\frac{1}{3}+\frac{2}{5}} = y^{-\frac{5}{15}+\frac{6}{15}} = y^{\frac{1}{15}}$

13. $\left(3a^{-3}b^2\right)\left(2a^2b^{-4}\right)$

$= 6a^{-3+2}b^{2+-4}$

$= 6a^{-1}b^{-2}$

$= \dfrac{6}{ab^2}$

14. $\left(4a^{-2}b^3\right)\left(-2a^4b^{-5}\right)$

$= -8a^{-2+4}b^{3+-5}$

$= -8a^2b^{-2}$

$= \dfrac{-8a^2}{b^2}$

15. $\left(\dfrac{2x^{-3}}{x^2}\right)^{-2} = \left(2x^{-3-2}\right)^{-2}$

$= \left(2x^{-5}\right)^{-2}$

$= (2)^{-2}\left(x^{-5}\right)^{-2}$

$= \dfrac{1}{2^2}x^{-5\cdot-2}$

$= \dfrac{1}{4}x^{10}$

$= \dfrac{x^{10}}{4}$

16. $\left(\dfrac{3y^{-4}}{2y^2}\right)^{-3} = \left(\dfrac{2y^2}{3y^{-4}}\right)^{3}$

$= \dfrac{\left(2y^2\right)^3}{\left(3y^{-4}\right)^3}$

$= \dfrac{8y^6}{27y^{-12}}$

$= \dfrac{8y^{6-(-12)}}{27}$

$= \dfrac{8y^{18}}{27}$

17. $\dfrac{28a^4b^{-3}}{-4a^6b^{-2}} = -7a^{4-6}b^{-3-(-2)}$

$= -7a^{-2}b^{-1}$

$= \dfrac{-7}{a^2b}$

18. $\dfrac{36x^5y^{-2}}{-6x^6y^{-4}} = -6x^{5-6}y^{-2-(-4)}$

$= -6x^{-1}y^2$

$= \dfrac{-6y^2}{x}$

19. 4.6×10^7

20. 8.62×10^{11}

21. 9.4×10^{-5}

22. 2.78×10^{-6}

23. $437,200$

24. $7,910,000$

25. 0.00056294

26. 0.0063478

27. $\left(6.25\times10^7\right)\left(5.933\times10^{-2}\right)$

$(6.25\times5.933)\times10^{7+-2}$

37.08125×10^5

Rewriting in scientific notation

3.708125×10^6

28. $\dfrac{2.961\times10^{-2}}{4.583\times10^{-4}}$

$\dfrac{2.961}{4.583}\times10^{-2-(-4)}$

0.6460833515×10^2

Rewriting in scientific notation

6.460833515×10^1

29. $x^{1/2}\cdot x^{5/6} = x^{1/2+5/6} = x^{3/6+5/6} = x^{8/6} = x^{4/3}$

30. $y^{2/5}\cdot y^{1/4} = y^{2/5+1/4} = y^{8/20+5/20} = y^{13/20}$

31. $\left(c^{2/3}\right)^{5/2} = c^{2/3 \cdot 5/2} = c^{5/3}$

32. $\left(x^{3/2}\right)^{3/4} = x^{3/2 \cdot 3/4} = x^{9/8}$

33. $\dfrac{x^{3/4}}{x^{1/2}} = x^{3/4 - 1/2} = x^{3/4 - 2/4} = x^{1/4}$

34. $\dfrac{y^{3/8}}{y^{1/4}} = y^{3/8 - 1/4} = y^{3/8 - 2/8} = y^{1/8}$

35. $\left(\dfrac{2b^{1/3}}{6c^{2/3}}\right)^3 = \left(\dfrac{b^{1/3}}{3c^{2/3}}\right)^3$

$= \dfrac{\left(b^{1/3}\right)^3}{\left(3c^{2/3}\right)^3}$

$= \dfrac{b}{27c^2}$

36. $\left(\dfrac{9x^{2/3}}{16y^{5/6}}\right)^{3/2} = \dfrac{\left(9x^{2/3}\right)^{3/2}}{\left(16y^{5/6}\right)^{3/2}}$

$= \dfrac{9^{3/2} x^{6/6}}{16^{3/2} y^{15/12}}$

$= \dfrac{27x}{64 y^{5/4}}$

37. $\left(x^{4/9} y^{2/3} z^{-1/3}\right)^{3/2} = x^{12/18} y^{6/6} z^{-3/6}$

$= x^{2/3} y z^{-1/2}$

$= \dfrac{x^{2/3} y}{z^{1/2}}$

38. $\left(\dfrac{-8x^9 y^{1/2}}{27z^{3/2}}\right)^{4/3} = \dfrac{\left(-8x^9 y^{1/2}\right)^{4/3}}{\left(27z^{3/2}\right)^{4/3}}$

$= \dfrac{(-8)^{4/3} x^{36/3} y^{4/6}}{27^{4/3} z^{12/6}}$

$= \dfrac{16x^{12} y^{2/3}}{81z^2}$

Section 5.1 Skills Check

1. Functions c), d), and e) represent exponential functions. They both fit the form $y = a^x$, where a is a constant, $a > 0$ and $a \neq 1$.

3. a.

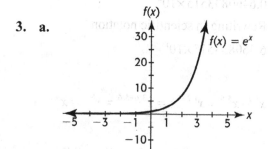

b. $f(1) = e^1 = e \approx 2.718$

$f(-1) = e^{-1} = \dfrac{1}{e} \approx 0.368$

$f(4) = e^4 \approx 54.598$

c. $y = 0$, the x-axis.

d. $(0,1)$ since $f(0) = 1$.

5.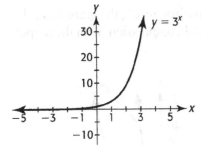

7.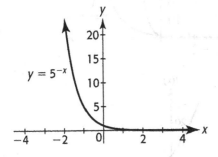

9.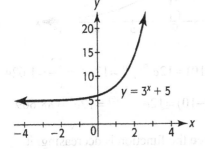

Notice that the y-intercept is $(0,6)$.

11.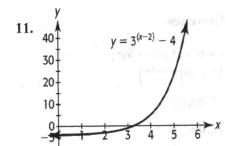

Notice that the y-intercept is $(0,-3.\overline{8})$.

13. The equation matches graph B.

15. The equation matches graph A.

17. The equation matches graph E.

19. In comparison to 4^x, the graph has the same shape but shifted 2 units up.

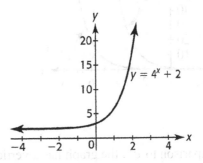

21. In comparison to 4^x, the graph has the same shape but is reflected across the y-axis.

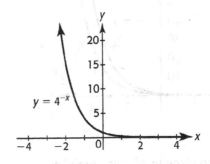

23. In comparison to 4^x, the graph has a vertical stretch by a factor of 3.

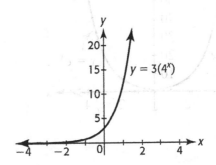

25. Both graphs have a vertical stretch by a factor of 3 in comparison with 4^x. Therefore, the graph in Exercise 24 has the same shape as the graph in Exercise 23, but it has a shift 2 units right and 3 units down.

27. In comparison to e^x, the graph has a horizontal shift 3 units right.

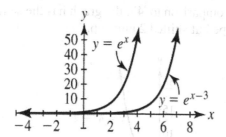

29. In comparison to e^x, the graph has a vertical compression with a factor of ½.

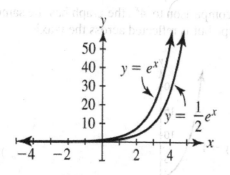

31. In comparison to e^x, the graph has a reflection across the y-axis.

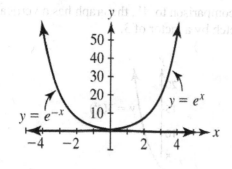

33. In comparison to e^x, the graph has a horizontal compression with the outputs cubed.

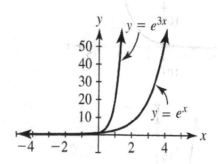

35. a.

b. $f(10) = 12e^{-0.2(10)} = 12e^{-2} = \dfrac{12}{e^2} \approx 1.624$

$f(-10) = 12e^{-0.2(-10)} = 12e^2 \approx 88.669$

c. Since the function is decreasing, it represents decay. Notice that the y-intercept is $(0,12)$.

Section 5.1 Exercises

37. a. Let $x = 0$ and solve for y.

$y = 12,000\left(2^{-0.08 \cdot 0}\right)$

$= 12,000\left(2^0\right)$

$= 12,000(1)$

$= 12,000$

At the end of the ad campaign, sales were $12,000 per week.

b. Let $x = 6$ and solve for y.

$$y = 12,000\left(2^{-0.08 \cdot 6}\right)$$
$$= 12,000\left(2^{-0.48}\right)$$
$$= 12,000(0.716977624)$$
$$\approx 8603.73$$

Six weeks after the end of the ad campaign, sales were $8603.73 per week.

c. No. Sales approach a level of zero but never actually reach that level. Consider the graph of the model below.

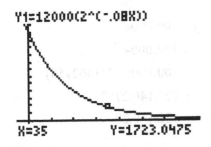

[−5, 75] by [−2000, 15,000]

39. a.

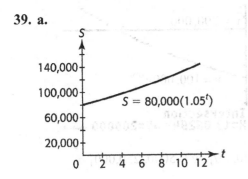

b. $S = 80,000\left(1.05^{10}\right)$

$$= 80,000(1.628894627)$$
$$\approx 130,311.57 \text{ after 10 years}$$

41. a.

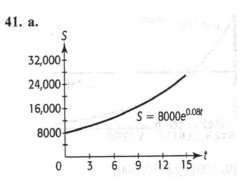

b. The future value will be $20,000 in approximately 11.45 years.

c.

t (Year)	S ($)
10	17,804.33
20	39,624.26
22	46,449.50

43. a. $A(10) = 500e^{-0.02828(10)}$

$$= 500e^{-0.2828}$$
$$= 500(0.7536705069)$$
$$\approx 376.84$$

Approximately 376.84 grams remain after 10 years.

b.

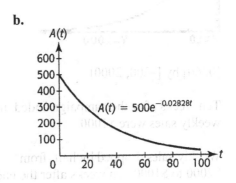

c.

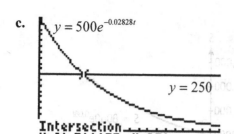

$y = 500e^{-0.02828t}$

$y = 250$

Intersection
X=24.510155 Y=250

[0, 100] by [−50, 500]

The half-life is approximately 24.5 years.

45. a.

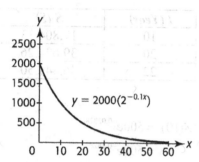

$y = 2000(2^{-0.1x})$

b.

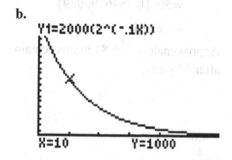

Y1=2000(2^(-.1X))

X=10   Y=1000

[0, 60] by [−200, 2000]

Ten weeks after the campaign ended, the weekly sales were $1000.

c. Weekly sales dropped by half, from $2000 to $1000, ten weeks after the end of the ad campaign. It is important for this company to advertise.

47. a. $P = 40,000(0.95^{20})$

$= 40,000(0.3584859224)$

$= 14,339.4369$

$\approx 14,339.44$

The purchasing power will be $14,339.44.

b. Since the purchasing power of $40,000 will decrease to $14,339 over the next twenty years, people who retire at age 50 should continue to save money to offset the decrease due to inflation. Answers to part b) could vary.

49. a. $y = 100,000e^{0.05(4)}$

$= 100,000e^{0.2}$

$= 100,000(1.221402758)$

$= 122,140.2758$

$\approx 122,140.28$

The value of this property after 4 years will be $122,140.28.

b.

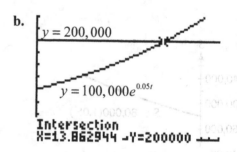

$y = 200,000$

$y = 100,000e^{0.05t}$

Intersection
X=13.862944 Y=200000

[0, 20] by [−5000, 250,000]

The value of this property doubles in 13.86 years or approximately 14 years.

51. a. Increasing. The exponent is positive for all values of $t \geq 0$.

b. $P(5) = 53,000e^{0.015(5)}$

$= 53,000e^{0.075}$

$= 53,000(1.077884151)$

$\approx 57,128$

The population was 57,128 in 2005.

c. $P(10) = 53,000e^{0.015(10)}$

$= 53,000e^{0.15}$

$= 53,000(1.161834243)$

$\approx 61,577$

The population was 61,577 in 2010.

d. $\dfrac{y_2 - y_1}{x_2 - x_1} = \dfrac{61,577 - 53,000}{10 - 0}$

$= \dfrac{8577}{10}$

$= 857.7$

The average rate of growth in population between 2000 and 2010 is approximately 858 people per year.

53. a. $y = 100e^{-0.00012097(1000)}$

$= 100e^{-0.12097}$

$= 100(0.886060541)$

≈ 88.61

Appriximately 88.61 grams remain after 1000 years.

b.
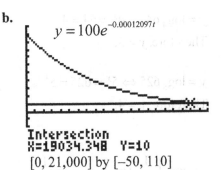
$y = 100e^{-0.00012097t}$

Intersection
X=19034.348   Y=10
[0, 21,000] by [−50, 110]

After approximately 19,034 years, 10 grams of Carbon-14 will remain.

55. a.
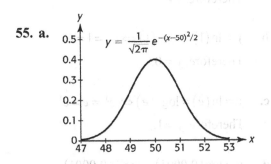
$y = \dfrac{1}{\sqrt{2\pi}}e^{-(x-50)^2/2}$

b. The average score is 50.

Section 5.2 Skills Check

1. $y = \log_3 x \Leftrightarrow 3^y = x$

3. $y = \ln(2x) = \log_e(2x) \Leftrightarrow e^y = 2x$

5. $x = 4^y \Leftrightarrow \log_4 x = y$

7. $32 = 2^5 \Leftrightarrow \log_2 32 = 5$

9. a. 0.845

b. 4.454

c. 4.806

11. a. $y = \log_2 32 \Leftrightarrow 2^y = 32 = 2^5$

Therefore, $y = 5$.

b. $y = \log_9 81 \Leftrightarrow 9^y = 81 = 9^2$

Therefore, $y = 2$.

c. $y = \log_3 27 \Leftrightarrow 3^y = 27 = 3^3$

Therefore, $y = 3$.

d. $y = \log_4 64 \Leftrightarrow 4^y = 64 = 4^3$

Therefore, $y = 3$.

e. $y = \log_5 625 \Leftrightarrow 5^y = 625 = 5^4$

Therefore, $y = 4$.

13. a. $y = \log_3\left(\dfrac{1}{27}\right) \Leftrightarrow 3^y = \dfrac{1}{27} = \dfrac{1}{3^3}$

$3^y = \dfrac{1}{3^3} = 3^{-3}$

Therefore, $y = -3$.

b. $y = \ln(1) = \log_e(1) \Leftrightarrow e^y = 1 = e^0$

Therefore, $y = 0$.

c. $y = \ln(e) = \log_e(e) \Leftrightarrow e^y = e = e^1$

Therefore, $y = 1$.

d. $y = \log(0.0001) = \log_{10}(0.0001)$

$y = \log_{10}(0.0001) \Leftrightarrow 10^y = 0.0001$

$10^y = \dfrac{1}{10,000}$

$10^y = \dfrac{1}{10^4} = 10^{-4}$

Therefore, $y = -4$.

15.

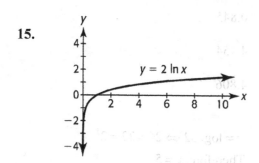

17.

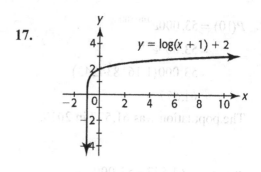

19. a. $y = 4^x$

$x = 4^y \Leftrightarrow \log_4 x = y$

Therefore, the inverse function is

$y = \log_4 x$.

b.

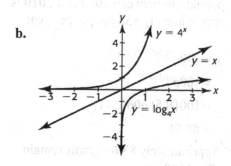

The graphs are symmetric about the line $y = x$.

21. $\log_a a = x \Leftrightarrow a^x = a = a^1$

If $a > 0$ and $a \neq 1$, then $x = 1$,

and therefore, $\log_a a = 1$.

23. $\log 10^{14} = \log_{10} 10^{14}$

$= 14 \log_{10} 10$

$= 14(1) = 14$

25. $10^{\log_{10} 12} = 12$

27. $\log_a(100) = \log_a(20 \cdot 5)$

$\qquad = \log_a(20) + \log_a(5)$

$\qquad = 1.4406 + 0.7740$

$\qquad = 2.2146$

29. $\log_a 5^3 = 3\log_a 5$

$\qquad = 3(0.7740)$

$\qquad = 2.322$

31. $\ln\left(\dfrac{3x-2}{x+1}\right) = \ln(3x-2) - \ln(x+1)$

33. $\log_3 \dfrac{\sqrt[3]{4x+1}}{4x^2}$

$\qquad = \log_3\left(\sqrt[3]{4x+1}\right) - \log_3\left(4x^2\right)$

$\qquad = \log_3\left[(4x+1)^{\frac{1}{3}}\right] - \left[\log_3(4) + \log_3\left(x^2\right)\right]$

$\qquad = \dfrac{1}{3}\log_3(4x+1) - \left[\log_3(4) + 2\log_3(x)\right]$

$\qquad = \dfrac{1}{3}\log_3(4x+1) - \log_3(4) - 2\log_3(x)$

35. $3\log_2 x + \log_2 y$

$\qquad = \log_2 x^3 + \log_2 y$

$\qquad = \log_2\left(x^3 y\right)$

37. $4\ln(2a) - \ln(b)$

$\qquad = \ln(2a)^4 - \ln(b)$

$\qquad = \ln\left(\dfrac{(2a)^4}{b}\right) = \ln\left(\dfrac{16a^4}{b}\right)$

Section 5.2 Exercises

39. a. In 1925, $x = 1925 - 1900 = 25$.

$\qquad f(25) = 11.027 + 14.304\ln(25)$

$\qquad f(25) = 57.0698$

In 1925, the expected life span is approximately 57 years.

In 2007, $x = 2007 - 1900 = 107$.

$\qquad f(107) = 11.027 + 14.304\ln(107)$

$\qquad f(107) = 77.8671$

In 2007, the expected life span is approximately 78 years.

b. Based on the model, life span increased tremendously between 1925 and 2007. The increase could be due to multiple factors, including improved healthcare and nutrition/better diet.

41. a. In 2025, $x = 2025 - 1960 = 65$.

$\qquad y = -35.700 + 27.063\ln(65)$

$\qquad y = 77.271$

In 2025, the number of women in the civilian workforce will be approximately 77.271 million.

In 2030, $x = 2030 - 1960 = 70$.

$\qquad y = -35.700 + 27.063\ln(70)$

$\qquad y = 79.277$

In 2030, the number of women in the civilian workforce will be approximately 79.277 million.

b. Based on the solutions to part a), the function seems to be increasing.

c.

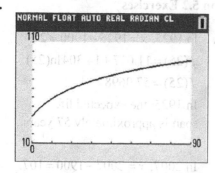

43. $p = 20 + 6\ln(2 \times 5200 + 1)$
$= \$75.50$

45. a.   In 2011, $x = 2011 - 1960 = 51$.
$f(51) = 27.4 + 5.02\ln(51)$
$f(51) = 47.138$
In 2011, the % of female workers in the work force will be 47.1%.

In 2015, $x = 2015 - 1960 = 55$.
$f(55) = 27.4 + 5.02\ln(55)$
$f(55) = 47.517$
In 2015, the % of female workers in the work force will be 47.5%.

b.   Based on part a), it appears the % is increasing.

47. $\dfrac{\ln 2}{0.10} \approx 6.9$ years

49. $n = \dfrac{\log 2}{0.0086}$
$n = 35.0035 \approx 35$

Since it takes approximately 35 quarters for an investment to double under this scenario, then in terms of years the time to double is approximately $\dfrac{35}{4} = 8.75$ years.

51. $t = \dfrac{\ln 2}{\ln(1 + .08)} = 9.0$ years

53. $R = \log\left(\dfrac{I}{I_0}\right)$

$R = \log\left(\dfrac{25{,}000 I_0}{I_0}\right)$

$R = \log(25{,}000) = 4.3979 \approx 4.4$

The earthquake measures 4.4 on the Richter scale.

55. $R = \log\left(\dfrac{I}{I_0}\right)$

$6.4 = \log\left(\dfrac{I}{I_0}\right)$

$10^{6.4} = \left(\dfrac{I}{I_0}\right) = 2{,}511{,}886.4$

Thus $I = 10^{6.4} I_0 = 2{,}511{,}886.4 I_0$.

57. $R = \log\left(\dfrac{I}{I_0}\right)$

$7.1 = \log\left(\dfrac{I}{I_0}\right)$

$10^{7.1} = \left(\dfrac{I}{I_0}\right)$

Thus $I = 10^{7.1} I_0 = 12{,}589{,}254 I_0$.

59. The difference in the Richter scale measurements is $8.25 - 7.1 = 1.15$.

Therefore, the intensity of the 1906 earthquake was $10^{1.15} \approx 14.13$ times as intense as that of the 1989 earthquake.

61. The difference in the Richter scale measurements is $9.0 - 6.8 = 2.2$.

Therefore, the intensity of the 2011 earthquake was $10^{2.2} \approx 158.5$ times as intense as that of the 2008 earthquake.

63. Suppose the intensity of one sound is AI_0, while the intensity of a second sound is $100AI_0$. Then,

$$L_1 = 10\log\left(\frac{AI_0}{I_0}\right) = 10\log(A)$$

$$L_2 = 10\log\left(\frac{100AI_0}{I_0}\right)$$
$$= 10\log(100A)$$
$$= 10(\log 100 + \log A)$$
$$= 20 + 10\log A$$
$$= 20 + L_1$$

As a decibel level, the higher intensity sound measures 20 more than the lower intensity sound.

65. $L = 10\log\left(\frac{I}{I_0}\right)$

$$140 = 10\log_{10}\left(\frac{I}{I_0}\right)$$

$$\log_{10}\left(\frac{I}{I_0}\right) = 14 \Leftrightarrow 10^{14} = \frac{I}{I_0}$$

$$\frac{I}{I_0} = 10^{14}$$

$$I = 10^{14}I_0 = 100{,}000{,}000{,}000{,}000I_0$$

67. $L = 10\log\left(\frac{I}{I_0}\right)$

Let $L = 140$.

$$140 = 10\log\left(\frac{I}{I_0}\right)$$

$$14 = \log\left(\frac{I}{I_0}\right) \Leftrightarrow 10^{14} = \frac{I}{I_0}$$

$$I = 10^{14}I_0$$

Let $L = 120$.

$$120 = 10\log\left(\frac{I}{I_0}\right)$$

$$12 = \log\left(\frac{I}{I_0}\right) \Leftrightarrow 10^{12} = \frac{I}{I_0}$$

$$I = 10^{12}I_0$$

Comparing the intensity levels:

$$\frac{10^{14}}{10^{12}} = 10^2 = 100$$

The decibel level of 140 is one hundred times as intense as a decibel level of 120.

69. $7.79 = -\log\left[H^+\right]$

multiply both sides by -1

$$-7.79 = \log_{10}\left[H^+\right] \Leftrightarrow 10^{-7.79} = \left[H^+\right]$$

$$\left[H^+\right] = 10^{-7.79} \approx 0.0000000162$$

71. If the pH of ketchup is 3.9, then $\left[H^+\right]$ for ketchup $= 10^{-3.9}$, and if the pH for peanut butter is 6.3, then $\left[H^+\right]$ for peanut butter $= 10^{-6.3}$.

Thus, $\dfrac{10^{-3.9}}{10^{-6.3}} = 10^{2.4} = 251.2$

Thus ketchup is 251.2 times as acidic as peanut butter.

Section 5.3 Skills Check

1. $1600 = 10^x$
    $x = \log(1600) \approx 3.204$

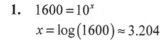
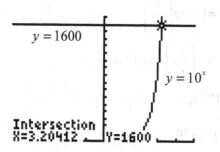

[−5, 5] by [−10, 1700]

3. $2500 = e^x$
    $x = \ln(2500) \approx 7.824$

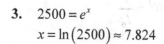
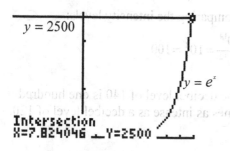

[−5, 8] by [−10, 2600]

5. $8900 = e^{5x}$
    $5x = \ln(8900)$
    $x = \dfrac{\ln(8900)}{5} \approx 1.819$

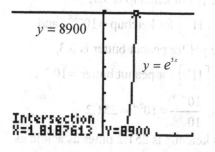

[−5, 5] by [−10, 9000]

7. $4000 = 200e^{8x}$
    $20 = e^{8x}$
    $8x = \ln(20)$
    $x = \dfrac{\ln(20)}{8} \approx 0.374$

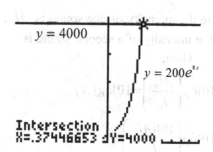

[−1, 1] by [−10, 4200]

9. $8000 = 500\left(10^x\right)$
    $16 = 10^x$
    $x = \log(16) \approx 1.204$

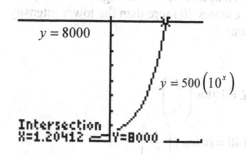

[−2, 2] by [−10, 8200]

11. $\log_6(18) = \dfrac{\ln(18)}{\ln(6)} = 1.6131$

    or

    $\log_6(18) = \dfrac{\log(18)}{\log(6)} = 1.6131$

13. $\log_8\left(\sqrt{2}\right) = \left(\dfrac{\ln\left(\sqrt{2}\right)}{\ln(8)}\right) = 0.1667$

or

$\log_8\left(\sqrt{2}\right) = \left(\dfrac{\log\left(\sqrt{2}\right)}{\log(8)}\right) = 0.1667$

15. $8^x = 1024$

$\left(2^3\right)^x = 2^{10}$

$2^{3x} = 2^{10}$

$3x = 10$

$x = 3.\overline{3} = \dfrac{10}{3}$

17. $2\left(5^{3x}\right) = 31{,}250$

$5^{3x} = 15{,}625 = 5^6$

$3x = 6$

$x = 2$

19. $5^{x-2} = 11.18$

$\ln\left(5^{x-2}\right) = \ln(11.18)$

$(x-2)\ln(5) = \ln(11.18)$

$x - 2 = \dfrac{\ln(11.18)}{\ln(5)}$

$x = \dfrac{\ln(11.18)}{\ln(5)} + 2$

$x \approx 3.5$

21. $18{,}000 = 30\left(2^{12x}\right)$

$600 = 2^{12x}$

$\log(600) = \log\left(2^{12x}\right)$

$12x\log(2) = \log(600)$

$x = \dfrac{\log(600)}{12\log(2)}$

$x \approx 0.769$

23. $\log_2 x = 3 \Leftrightarrow 2^3 = x$

$x = 8$

25. $5 + 2\ln x = 8$

$2\ln x = 3$

$\ln x = \dfrac{3}{2}$

$\log_e x = \dfrac{3}{2} \Leftrightarrow e^{\frac{3}{2}} = x$

$x = e^{\frac{3}{2}} \approx 4.482$

27.

$5 + \ln(8x) = 23 - 2\ln(x)$

$\ln(8x) + 2\ln(x) = 23 - 5$

$\ln(8x) + \ln(x)^2 = 18$

$\ln\left(8x \cdot x^2\right) = 18$

$\ln\left(8x^3\right) = 18$

$8x^3 = e^{18}$

$x = \sqrt[3]{\dfrac{e^{18}}{8}}$

$x = \dfrac{e^6}{2} \approx 201.7$

29.

$2\log(x) - 2 = \log(x - 25)$

$\log\left(x^2\right) - \log(x - 25) = 2$

$\log\left(\dfrac{x^2}{x - 25}\right) = 2$

$\dfrac{x^2}{x - 25} = 10^2 = 100$

$x^2 = 100(x - 25)$

$x^2 = 100x - 2500$

$x^2 - 100x + 2500 = 0$

$(x - 50)(x - 50) = 0$

$x = 50$

31. $\log_3 x + \log_3 9 = 1$

$\log_3 (9 \cdot x) = 1$

$9 \cdot x = 3^1$

$x = 3/9 = 1/3$

33. $\log_2 x = \log_2 5 + 3$

$\log_2 x - \log_2 5 = 3$

$\log_2 \dfrac{x}{5} = 3$

$\dfrac{x}{5} = 2^3 = 8$

$x = 40$

35.

$\log 3x + \log 2x = \log 150$

$\log(2x \cdot 3x) = \log 150$

$\log(6x^2) = \log 150$

$6x^2 = 150$

$x^2 = 25$

$x = \pm 5$, but $x = -5$

does not check in the original

equation since $\log(-10)$ nor

$\log(-15)$ is undefined.

37. $\qquad 3^x < 243$

$\ln(3^x) < \ln(243)$

$x \ln(3) < \ln(243)$

$x < \dfrac{\ln(243)}{\ln(3)}$

$x < 5$

39. $5(2^x) \geq 2560$

$\ln(2^x) \geq \ln\left(\dfrac{2560}{5}\right)$

$x \ln(2) \geq \ln(512)$

$x \geq \dfrac{\ln(512)}{\ln(2)}$

$x \geq 9$

Section 5.3 Exercises

41. $10,880 = 340(2^q)$

$2^q = \dfrac{10,880}{340}$

$2^q = 32$

$\ln(2^q) = \ln(32)$

$q \ln(2) = \ln(32)$

$q = \dfrac{\ln(32)}{\ln(2)}$

$q = 5$

When the price is \$10,880, the quantity
supplied is 5.

43. a. $S = 25,000e^{-0.072x}$

$\dfrac{S}{25,000} = e^{-0.072x}$

$\Leftrightarrow \ln\left(\dfrac{S}{25,000}\right) = -0.072x$

 b. $\ln\left(\dfrac{16,230}{25,000}\right) = -0.072x$

$x = \dfrac{\ln\left(\dfrac{16,230}{25,000}\right)}{-0.072}$

$x = 6$

Six weeks after the completion of the
campaign, the weekly sales fell to
\$16,230.

45. a. $S = 3200e^{-0.08(0)} = 3200e^0 = 3200$

At the end of the ad campaign, daily
sales were \$3200.

b.

$$S = 3200e^{-0.08x}$$
$$1600 = 3200e^{-0.08x}$$
$$\frac{1}{2} = e^{-0.08x}$$
$$-0.08x = \ln\left(\frac{1}{2}\right)$$
$$x = \frac{\ln\left(\frac{1}{2}\right)}{-0.08} = 8.664$$

Approximately 9 days after the completion of the ad campaign, daily sales dropped below half of what they were at the end of the campaign.

47. a. $y = 0.0000966(1.101^x)$

When $x = 100$, $(2000 - 1900)$

$$y = 0.0000966(1.101^{100})$$
$$y = 1.457837 \text{ million}$$

Based on the model, in 2000, the cost of a 30-second Super Bowl ad was $1,457,837.

b. $y = 0.0000966(1.101^x)$

When $x = 115$, $(2015 - 1900)$

$$y = 0.0000966(1.101^{115})$$
$$y = 6.2 \text{ million}$$

Based on the model, in 2015, the cost of a 30-second Super Bowl ad should have been $6.2 million therefore, the model overestimated the cost.

49.
$$20,000 = 40,000e^{-0.05t}$$
$$e^{-0.05t} = 0.5$$
$$\ln\left(e^{-0.05t}\right) = \ln(0.5)$$
$$-0.05t = \ln(0.5)$$
$$t = \frac{\ln(0.5)}{-0.05}$$
$$t = 13.86294361$$

It will take approximately 13.86 years for the $40,000 pension to decrease to $20,000 in purchasing power.

51.
$$200,000 = 100,000e^{0.03t}$$
$$2 = e^{0.03t}$$
$$\ln(2) = \ln\left(e^{0.03t}\right)$$
$$\ln(2) = 0.03t$$
$$t = \frac{\ln(2)}{0.03}$$
$$t = 23.1049$$

It will take approximately 23.1 years for the value of the property to double.

53. a. $A(0) = 500e^{-0.02828(0)} = 500e^0 = 500$
The initial quantity is 500 grams.

b.
$$250 = 500e^{-0.02828t}$$
$$0.5 = e^{-0.02828t}$$
$$\ln(0.5) = \ln\left(e^{-0.02828t}\right)$$
$$-0.02828t = \ln(0.5)$$
$$t = \frac{\ln(0.5)}{-0.02828} = 24.51$$

The half-life, the time it takes the initial quantity to become half, is approximately 24.51 years.

55.

X	Y1
-1	93.967
0	91.759
1	88.741
2	84.618
3	78.986
4	71.292
5	60.781

X=3

The concentration reaches 79% in about 3 hours.

Solving algebraically:

$$79 = 100\left(1 - e^{-0.312(8-t)}\right)$$

$$\frac{79}{100} = 1 - e^{-0.312(8-t)}$$

$$e^{-0.312(8-t)} = 1 - .79 = .21$$

$$\ln\left(e^{-0.312(8-t)}\right) = \ln(.21) = -1.5606$$

$$-0.312(8-t) = -1.5606$$

$$8 - t = \frac{-1.5606}{-0.312} = 5$$

$$t = 3$$

57.

$$335 = 172(1.04^x)$$

$$1.04^x = \frac{335}{172}$$

$$\ln\left(1.04^x\right) = \ln\left(\frac{335}{172}\right)$$

$$x \ln(1.04) = \ln\left(\frac{335}{172}\right)$$

$$t = \frac{\ln\left(\frac{335}{172}\right)}{\ln(1.04)}$$

$$t \approx 17$$

The number of passengers using international air traffic into and out of the U.S. will be 355 million 17 years after 2012 or 2029.

59.

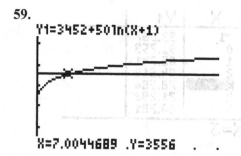

[0, 40] by [3200, 3800]

When the cost is $3556, approximately 7 units are produced.

61. $S = P(1.07)^t$

Note that the initial investment is P and that double the initial investment is $2P$.

$$2P = P(1.07)^t$$

$$2 = 1.07^t$$

$$\ln(2) = \ln\left(1.07^t\right)$$

$$t = \frac{\ln(2)}{\ln(1.07)}$$

The time to double is $\ln(2)$ divided by $\ln(1.07)$.

63.

$$S = 20,000(1 + .07)^t$$

$$48,196.90 = 20,000(1 + .07)^t$$

$$\frac{48,196.90}{20,000} = (1.07)^t$$

$$2.409845 = (1.07)^t$$

$$t = \log_{1.07}(2.409845)$$

$$t = 13$$

65.

$$S = 40,000(1 + .10)^t$$

$$64,420.40 = 40,000(1 + .10)^t$$

$$\frac{64,420.40}{40,000} = (1.10)^t$$

$$1.61051 = (1.10)^t$$

$$t = \log_{1.10}(1.61051)$$

$$t = 5 \text{ years}$$

67. a.

$$f(x) = 11.027 + 14.304 \ln x$$

For $f(x) = 78$,

$$78 = 11.027 + 14.304 \ln x$$

$$\frac{78 - 11.027}{14.304} = \ln x$$

$$4.682 = \ln x$$

$$x = e^{4.682} = 107.99 \approx 108 \text{ years}$$

Thus, for an expected life span of 78 years, the birth year is $1900 + 108 = 2008$.

b.

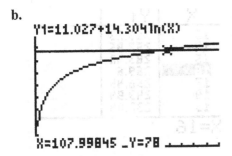

Y1=11.027+14.304ln(X)

X=107.99845 _Y=78

[0, 150] by [0, 100]
Yes, it agrees with part a).

69.

$$8 = 3.294(1.025^x)$$

$$1.025^x = \frac{8}{3.294}$$

$$\ln\left(1.025^x\right) = \ln\left(\frac{8}{3.294}\right)$$

$$x \ln\left(1.025\right) = \ln\left(\frac{8}{3.294}\right)$$

$$t = \frac{\ln\left(\dfrac{8}{3.294}\right)}{\ln(1.025)}$$

$$t \approx 36$$

The number of U.S. citizens with Alzheimer's disease will be 8 million 36 years after 1990 or 2026.

71. $y = 0.0131(1.725^x)$

$1.77 = 0.0131(1.725^x)$

$135.115 = 1.725^x$

$x = \log_{1.725}(135.115) = 9$

Thus $2010 + 9 = 2019$

73. a.

Annual Interest Rate	Rule of 72 Years	Exact Years
2%	36	34.66
3%	24	23.10
4%	18	17.33
5%	14.4	13.86
6%	12	11.55
7%	10.29	9.90
8%	9	8.66
9%	8	7.70
10%	7.2	6.93
11%	6.55	6.30

b. The differences between the two sets of outputs are: 1.34, 0.90, 0.67, 0.54, 0.45, 0.39, 0.34, 0.30, 0.27, and 0.25.

c. As interest rate increases, the estimate gets closer to actual value.

75. a. $n = \log_{1.06} 2 = \dfrac{\ln 2}{\ln 1.06} \approx 11.9$

b. Since the compounding is semi-annual, 11.9 compounding periods correspond to approximately 6 years.

77. $t = \log_{1.08} 3.4$

$$t = \frac{\log 3.4}{\log 1.08}$$

$t = 15.9012328$

$t \approx 15.9$

The future value will be $30,000 in approximately 16 years.

79. $m = 20 \ \ln \dfrac{50}{50 - x}$, for $x < 50$

$$m = 20 \ \ln \dfrac{50}{50 - 45}$$

$$m = 20 \ \ln \dfrac{50}{5} = 20 \ \ln(10)$$

$$m = 20(2.303) = 46.052 \text{ months}$$

Thus, the market share is more than 45% for approximately 46 months.

81. Applying the intersection of graphs method:

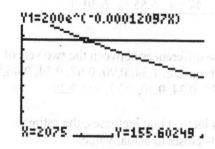

[0, 6000] by [0, 200]

After approximately 2075 years, 155.6 grams of carbon-14 remain.

83. a.

$$S = 600e^{-0.05x}$$

$$269.60 = 600e^{-0.05x}$$

$$e^{-0.05x} = \dfrac{269.60}{600}$$

$$-0.05x = \ln\left(\dfrac{269.60}{600}\right)$$

$$x = \dfrac{\ln\left(\dfrac{269.60}{600}\right)}{-0.05}$$

$$x \approx 16$$

Sixteen weeks after the end of the campaign, sales dropped below $269.60 thousand.

b.

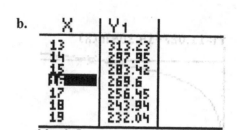

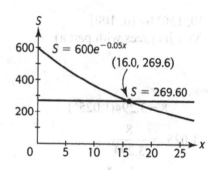

Section 5.4 Skills Check

1.

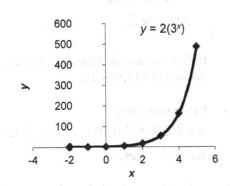

$y = 2(3^x)$

3.

x	$g(x)$	First Differences	Percent Change
1	2.5		
2	6	3.5	140.00%
3	8.5	2.5	41.67%
4	10	1.5	17.65%
5	8	−2	−20.00%
6	6	−2	−25.00%

Since the percent change in the table is both positive and negative, $g(x)$ is not exponential.

5.

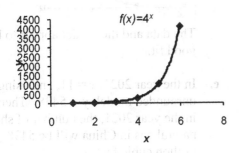

$f(x) = 4^x$

7. a.

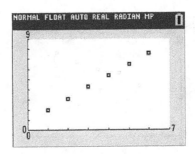

b. Considering the scatter plot from part a), a linear model fits the data very well.

9.

x	y	First Differences	Percent Change
1	2		
2	6	4	200.00%
3	14	8	133.33%
4	34	20	142.86%
5	81	47	138.24%

Since the percent change is approximately constant and the first differences vary, an exponential function will fit the data best. Also, since the first differences are not constant, it cannot be linear.

11. Using technology yields, $y = 0.876(2.494^x)$.

13. a.

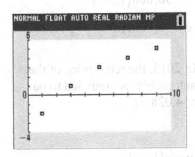

b. Based on the scatter plot, it appears that a logarithmic model fits the data better.

15. a.

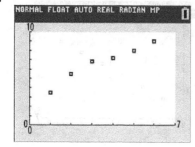

b. Using technology, $y = 3.671x^{0.505}$ is a power function that models the data.

c. Using technology,
$y = -0.125x^2 + 1.886x + 1.960$ is a quadratic function that models the data.

d. Using technology,
$y = 3.468 + 2.917 \ln x$ is a logarithmic function that models the data.

Section 5.4 Exercises

17. a. $y = a(1+r)^x$
$y = 30,000(1+0.04)^t$
$y = 30,000(1.04^t)$

b. $y = 30,000(1.04^t)$
$y = 30,000(1.04^{15}) \approx 54,028.31$

In 2015, the retail price of the automobile is predicted to be $54,028.31.

19. a. $y = a(1+r)^x$
$y = 20,000(1-0.02)^x$
$y = 20,000(0.98^x)$

b. $y = 20,000(0.98^t)$
$= 20,000(0.98^5)$
$= 18,078.42$

In five weeks, the sales are predicted to decline to $18,078.42.

21. a. Using technology,
$y = 492.439(1.070^x)$, correct to three decimal places.

b. Using the unrounded model
for the year 2015, $2015 - 1960 = 55$
$y = \$20,100.80$ billion

c. Using the unrounded model, it will take approximately 54 years to reach 19 trillion. Thus $1960 + 54 =$ the year 2014.

23. a. Using technology, $y = 11.682(1.748^x)$

b.

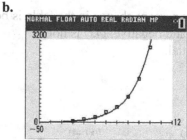

The data and the model appear to be a good fit.

c. In the year 2021, $x = 11$, and using the unrounded model, y = 5438. Therefore, in the year 2021, the number of shale natural gas in China will be 5438 million cubic feet.

d. For y = 9506, x will equal 12 years. In the year 2022 (2010 + 12), the number of shale natural gas in China will be 9506 million cubic feet.

25. a. Using technology, $y = 13.410(1.040^x)$

b.

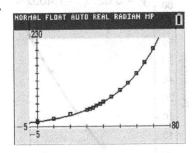

c. In the year 2040, $x = 60$, and using the unrounded model, $y = 137.2$. In the year 2040, the average annual wage will be $137.2 thousand. This is interpolation because the year is included within the data.

d. Using the unrounded model, for $y = 225$, x will equal 72.75 years. In the year 2053 (1980 + 73), the average annual wage will reach $225,000.

27. a. $y = 11.027 + 14.304 \ \ln x$

b. $y = -0.0018x^2 + 0.4880x + 46.2495$

c.

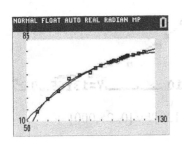

Based on the graph, it appears that the logarithmic model is the better fit.

d. In 2016, $x = 116$.
Using the unrounded model of the logarithmic function, $y = 79.0$.

Using the unrounded model of the quadratic function, $y = 78.6$.

29. a. Using technology,
$y = 3.885 - 0.733 \ \ln x$

b. Using technology, $y = 2.572(0.982^x)$

c. Logarithmic Function

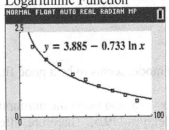

Exponential Function

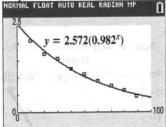

The exponential model seems to fit better.

31. a.

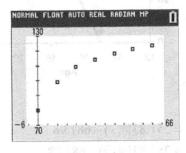

b. Using technology,
$y = 68.371 + 13.424 \ln x$.

The first data point must be omitted because the input cannot be 0 for a logarithmic function.

c.

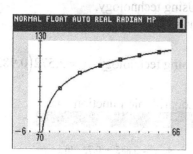

The model seems to be a good fit.

d. Let $x = 45$, and using the unrounded model, y = 119.5. Therefore, after 45 minutes, the interior air temperature will be approximately 119.5 degrees Fahrenheit.

33. a.

Sexually Active Girls, Logarithmic Model

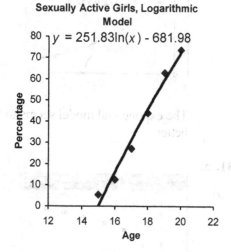

b. $y = 251.83\ln(x) - 681.98$

$y = 251.83\ln(17) - 681.98$

Substituting into the unrounded model yields $y \approx 31.5$.

The percentage of girls 17 or younger who have been sexually active is 31.5%.

c.

Sexually Active Girls, Quadratic Model

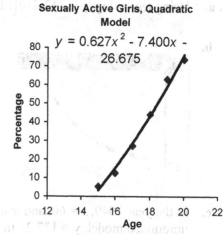

d. Based on the graphs in parts a) and c), the quadratic function seems to be the better fit.

35. a. Based on the figure shown, the data should be modeled by an exponential decay function.

b. Using technology,
$y = 16,278.587(0.979^x)$

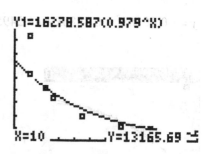

[0, 120] by [0, 25,000]

c. Using technology,
$$y = 210,002.816x^{-1}$$

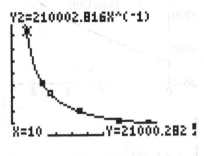

[0, 120] by [0, 25,000]

d. The power function appears to be the better fit to the data.

e. For x = a fuel economy of 100 mpg, according the power model, its lifetime gasoline use would be 2100 gallons.

X	Y2
97	2165
98	2142.9
99	2121.2
100	2100
101	2079.2
102	2058.9
103	2038.9

X=100

Section 5.5 Skills Check

1. $15,000e^{0.06(20)}$
$$= 15,000e^{1.2}$$
$$= 15,000(3.320116923)$$
$$= 49,801.75$$

3. $3000(1.06)^{30}$
$$= 3000(5.743491173)$$
$$= 17,230.47$$

5. $12,000\left(1 + \dfrac{0.10}{4}\right)^{(4)(8)}$
$$= 12,000(1+.025)^{32}$$
$$= 12,000(1.025)^{32}$$
$$= 12,000(2.203756938)$$
$$= 26,445.08$$

7. $P\left(1 + \dfrac{r}{k}\right)^{kn}$
$$= 3000\left(1 + \dfrac{0.08}{2}\right)^{(2)(18)}$$
$$= 3000(1.04)^{36}$$
$$= 12,311.80$$

9. $300\left[\dfrac{1.02^{240} - 1}{0.02}\right]$
$$= 300\left[\dfrac{115.8887352 - 1}{0.02}\right]$$
$$= 300\left[\dfrac{114.8887352}{0.02}\right]$$
$$= 300[5744.436758]$$
$$= 1,723,331.03$$

11. $g(2.5) = 1123.60$
$$g(3) = 1191.00$$
$$g(3.5) = 1191.00$$

13.

$$S = P\left(1+\frac{r}{k}\right)^{kn}$$

$$P\left(1+\frac{r}{k}\right)^{kn} = S$$

$$P = \frac{S}{\left(1+\frac{r}{k}\right)^{kn}}$$

$$P = S\left(1+\frac{r}{k}\right)^{-kn}$$

Section 5.5 Exercises

15. a. $S = P\left(1+\frac{r}{k}\right)^{kt}$

$P = 8800,\ r = 0.08,\ k = 1,\ t = 8$

$S = 8800\left(1+\dfrac{0.08}{1}\right)^{(1)(8)}$

$S = 8800(1.08)^{8}$

$S = 16,288.19$

The future value is $16,288.19.

b. $S = P\left(1+\frac{r}{k}\right)^{kt}$

$P = 8800,\ r = 0.08,\ k = 1,\ t = 30$

$S = 8800\left(1+\dfrac{0.08}{1}\right)^{(1)(30)}$

$S = 8800(1.08)^{30}$

$S = 88,551.38$

The future value is $88,551.38.

17. a.

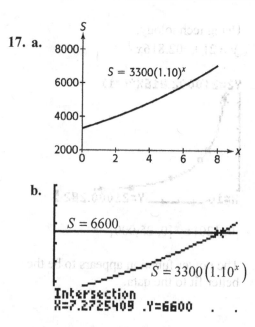

b.

[0, 8] by [2500, 9000]

The initial investment doubles in approximately 7.3 years. After 8 years compounded annually, the initial investment will be more than doubled.

19. $S = P\left(1+\frac{r}{k}\right)^{kt}$

$P = 10,000,\ r = 0.12,\ k = 4,\ t = 10$

$S = 10,000\left(1+\dfrac{0.12}{4}\right)^{(4)(10)}$

$S = 10,000(1.03)^{40}$

$S = 32,620.38$

The future value is $32,620.38.

21. a. $S = P\left(1+\frac{r}{k}\right)^{kt}$

$P = 10,000,\ r = 0.12,\ k = 365,\ t = 10$

$S = 10,000\left(1+\dfrac{0.12}{365}\right)^{(365)(10)}$

$S = 10,000(1.0003287671233)^{3650}$

$S = 33,194.62$

The future value is $33,194.62.

b. Since the compounding occurs more frequently in Exercise 21 than in Exercise 19, the future value in Exercise 21 is greater.

23. $S = P\left(1 + \dfrac{r}{k}\right)^{kt}$

$P = 10,000,\ r = 0.12,\ k = 12,\ t = 15$

$S = 10,000\left(1 + \dfrac{0.12}{12}\right)^{(12)(15)}$

$S = 10,000(1.01)^{180}$

$S = 59,958.02$

The future value is $59,958.02. The interest earned is the future value minus the original investment. In this case, $59,958.02 − $10,000 = $49,958.02.

25. a. $S = Pe^{rt}$

$P = 10,000,\ r = 0.06,\ t = 12$

$S = 10,000e^{(0.06)(12)}$

$S = 10,000e^{0.72}$

$S = 20,544.33$

The future value is $20,544.33.

b. $S = Pe^{rt}$

$P = 10,000,\ r = 0.06,\ t = 18$

$S = 10,000e^{(0.06)(18)}$

$S = 10,000e^{1.08}$

$S = 29,446.80$

The future value is $29,446.80.

27. a. $S = P\left(1 + \dfrac{r}{k}\right)^{kt}$

$P = 10,000,\ r = 0.06,\ k = 1,\ t = 18$

$S = 10,000\left(1 + \dfrac{0.06}{1}\right)^{(1)(18)}$

$S = 10,000(1.06)^{18}$

$S \approx 28,543.39$

The future value is $28,543.39.

b. Continuous compounding yields a higher future value, $29,446.80 − $28,543.39 = $903.41 additional dollars.

29. a. $S = P\left(1 + \dfrac{r}{k}\right)^{kt}$

Doubling the investment implies $S = 2P$.

$2P = P\left(1 + \dfrac{r}{k}\right)^{kt}$

$\dfrac{2P}{P} = \dfrac{P\left(1 + \dfrac{r}{k}\right)^{kt}}{P}$

$2 = \left(1 + \dfrac{r}{k}\right)^{kt}$

$k = 1,\ r = 0.10$

$2 = \left(1 + \dfrac{0.10}{1}\right)^{(1)t}$

$2 = (1.10)^{t}$

Applying the intersection of graphs method:

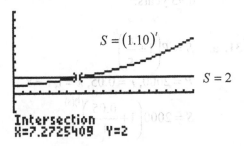

[0, 20] by [−5, 10]
The time to double is approximately 7.27 years.

b. $S = Pe^{rt}$

Doubling the investment implies
$S = 2P$.

$2P = Pe^{rt}$

$\dfrac{2P}{P} = \dfrac{Pe^{rt}}{P}$

$2 = e^{rt}$

$r = 0.10$

$2 = e^{0.10t}$

Applying the intersection of graphs method:

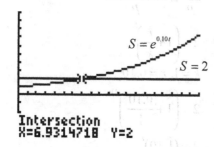

$S = e^{0.10t}$

$S = 2$

Intersection
X=6.9314718   Y=2

[0, 20] by [−5, 10]

The time to double is approximately 6.93 years.

31. a. $S = P\left(1 + \dfrac{r}{k}\right)^{kt}$

$P = 2000,\ r = 0.05,\ k = 1,\ t = 8$

$S = 2000\left(1 + \dfrac{0.05}{1}\right)^{(1)(8)}$

$S = 2000(1.05)^{8}$

$S = 2954.91$

The future value is $2954.91.

b. $S = P\left(1 + \dfrac{r}{k}\right)^{kt}$

$P = 2000,\ r = 0.05,\ k = 1,\ t = 18$

$S = 2000\left(1 + \dfrac{0.05}{1}\right)^{(1)(18)}$

$S = 2000(1.05)^{18}$

$S = 4813.24$

The future value is $4813.24.

33. $S = P\left(1 + \dfrac{r}{k}\right)^{kt}$

$P = 3000,\ r = 0.06,\ k = 12,\ t = 12$

$S = 3000\left(1 + \dfrac{0.06}{12}\right)^{(12)(12)}$

$S = 3000(1.005)^{144}$

$S = 6152.25$

The future value is $6152.25.

35. a.

Years	Future Value
0	1000
7	2000
14	4000
21	8000
28	16,000

b. $S = 1000\left(1 + \dfrac{0.10}{4}\right)^{4t}$

$S = 1000(1.025)^{4t}$

$S = 1000\left((1.025)^{4}\right)^{t}$

$S = 1000(1.104)^{t}$

c.   After five years, the investment is worth
$S = 1000(1.104)^{5} = \$1640.01$.

After 10.5 years, the investment is worth
$S = 1000(1.104)^{10.5} = \$2826.02$.

37.

$$S = P\left(1+\frac{r}{k}\right)^{kt}$$

$$65,000 = P\left(1+\frac{0.10}{12}\right)^{(12)(8)}$$

$$65,000 = P\left(1.008\overline{3}\right)^{96}$$

$$P = \frac{65,000}{\left(1.008\overline{3}\right)^{96}}$$

$$P = \$29,303.36$$

39.

$$S = P\left(1+\frac{r}{k}\right)^{kt}$$

$$10,000 = P\left(1+\frac{0.06}{1}\right)^{(1)(10)}$$

$$10,000 = P\left(1.06\right)^{10}$$

$$P = \frac{10,000}{\left(1.06\right)^{10}}$$

$$P = \$5,583.95$$

An initial amount of \$5583.95 will grow to \$10,000 in 10 years if invested at 6% compounded annually.

41.

$$S = P\left(1+\frac{r}{k}\right)^{kt}$$

$$30,000 = P\left(1+\frac{0.10}{12}\right)^{(12)(18)}$$

$$30,000 = P\left(1.008\overline{3}\right)^{216}$$

$$P = \frac{30,000}{\left(1.008\overline{3}\right)^{216}}$$

$$P = \$4,996.09$$

An initial amount of \$4996.09 will grow to \$30,000 in 18 years if invested at 10% compounded monthly.

43.

$$S = P\left(1+\frac{r}{k}\right)^{kt}$$

$$80,000 = P\left(1+\frac{0.10}{12}\right)^{(12)(12)}$$

$$80,000 = P\left(1.008\overline{3}\right)^{144}$$

$$P = \frac{80,000}{\left(1.008\overline{3}\right)^{144}}$$

$$P = \$24,215.65$$

An initial amount of \$24,215.65 will grow to \$80,000 in 12 years if invested at 10% compounded monthly.

45.

$$40,000 = 10,000\left(1+\frac{0.08}{12}\right)^{12t}$$

$$4 = \left(1+\frac{0.08}{12}\right)^{12t}$$

$$\ln(4) = \ln\left[\left(1+\frac{0.08}{12}\right)^{12t}\right]$$

$$12t\ln\left(1.00\overline{6}\right) = \ln(4)$$

$$t = \frac{\ln(4)}{12\ln\left(1.00\overline{6}\right)}$$

$$t \approx 17.3864$$

It will take approximately 17.39 years, or 17 years and 5 months, for the initial investment of \$10,000 to grow to \$40,000.

47. Applying the intersection of graphs method:

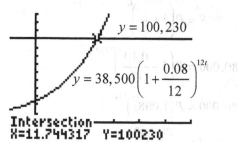

$y = 100,230$

$y = 38,500\left(1 + \dfrac{0.08}{12}\right)^{12t}$

Intersection
X=11.744317   Y=100230

[−5, 30] by [−20,000, 130,000]
After 12 years, the future value of the investment will be greater than $100,230.

49. $140,900 = 15,000e^{28r}$

$$\dfrac{140,900}{15,000} = e^{28r}$$

$$\ln\left(\tfrac{140,900}{15,000}\right) = \ln\left[e^{28r}\right]$$

$$\ln\left(\tfrac{140,900}{15,000}\right) = 28r$$

$$r = \dfrac{\ln\left(\tfrac{140,900}{15,000}\right)}{28}$$

$$r \approx 0.08$$

An interest rate of 8% will allow an initial investment of $15,000 to grow to $140,900 in 28 years.

51. $81,104 = 20,000e^{28r}$

$$\dfrac{81,104}{20,000} = e^{28r}$$

$$\ln\left(\tfrac{81,104}{20,000}\right) = \ln\left[e^{28r}\right]$$

$$\ln\left(\tfrac{81,104}{20,000}\right) = 28r$$

$$r = \dfrac{\ln\left(\tfrac{81,104}{20,000}\right)}{28}$$

$$r \approx 0.05$$

An interest rate of 5% will allow an initial investment of $20,000 to grow to $81,104 in 28 years.

Section 5.6 Skills Check

1. $S = P(1+i)^n$

$$\dfrac{S}{(1+i)^n} = \dfrac{P(1+i)^n}{(1+i)^n}$$

$$P = \dfrac{S}{(1+i)^n}$$

3. $Ai = R\left[1 - (1+i)^{-n}\right]$

$$\dfrac{Ai}{i} = \dfrac{R\left[1 - (1+i)^{-n}\right]}{i}$$

$$A = R\left[\dfrac{1 - (1+i)^{-n}}{i}\right]$$

5.

$$A = R\left[\frac{1-(1+i)^{-n}}{i}\right]$$

$$i\ A = i\left(R\left[\frac{1-(1+i)^{-n}}{i}\right]\right)$$

$$iA = R\left[1-(1+i)^{-n}\right]$$

$$\frac{iA}{\left[1-(1+i)^{-n}\right]} = \frac{R\left[1-(1+i)^{-n}\right]}{\left[1-(1+i)^{-n}\right]}$$

$$R = \frac{iA}{\left[1-(1+i)^{-n}\right]} = A\left[\frac{i}{1-(1+i)^{-n}}\right]$$

Section 5.6 Exercises

7. $A = R\left[\frac{(1+i)^{n}-1}{i}\right]$

$$A = 4000\left[\frac{(1+.06)^{10}-1}{.06}\right] = 52{,}723.18$$

9. $A = R\left[\frac{(1+i)^{n}-1}{i}\right]$

$$A = 1000\left[\frac{(1+.08/2)^{2(8)}-1}{.04}\right] = 21{,}824.53$$

11. $A = R\left[\frac{(1+i)^{n}-1}{i}\right]$

$$A = 600\left[\frac{(1+.07/12)^{12(25)}-1}{.00583}\right] = 486{,}043.02$$

13. $A = R\left[\frac{(1+i)^{n}-1}{i}\right]$

$$A = 1000\left[\frac{(1+.10/2)^{2(4)}-1}{.05}\right] = 9549.11$$

15. $A = R\left[\frac{1-(1+i)^{-n}}{i}\right]$

$$A = 1000\left[\frac{1-(1+0.07)^{-10}}{0.07}\right]$$

$$A = 1000\left[\frac{1-(1.07)^{-10}}{0.07}\right]$$

$$A = 1000\left[\frac{1-(0.5083492921)}{0.07}\right]$$

$$A = 1000[7.023581541]$$

$$A = 7023.58$$

Investing \$7023.58 initially will produce an income of \$1000 per year for 10 years if the interest rate is 7% compounded annually.

17. $A = R\left[\frac{1-(1+i)^{-n}}{i}\right]$

$$A = 50{,}000\left[\frac{1-(1+0.08)^{-19}}{0.08}\right]$$

$$A = 50{,}000\left[\frac{1-(1.08)^{-19}}{0.08}\right]$$

$$A = 50{,}000\left[\frac{1-(0.231712064)}{0.08}\right]$$

$$A = 50{,}000[9.6035992]$$

$$A = 480{,}179.96$$

The formula above calculates the present value of the annuity given the payment made at the end of each period. Twenty total payments were made, but only nineteen occurred at the end of a compounding period. The first payment of $50,000 was made up front. Therefore, the total value of the lottery winnings is
$$\$50,000 + \$480,179.96 = \$530,179.96.$$

19. $A = R\left[\dfrac{1-(1+i)^{-n}}{i}\right]$

$$A = 3000\left[\dfrac{1-\left(1+\dfrac{0.09}{12}\right)^{-(30)(12)}}{\dfrac{0.09}{12}}\right]$$

$$A = 3000\left[\dfrac{1-(1.0075)^{-360}}{0.0075}\right]$$

$$A = 3000\left[\dfrac{1-0.0678860074}{0.0075}\right]$$

$$A = 3000[124.2818657]$$

$$A = 372,845.60$$

The disabled man should seek a lump sum payment of $372,845.60.

21. a.

$$A = R\left[\dfrac{1-(1+i)^{-n}}{i}\right]$$

$$A = 122,000\left[\dfrac{1-(1+0.10)^{-9}}{0.10}\right]$$

$$A = 122,000\left[\dfrac{1-(1.10)^{-9}}{0.10}\right]$$

$$A = 122,000\left[\dfrac{1-0.4240976184}{0.10}\right]$$

$$A = 122,000[5.759023816]$$

$$A = \$702,600.91$$

b.

$$R = A\left[\dfrac{i}{1-(1+i)^{-n}}\right]$$

$$R = 700,000\left[\dfrac{0.10}{1-(1+0.10)^{-9}}\right]$$

$$R = 700,000\left[\dfrac{0.10}{1-(1.10)^{-9}}\right]$$

$$R = 700,000\left[\dfrac{0.10}{1-0.4240976184}\right]$$

$$R = 700,000[0.1736405391]$$

$$R = \$121,548.38$$

The annuity payment is $121,548.38.

c. The $100,000 plus the annuity yields a higher present value and therefore would be the better choice. Over the nine year annuity period, the $100,000 cash plus $122,000 annuity yields $452 more per year than investing $700,000 in cash.

23. a.

$$A = R\left[\dfrac{1-(1+i)^{-n}}{i}\right]$$

$$A = 1600\left[\dfrac{1-\left(1+\dfrac{0.09}{12}\right)^{-(30)(12)}}{\dfrac{0.09}{12}}\right]$$

$$A = 1600\left[\dfrac{1-(1.0075)^{-360}}{0.0075}\right]$$

$$A = 1600\left[\dfrac{1-0.0678860074}{0.0075}\right]$$

$$A = 1600[124.2818657]$$

$$A = \$198,850.99$$

The couple can afford to pay $198,850.99 for a house.

b. ($1600 per month)×(12 months)

  ×(30 years) = $576,000

c. $576,000 − $198,850.99 = $377,149.01

25. a. $\dfrac{8}{4} = 2\%$

b. (4 years)×(4 payments per year)

  = 16 payments

c.

$$R = A\left[\dfrac{i}{1-(1+i)^{-n}}\right]$$

$$R = 10,000\left[\dfrac{0.02}{1-(1+0.02)^{-16}}\right]$$

$$R = 10,000\left[\dfrac{0.02}{1-(1.02)^{-16}}\right]$$

$$R = 10,000\left[\dfrac{0.02}{1-0.7284458137}\right]$$

$$R = 10,000[0.0736501259]$$

$$R = \$736.50$$

The quarterly payment is $736.50.

27. a.

$$i = \dfrac{0.06}{12} = 0.005,\ n = 360$$

$$R = A\left[\dfrac{i}{1-(1+i)^{-n}}\right]$$

$$R = 250,000\left[\dfrac{0.005}{1-(1+0.005)^{-360}}\right]$$

$$R = 250,000\left[\dfrac{0.005}{1-0.166041928}\right]$$

$$R = 250,000\left[\dfrac{0.005}{0.833958072}\right]$$

$$R = 250,000[0.0059955053]$$

$$R = \$1498.88$$

The monthly mortgage payment is $1498.88.

b. (30 years)×(12 payments per year)

  ×($1498.88) = $539,596.80

Including the down payment, the total cost of the house is $639,596.80.

c. $639,596.80 − $350,000 = $289,596.80

Section 5.7 Skills Check

1.
$$\frac{79.514}{1+0.835e^{-0.0298(80)}}$$

$$=\frac{79.514}{1+0.835e^{-2.384}}$$

$$=\frac{79.514}{1+0.835(0.0921811146)}$$

$$=\frac{79.514}{1.076971231}$$

$$=73.83112727$$

$$\approx 73.83$$

3. $1000(0.06)^{0.2^t}$

Let $t = 4$.
$$1000(0.06)^{0.2^4} = 1000(0.06)^{0.0016}$$
$$= 1000(0.9955086592)$$
$$\approx 995.51$$

Let $t = 6$.
$$1000(0.06)^{0.2^6} = 1000(0.06)^{0.000064}$$
$$= 1000(0.9998199579)$$
$$\approx 999.82$$

5. a.

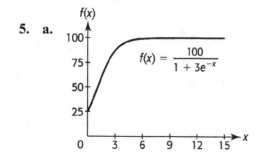

b.

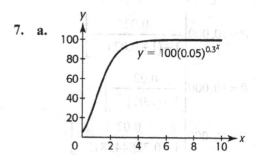

$$f(0) = 25$$
$$f(10) = 99.986 = 99.99$$

c. The graph is increasing.

d. Based on the graph, the y-values of the function approach 100. Therefore the limiting value of the function is 100. $y = c = 100$ is a horizontal asymptote of the function.

7. a.

$y = 100(0.05)^{0.3^x}$

b. Let $x = 0$, and solve for y.
$$y = 100(0.05)^{0.3^0} = 100(0.05)^1 = 5$$
The initial value is 5.

c. The limiting value is c. In this case, $c = 100$.

Section 5.7 Exercises

9. a.

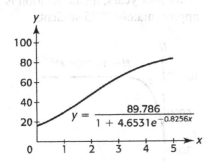

$$y = \frac{5000}{1 + 999e^{-0.8x}}$$

b. At $x = 0$, the number of infected students is the value of the y-intercept of the function. The y-intercept is

$$\frac{5000}{1 + 999e^{-0.8(0)}} = \frac{5000}{1 + 999} = 5.$$

c. The upper limit is $c = 5000$ students.

11. a.

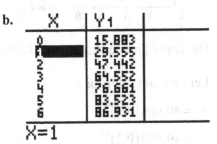

$$y = \frac{89.786}{1 + 4.6531e^{-0.8256x}}$$

b.

X	Y1
0	15.883
1	29.555
2	47.442
3	64.552
4	76.661
5	83.523
6	86.931

X=1

The model indicates that 29.56% of 16-year old boys have been sexually active

c. Consider the table in part b) above.

The model indicates that 86.93% of 21-year old boys have been sexually active

d. The upper limit is $c = 89.786\%$.

13. a. Let $t = 1$ and solve for N.

$$N = \frac{10,000}{1 + 100e^{-0.8(1)}}$$
$$= \frac{10,000}{1 + 100e^{-0.8}}$$
$$= \frac{10,000}{45.393289641}$$
$$\approx 218$$

Approximately 218 people have heard the rumor by the end of the first day.

b. Let $t = 4$ and solve for N.

$$N = \frac{10,000}{1 + 100e^{-0.8(4)}}$$
$$= \frac{10,000}{1 + 100e^{-3.2}}$$
$$= \frac{10,000}{5.076220398}$$
$$\approx 1970$$

Approximately 1970 people have heard the rumor by the end of the fourth day.

c.

X	Y1
3	992.87
4	1970
5	3531.6
6	5485.5
7	7300.4
8	8575.2
9	9305.3

X=7

By the end of the seventh day, 7300 people have heard the rumor.

15. a. $y = \dfrac{89.786}{1 + 4.6531e^{-0.8256x}}$

b. Yes, the models are the same.

c. $y = 14.137x + 17.624$

d.

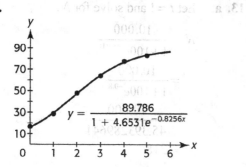

$$y = \frac{89.786}{1 + 4.6531e^{-0.8256x}}$$

The logistic model is a better fit.

17. a. $y = \dfrac{10.641}{1 + 53.050e^{-2.201x}}$

b.

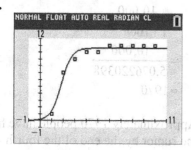

c. The model predicts the maximum will be 10.641 million. This value is close but less than the projections given in the table.

19. a. $y = \dfrac{82.488}{1 + 0.816e^{-0.024x}}$

b. The expected life span for a person born in 1955 was 67.9 years, and in 2006, it was 77.7 years.

c. The upper limit for a person's life span is 82.5 years.

21. a. Let $t = 0$ and solve for N.

$$\begin{aligned} N &= 10,000(0.4)^{0.2^{0}} \\ &= 10,000(0.4)^{1} \\ &= 10,000(0.4) \\ &= 4000 \end{aligned}$$

The initial population size is 4000 students.

b. Let $t = 4$ and solve for N.

$$\begin{aligned} N &= 10,000(0.4)^{0.2^{4}} \\ &= 10,000(0.4)^{0.0016} \\ &= 10,000(0.998535009) \\ &= 9985.35009 \\ &\approx 9985 \end{aligned}$$

After four years, the population is approximately 9985 students.

c.

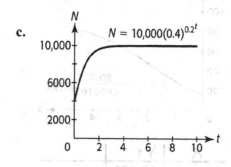

The upper limit appears to be 10,000.

23. a. Let $t = 1$ and solve for N.

$$\begin{aligned} N &= 40,000(0.2)^{0.4^{1}} \\ &= 40,000(0.2)^{0.4} \\ &= 40,000(0.5253055609) \\ &= 21,012.22244 \\ &\approx 21,012 \end{aligned}$$

After one month, the approximately 21,012 units will be sold.

b.
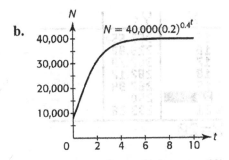

c. The upper limit appears to be 40,000.

25. a. Let $t = 0$ and solve for N.

$$N = 1000(0.01)^{0.5^0}$$
$$= 1000(0.01)^1$$
$$= 1000(0.01)$$
$$= 10$$

Initially the company had 10 employees.

b. Let $t = 1$ and solve for N.

$$N = 1000(0.01)^{0.5^1}$$
$$= 1000(0.01)^{0.5}$$
$$= 1000(0.1)$$
$$= 100$$

After one year, the company had 100 employees.

c. The upper limit is 1000 employees.

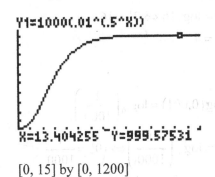

[0, 15] by [0, 1200]

d.
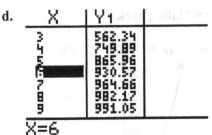

In the sixth year 930 people were employed by the company.

27.
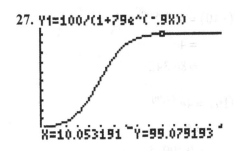

[0, 15] by [0, 120]

After 10 days, 99 people are infected.

29.
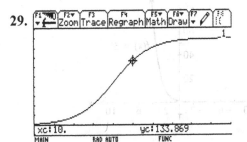

[0, 20] by [−20, 200]

In approximately 11 years, the deer population reaches a level of 150.

Chapter 5 Skills Check

1. **a.**

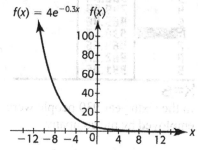

$f(x) = 4e^{-0.3x}$

b. $f(-10) = 4e^{-0.3(-10)}$

$= 4e^{3}$

≈ 80.342

$f(10) = 4e^{-0.3(10)}$

$= 4e^{-3}$

≈ 0.19915

3.

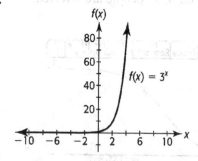

$f(x) = 3^{x}$

5. The graph in Exercise 4 is shifted right one unit and up four units in comparison with the graph in Exercise 3.

7. **a.** $y = 1000(2)^{-0.1x}$

$= 1000(2)^{-0.1(10)}$

$= 1000(2)^{-1}$

$= 500$

b.

X	Y1
15	353.55
16	329.88
17	307.79
18	287.17
19	267.94
20	250
21	233.26

X=20

When $y = 250$, $x = 20$.

9. $y = 7^{3x} \Leftrightarrow \log_7 y = 3x$

11. $y = \log(x) = \log_{10} x$

$y = \log_{10} x \Leftrightarrow x = 10^{y}$

13. $y = 4^{x}$

$x = 4^{y}$

$x = 4^{y} \Leftrightarrow \log_4 x = y$

Therefore, the inverse function is

$y = \log_4 x$.

15. $\ln 56 = \log_e 56 = 4.0254$

17. $\log_2 16$

$y = \log_2 16 \Leftrightarrow 2^{y} = 16$

$y = 4$

19. $\log(0.001) = \log_{10}\left(\dfrac{1}{1000}\right)$

$y = \log_{10}\left(\dfrac{1}{1000}\right) \Leftrightarrow 10^{y} = \dfrac{1}{1000}$

$10^{y} = \dfrac{1}{1000} = \dfrac{1}{10^{3}} = 10^{-3}$

$y = -3$

21. $\log_8(56) = \dfrac{\ln(56)}{\ln(8)} = 1.9358$

23.

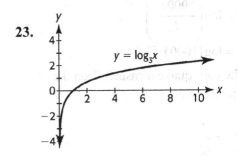

25. $1500 = 300e^{8x}$

$\dfrac{1500}{300} = \dfrac{300e^{8x}}{300}$

$5 = e^{8x}$

$8x = \ln(5)$

$x = \dfrac{\ln(5)}{8}$

$x \approx 0.2012$

27.

$4(3^x) = 36$

$3^x = 9$

$\log(3^x) = \log(9)$

$x\log(3) = \log(9)$

$x = \dfrac{\log(9)}{\log(3)}$

$x = 2$, or since

$3^x = 9 = 3^2$

$x = 2$

29. $6\log_4 x - 2\log_4 y$

$= \log_4 x^6 - \log_4 y^2$

$= \log_4\left(\dfrac{x^6}{y^2}\right)$

31. $P\left(1 + \dfrac{r}{k}\right)^{kn}$

$= 1000\left(1 + \dfrac{0.08}{12}\right)^{(12)(20)}$

$= 1000\left(1.00\overline{6}\right)^{240}$

≈ 4926.80

33. a.

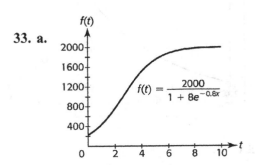

b. $f(0) = \dfrac{2000}{1 + 8e^{-0.8(0)}}$

$= \dfrac{2000}{1 + 8e^0}$

$= \dfrac{2000}{9}$

≈ 222.22

$f(8) = \dfrac{2000}{1 + 8e^{-0.8(8)}}$

$= \dfrac{2000}{1 + 8e^{-6.4}}$

$= \dfrac{2000}{1.013292458}$

≈ 1973.76

c. The limiting value of the function is 2000.

Chapter 5 Review Exercises

35. a. $P = 2969e^{0.051t}$
This model is exponential growth since the base (e) is > 1, and the exponent has a positive coefficient.

b.

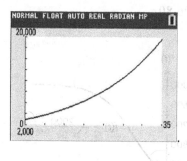

36. Let $x = 4$.

$$y = 2000(2)^{-0.1(4)}$$
$$= 2000(2)^{-0.4}$$
$$= 2000(0.7578582833)$$
$$\approx 1515.72$$

Four weeks after the end of the advertising campaign, the daily sales in dollars will be $1515.72.

37.

$$1000 = 2000(2)^{-0.1x}$$
$$0.5 = 2^{-0.1x}$$
$$\ln(0.5) = \ln\left(2^{-0.1x}\right)$$
$$\ln(0.5) = -0.1x \ln 2$$
$$x = \frac{\ln 0.5}{-0.1 \ln 2}$$
$$x = 10$$

In 10 weeks, sales will decay by half).

38. a. $R = \log\left(\dfrac{I}{I_0}\right)$

$$R = \log\left(\frac{1000 I_0}{I_0}\right)$$
$$R = \log(1000) = 3$$

The earthquake measures 3 on the Richter scale.

b. $10^R = \dfrac{I}{I_0}$

$$I = 10^R I_0$$
$$I = 10^{6.5} I_0$$
$$I = 3,162,277.66 I_0$$

39. The difference in the Richter scale measurements is $7.9 - 4.8 = 3.1$. Therefore the intensity of the Indian earthquake was $10^{3.1} \approx 1259$ times as intense as the American earthquake.

40. $t = \log_{1.12} 3 = \dfrac{\ln 3}{\ln 1.12} \approx 9.69$

The investment will triple in approximately 10 years.

41. a. $S = 1000(2)^{\left(\frac{x}{7}\right)}$

$$\frac{S}{1000} = (2)^{\left(\frac{x}{7}\right)} \Leftrightarrow \log_2\left(\frac{S}{1000}\right) = \frac{x}{7}$$
$$x = 7\log_2\left(\frac{S}{1000}\right)$$

b. $x = 7\log_2\left(\dfrac{19,504}{1000}\right)$

$$= 7\log_2(19.504)$$
$$= 7\left(\frac{\ln 19.504}{\ln 2}\right)$$
$$\approx 29.99989$$

In about 30 years, the future value will be $19,504.

42. $15.66 = 3.196 + 3.975 \ln x$

$12.464 = 3.975 \ln x$

$\frac{12.464}{3.975} = \ln x$

$e^{\frac{12.464}{3.975}} = x$

$23 = x$

In the year 2023 (2000 +23), the annual increase is projected to be 15.66 million.

43. $L = 10 \log\left(\frac{I}{I_0}\right)$

Let $L = 90$.

$90 = 10 \log\left(\frac{I}{I_0}\right)$

$9 = \log\left(\frac{I}{I_0}\right) \Leftrightarrow 10^9 = \frac{I}{I_0}$

$I = 10^9 I_0$

Let $L = 30$.

$30 = 10 \log\left(\frac{I}{I_0}\right)$

$3 = \log\left(\frac{I}{I_0}\right) \Leftrightarrow 10^3 = \frac{I}{I_0}$

$I = 10^3 I_0$

Comparing the intensity levels:

$\frac{10^9}{10^3} = 10^6 = 1,000,000$

The decibel level of 90 is one million times as loud as a decibel level of 30.

44. a. $P(x) = R(x) - C(x)$

$P(x) = 10(1.26^x) - (2x + 50)$

$= 10(1.26^x) - 2x - 50$

b. Applying the intersection of graphs method $P(x)$ is in thousands:

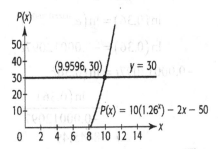

Selling at least 10 mobile homes produces a profit of at least \$30,000.

45. Applying the intersection of graphs method:

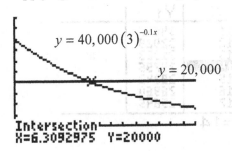

$[0, 15]$ by $[-7500, 50,000]$

After seven weeks, sales will be less than half.

46. a. $y = 100e^{-0.00012097(5000)}$

$= 100e^{-0.60485}$

$= 100(0.5461563439)$

≈ 54.62

After 5000 years, approximately 54.62 grams of carbon-14 remains.

b.

$$36\% y_0 = y_0 e^{-0.00012097t}$$

$$0.36 = e^{-0.00012097t}$$

$$\ln(0.36) = \ln\left(e^{-0.00012097t}\right)$$

$$\ln(0.36) = -0.00012097t$$

$$-0.00012097t = \ln(0.36)$$

$$t = \frac{\ln(0.36)}{-0.00012097}$$

$$t \approx 8445.49$$

The wood was cut approximately 8445 years ago.

47. $P = 60000e^{-0.05t}$

X	Y1
11	34617
12	32929
13	31323
14	29795
15	28342
16	26960
17	25645

X=14

After approximately 14 years, the purchasing power will be less than half of the original $60,000 income.

48. $S = 2000e^{0.08(10)} = 2000e^{0.8} \approx 4451.08$

The future value is approximately $4451.08 after 10 years.

49.

$$13,784.92 = 3300(1.10)^x$$

$$(1.10)^x = 4.177248485$$

$$\ln\left[(1.10)^x\right] = \ln[4.177248485]$$

$$x\ln(1.10) = \ln(4.177248485)$$

$$x = \frac{\ln(4.177248485)}{\ln(1.10)}$$

$$x \approx 15$$

The investment reaches the indicated value in 15 years.

50. a. Using technology, $y = 2.7(1.057^x)$

b. Using the unrounded model for $x = 10$, the national health care spending in 2021 is projected to be $4.7 trillion.

51. a. Using technology, $y = 172(1.04^x)$

b. Using the unrounded model for $x = 20$, the number of passengers using international air traffic into and out of the U.S. in 2032 is projected to be 376.9 million.

c. Using technology and the unrounded model for $y = 254.6$, $x = 10$. Therefore, the number of passengers using international air traffic into and out of the U.S. is projected to be 254.6 million in the year 2022 (2012 + 10).

52. a. Using technology, $y = 81.041(0.981^x)$. This function would be decreasing.

b.

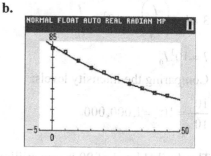

c. Using the unrounded model for $x = 42$, the energy use in 2032 is projected to be approximately 36% of GDP.

53. a. Using technology,
$y = 133.240 + 9.373 \ln x$

b.

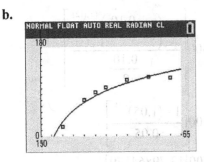

c. Using the unrounded model for $x = 29$, the number of White, non-Hispanic individuals 16 years and older in the U.S. civilian non-institutional population in 2019 is projected to be 164.8 million.

54. Using technology, $y = 4.337 + 40.890 \ln x$

55. a. Using technology, $y = \dfrac{120}{1 + 5.26e^{-0.0639x}}$

b. Using the unrounded model for $x = 28$, the number of U.S. adults with diabetes in 2028 is projected to be 63.9 million.

c. According to the model, the maximum number of U.S. adults with diabetes will be 120 million.

d. Using the unrounded model for $y = 80$, $x \approx 36.8$. Therefore, the number of U.S. adults with diabetes is projected to be 80 million in the year 2037 (2000 + 37).

56. a. Using technology, $y = \dfrac{21,240}{1 + 0.913e^{-0.2882x}}$

b. Using the unrounded model for $x = 15$, the number of Starbuck stores in 2020 is projected to be 20,986.

57. a. Using technology,
$y = -823.535 + 558.720 \ln x$

b. Using the unrounded model for $x = 37$, the number of endangered species in 2017 is projected to be 1194.

c. Using the unrounded model for $y = 1303$, $x = 45$. Therefore, the number of endangered species is projected to be 1303 in the year 2025 (1980 + 45).

58. $S = P\left(1 + \dfrac{r}{k}\right)^{kt}$

$S = 20,000\left(1 + \dfrac{0.06}{1}\right)^{(1)(7)}$

$S = 20,000(1.06)^7 \approx 30,072.61$

The future value is $30,072.61.

59.

$S = R\left[\dfrac{(1+i)^n - 1}{i}\right]$

$S = 1000\left[\dfrac{\left(1 + \dfrac{0.12}{4}\right)^{(4)(6)} - 1}{\dfrac{0.12}{4}}\right]$

$S = 1000\left[\dfrac{(1.03)^{24} - 1}{0.03}\right]$

$S = 1000(34.42647022) \approx 34,426.47$

The future value is $34,426.47.

60. $S = R\left[\dfrac{(1+i)^n - 1}{i}\right]$

$$S = 1500\left[\dfrac{\left(1 + \dfrac{0.08}{12}\right)^{(12)(10)} - 1}{\dfrac{0.08}{12}}\right]$$

$$S = 1500\left[\dfrac{(1.00\overline{6})^{120} - 1}{0.00\overline{6}}\right]$$

$$S = 1500(182.9460352)$$

$$S \approx 274,419.05$$

The future value is \$274,419.05.

61. $A = R\left[\dfrac{1-(1+i)^{-n}}{i}\right]$

$$A = 2000\left[\dfrac{1 - \left(1 + \dfrac{0.08}{12}\right)^{-(12)(15)}}{\dfrac{0.08}{12}}\right]$$

$$A = 2000\left[\dfrac{1 - (1.00\overline{6})^{-180}}{0.00\overline{6}}\right]$$

$$A = 2000[104.6405922] \approx 209,281.18$$

The formula above calculates the present value of the annuity given the payment made at the end of each period. The present value is \$209,218.18.

62. $A = R\left[\dfrac{1-(1+i)^{-n}}{i}\right]$

$$A = 500\left[\dfrac{1 - \left(1 + \dfrac{0.10}{2}\right)^{-(2)(12)}}{\dfrac{0.10}{2}}\right]$$

$$A = 500\left[\dfrac{1 - (1.05)^{-24}}{0.05}\right]$$

$$A = 500[13.79864179]$$

$$A \approx 6899.32$$

The formula above calculates the present value of the annuity given the payment made at the end of each period. The present value is \$6899.32.

63. $R = A\left[\dfrac{i}{1-(1+i)^{-n}}\right]$

$$R = 2000\left[\dfrac{\dfrac{0.12}{12}}{1 - \left(1 + \dfrac{0.12}{12}\right)^{-36}}\right]$$

$$R = 2000\left[\dfrac{0.01}{1 - (1.01)^{-36}}\right]$$

$$R = 2000[0.0332143098]$$

$$R \approx 66.43$$

The monthly payment is \$66.43.

64. $R = A\left[\dfrac{i}{1-(1+i)^{-n}}\right]$

$R = 120{,}000\left[\dfrac{\dfrac{0.06}{12}}{1-\left(1+\dfrac{0.06}{12}\right)^{-(12)(25)}}\right]$

$R = 120{,}000\left[\dfrac{0.005}{1-(1.005)^{-300}}\right]$

$R = 120{,}000[0.006443014]$

$R \approx 773.16$

The monthly payment is \$773.16.

65. a. In 1990, $x = 1990 - 1960 = 30$.

$y = \dfrac{44.472}{1+6.870e^{-0.0782(30)}}$

$= \dfrac{44.472}{1+6.870e^{-2.346}}$

$= \dfrac{44.472}{1+0.6578121375}$

≈ 26.989

Based on the model, the percentage of live births to unmarried mothers in 1990 was 26.989%.

In 1996, $x = 1996 - 1960 = 36$.

$y = \dfrac{44.742}{1+6.870e^{-0.0782(36)}}$

$= \dfrac{44.742}{1+6.870e^{-2.8152}}$

$= \dfrac{44.742}{1+0.4114631169}$

≈ 31.699

Based on the model, the percentage of live births to unmarried mothers in 1996 was 31.699%.

b. The upper limit on the percentage of live births to unmarried mothers is 44.742%.

66. a. Let $x = 14$.

$y = \dfrac{1400}{1+200e^{-0.5(14)}}$

$= \dfrac{1400}{1+200e^{-7}}$

$= \dfrac{1400}{1+0.1823763931}$

≈ 1184.06

After 14 days, approximately 1184 students are infected.

b.

X	Y1
13	1076.4
14	1184.1
15	1260.6
16	1312
17	1345.3
18	1366.3
19	1379.4

X=16

After 16 days, 1312 students are infected.

67. a. $N = 4000(0.06)^{0.4^{(2-1)}}$

$= 4000(0.06)^{0.4^{1}}$

$= 4000(0.06)^{0.4}$

≈ 1298.13

At the beginning of the second year, the enrollment will be approximately 1298 students.

b. $N = 4000(0.06)^{0.4^{(10-1)}}$

$= 4000(0.06)^{0.4^{9}}$

$= 4000(0.06)^{0.000262144}$

≈ 3997.05

At the beginning of the tenth year, the enrollment will be approximately 3997 students.

c. The upper limit on the number of students is 4000.

68. a. $N = 18,000(0.03)^{0.4^{10}}$

$$= 18,000(0.03)^{0.0001048576}$$

$$\approx 17,993.38$$

After ten months, the number of units sold in a month will be approximately 17,993.

b. The upper limit on the number of units sold per month is 18,000.

69. $202,760 = 50,000e^{0.07t}$

$$\frac{202,760}{50,000} = e^{0.07t}$$

$$\ln\left(\frac{202,760}{50,000}\right) = \ln\left[e^{0.07t}\right]$$

$$\ln\left(\frac{202,760}{50,000}\right) = 0.07t$$

$$t = \frac{\ln\left(\frac{202,760}{50,000}\right)}{0.07}$$

$$t \approx 20$$

It will take 20 years for an initial investment of $50,000 to grow to $202,760 at an interest rate of 7%.

70. $121,656 = 30,000e^{28r}$

$$\frac{121,656}{30,000} = e^{28r}$$

$$\ln\left(\frac{121,656}{30,000}\right) = \ln\left[e^{28r}\right]$$

$$\ln\left(\frac{121,656}{30,000}\right) = 28r$$

$$r = \frac{\ln\left(\frac{121,656}{30,000}\right)}{28}$$

$$r \approx 0.05$$

An interest rate of 5% will allow an initial investment of $30,000 to grow to $121,656 in 28 years.

71. $45,552.97 = 25,000e^{15r}$

$$\frac{45,552.97}{25,000} = e^{15r}$$

$$\ln\left(\frac{45,552.97}{25,000}\right) = \ln\left[e^{15r}\right]$$

$$\ln\left(\frac{45,552.97}{25,000}\right) = 15r$$

$$r = \frac{\ln\left(\frac{45,552.97}{25,000}\right)}{15}$$

$$r \approx 0.04$$

An interest rate of 4% will allow an initial investment of $25,000 to grow to $45,552.97 in 15 years.

Group Activity/Extended Applications

1. The first person on the list receives $36. Each of the original six people on the list sends their letter to six people. Therefore, 36 people receive letters with the original six names, and each of the 36 forwards a dollar to the first person on the original list.

2. The 36 people receiving the first letter place their name on the bottom of the list, shift up the second person to first place. The 36 people send out six letters each, for a total of $36 \times 6 = 216$ letters. Therefore the second person on the original list receives $216.

3.

Cycle Number	Money Sent to the Person on Top of the List
1	$6^2 = 36$
2	$6^3 = 216$
3	$6^4 = 1296$
4	$6^5 = 7776$
5	$6^6 = 46,656$

4. Position 5 generates the most money!

5. QuadReg
    y=ax²+bx+c
    a=5914.285714
    b=-25405.71429
    c=22356

    PwrReg
    y=a*x^b
    a=20.33965715
    b=4.338874682

    ExpReg
    y=a*b^x
    a=6
    b=6

The exponential model, $y = 6(6)^x = 6^{x+1}$, fits the data exactly.

6. $y = 6^{6+1} = 6^7 = 279,936$
    The sixth person on the original list receives $279,936.

7. The total number of responses on the sixth cycle would be
    $6 + 36 + 216 + 1296 + 7776 + 46,656 + 279,936 = 335,922$

8. $y = 6^{10+1} = 6^{11} = 362,797,056$
    On the tenth cycle 362,797,056 people receive the chain letter and are supposed to respond with $1.00 to the first name on the list.

9. The answer to problem 8 is larger than the U.S. population. There is no unsolicited person in the U.S. to whom to send the letter.

10. Chain letters are illegal since people entering lower on the chain have a very small chance of earning money from the scheme.

Chapter 6
Higher-Degree Polynomial and Rational
Functions

Toolbox Exercises

1.   **a.**   The polynomial is 4th degree.

     b.   The leading coefficient is 3.

2.   **a.**   The polynomial is 3rd degree.

     b.   The leading coefficient is 5.

3.   **a.**   The polynomial is 5th degree.

     b.   The leading coefficient is -14.

4.   **a.**   The polynomial is 6th degree.

     b.   The leading coefficient is -8.

5. $4x^3 - 8x^2 - 140x$
$$= 4x\left(x^2 - 2x - 35\right)$$
$$= 4x(x-7)(x+5)$$

6. $4x^2 + 7x^3 - 2x^4$
$$= -2x^4 + 7x^3 + 4x^2$$
$$= -1x^2\left(2x^2 - 7x - 4\right)$$
$$= -x^2(2x+1)(x-4)$$

7. $x^4 - 13x^2 + 36$
$$= \left(x^2 - 9\right)\left(x^2 - 4\right)$$
$$= (x+3)(x-3)(x+2)(x-2)$$

8. $x^4 - 21x^2 + 80$
$$= \left(x^2 - 16\right)\left(x^2 - 5\right)$$
$$= (x+4)(x-4)\left(x^2 - 5\right)$$

9. $2x^4 - 8x^2 + 8$
$$= 2\left(x^4 - 4x^2 + 4\right)$$
$$= 2\left(x^2 - 2\right)\left(x^2 - 2\right)$$
$$= 2\left(x^2 - 2\right)^2$$

10. $3x^5 - 24x^3 + 48x$
$$= 3x\left(x^4 - 8x^2 + 16\right)$$
$$= 3x\left(x^2 - 4\right)\left(x^2 - 4\right)$$
$$= 3x(x+2)(x-2)(x+2)(x-2)$$
$$= 3x(x+2)^2(x-2)^2$$

11. $\dfrac{x - 3y}{3x - 9y} = \dfrac{x - 3y}{3(x - 3y)} = \dfrac{1}{3}$

12. $\dfrac{x^2 - 9}{4x + 12} = \dfrac{(x+3)(x-3)}{4(x+3)} = \dfrac{x-3}{4}$

13. $\dfrac{2y^3 - 2y}{y^2 - y}$
$$= \dfrac{2y\left(y^2 - 1\right)}{y(y-1)}$$
$$= \dfrac{2y(y+1)(y-1)}{y(y-1)}$$
$$= 2(y+1) = 2y + 2$$

14. $\dfrac{4x^3-3x}{x^2-x}$

$=\dfrac{x(4x^2-3)}{x(x-1)}$

$=\dfrac{4x^2-3}{x-1}$

15. $\dfrac{x^2-6x+8}{x^2-16}$

$=\dfrac{(x-4)(x-2)}{(x+4)(x-4)}$

$=\dfrac{x-2}{x+4}$

16. $\dfrac{3x^2-7x-6}{x^2-4x+3}$

$=\dfrac{(3x+2)(x-3)}{(x-3)(x-1)}$

$=\dfrac{3x+2}{x-1}$

17. $\dfrac{6x^3}{8y^3}\cdot\dfrac{16x}{9y^2}\cdot\dfrac{15y^4}{x^3}$

$=\dfrac{1440x^4y^4}{72x^3y^5}$

$=\dfrac{20x}{y}$

18. $\dfrac{x-3}{x^3}\cdot\dfrac{x(x-4)}{(x-4)(x-3)}$

$=\dfrac{x}{x^3}$

$=\dfrac{1}{x^2}$

19. $(x+2)(x-2)\left(\dfrac{2x-3}{x+2}\right)$

$=(x-2)(2x-3)$

$=2x^2-7x+6$

20. $\dfrac{x^2-x-6}{1}\div\dfrac{9-x^2}{x^2+3x}$

$=\dfrac{x^2-x-6}{1}\cdot\dfrac{x^2+3x}{9-x^2}$

$=\dfrac{(x-3)(x+2)}{1}\cdot\dfrac{x(x+3)}{-(x+3)(x-3)}$

$=-x(x+2)$

21. $\dfrac{4x+4}{x-4}\div\dfrac{8x^2+8x}{x^2-6x+8}$

$=\dfrac{4x+4}{x-4}\cdot\dfrac{x^2-6x+8}{8x^2+8x}$

$=\dfrac{4(x+1)}{x-4}\cdot\dfrac{(x-2)(x-4)}{8x(x+1)}$

$=\dfrac{x-2}{2x}$

22. $\dfrac{6x^2}{4x^2y-12xy}\div\dfrac{3x^2+12x}{x^2+x-12}$

$=\dfrac{6x^2}{4x^2y-12xy}\cdot\dfrac{x^2+x-12}{3x^2+12x}$

$=\dfrac{6x^2}{4xy(x-3)}\cdot\dfrac{(x+4)(x-3)}{3x(x+4)}$

$=\dfrac{1}{2y}$

23. $\dfrac{x^2+x}{x^2-5x+6} \cdot \dfrac{x^2-2x-3}{2x+4} \div \dfrac{x^3-3x^2}{4-x^2}$

$=\dfrac{x^2+x}{x^2-5x+6} \cdot \dfrac{x^2-2x-3}{2x+4} \cdot \dfrac{4-x^2}{x^3-3x^2}$

$=\dfrac{-x(x+1)(x-3)(x+1)(x-2)(x+2)}{(x-3)(x-2)2(x+2)x^2(x-3)}$

$=\dfrac{-(x+1)^2}{2x(x-3)}$

24. $\dfrac{6x-2}{3xy}+\dfrac{3x+2}{3xy}$ LCD: $3xy$

$=\dfrac{9x}{3xy}$

$=\dfrac{3}{y}$

25. $\dfrac{2x+3}{x^2-1}+\dfrac{4x+3}{x^2-1}$ LCD: x^2-1

$=\dfrac{6x+6}{x^2-1}$

$=\dfrac{6(x+1)}{(x+1)(x-1)}$

$=\dfrac{6}{x-1}$

26. $3+\dfrac{1}{x^2}-\dfrac{2}{x^3}$ LCD: x^3

$=\dfrac{3x^3}{x^3}+\dfrac{x}{x^3}-\dfrac{2}{x^3}$

$=\dfrac{3x^3+x-2}{x^3}$

27.

$\dfrac{5}{x}-\dfrac{x-2}{x^2}+\dfrac{4}{x^3}$ LCD: x^3

$=\dfrac{5x^2}{x^3}-\dfrac{x(x-2)}{x^3}+\dfrac{4}{x^3}$

$=\dfrac{5x^2-(x^2-2x)+4}{x^3}$

$=\dfrac{5x^2-x^2+2x+4}{x^3}$

$=\dfrac{4x^2+2x+4}{x^3}$

28.

$$\frac{a}{a^2-2a}-\frac{a-2}{a^2}=\frac{a}{a(a-2)}-\frac{a-2}{a^2} \qquad \text{LCD: } a^2(a-2)$$

$$=\frac{a(a)}{a(a)(a-2)}-\frac{(a-2)(a-2)}{a^2(a-2)}$$

$$=\frac{a^2}{a^2(a-2)}-\frac{a^2-4a+4}{a^2(a-2)}$$

$$=\frac{a^2-(a^2-4a+4)}{a^2(a-2)}$$

$$=\frac{a^2-a^2+4a-4}{a^2(a-2)}$$

$$=\frac{4a-4}{a^2(a-2)}$$

$$=\frac{4(a-1)}{a^2(a-2)}$$

$$=\frac{4a-4}{a^3-2a^2}$$

29.

$$\frac{5x}{x^4-16}+\frac{8x}{x+2}=\frac{5x}{(x^2+4)(x^2-4)}+\frac{8x}{x+2}$$

$$=\frac{5x}{(x^2+4)(x+2)(x-2)}+\frac{8x}{x+2} \qquad \text{LCD: } (x^2+4)(x+2)(x-2)$$

$$=\frac{5x}{(x^2+4)(x+2)(x-2)}+\frac{8x(x^2+4)(x-2)}{(x^2+4)(x+2)(x-2)}$$

$$=\frac{5x+8x(x^3-2x^2+4x-8)}{(x^2+4)(x+2)(x-2)}$$

$$=\frac{5x+8x^4-16x^3+32x^2-64x}{(x^2+4)(x+2)(x-2)}$$

$$=\frac{8x^4-16x^3+32x^2-59x}{(x^2+4)(x+2)(x-2)}$$

$$=\frac{8x^4-16x^3+32x^2-59x}{x^4-16}$$

30.

$$\frac{x-1}{x+1} - \frac{2}{x(x+1)}$$

$$\{\text{LCD: } x(x+1)\}$$

$$= \frac{x(x-1)}{x(x+1)} - \frac{2}{x(x+1)}$$

$$= \frac{x^2 - x - 2}{x(x+1)}$$

$$= \frac{(x-2)(x+1)}{x(x+1)}$$

$$= \frac{x-2}{x}$$

31.

$$1 + \frac{1}{x-2} - \frac{2}{x^2}$$

$$\{\text{LCD: } x^2(x-2)\}$$

$$= \frac{x^2(x-2)}{x^2(x-2)} + \frac{x^2}{x^2(x-2)} - \frac{2(x-2)}{x^2(x-2)}$$

$$= \frac{x^3 - x^2 - 2x + 4}{x^2(x-2)}$$

32.

$$\frac{x-7}{x^2 - 9x + 20} + \frac{x+2}{x^2 - 5x + 4}$$

$$\{\text{LCD: } (x-5)(x-4)(x-1)\}$$

$$= \frac{(x-7)(x-1)}{(x-5)(x-4)(x-1)} + \frac{(x+2)(x-5)}{(x-5)(x-4)(x-1)}$$

$$= \frac{2x^2 - 11x - 3}{(x-5)(x-4)(x-1)}$$

33.

$$\frac{2x+1}{2(2x-1)} + \frac{5}{2x} - \frac{x+1}{x(2x-1)}$$

$$\{\text{LCD: } 2x(2x-1)\}$$

$$= \frac{x(2x+1)}{2x(2x-1)} + \frac{5(2x-1)}{2x(2x-1)} - \frac{2(x+1)}{2x(2x-1)}$$

$$= \frac{2x^2 + x + (10x - 5) - (2x + 2)}{2x(2x-1)}$$

$$= \frac{2x^2 + x + 10x - 5 - 2x - 2}{2x(2x-1)}$$

$$= \frac{2x^2 + 9x - 7}{2x(2x-1)} = \frac{2x^2 + 9x - 7}{4x^2 - 2x}$$

34.

$$\frac{\dfrac{1}{x} + \dfrac{1}{y}}{\dfrac{1}{x} - \dfrac{1}{y}}$$

$$\{\text{LCD: } xy\}$$

$$= \frac{y+x}{y-x}$$

35.

$$\frac{\dfrac{5}{2y} + \dfrac{3}{y}}{\dfrac{1}{4} + \dfrac{1}{3y}}$$

$$\{\text{LCD: } 12y\}$$

$$= \frac{30 + 36}{3y + 4}$$

$$= \frac{66}{3y + 4}$$

36.

$$\cfrac{\dfrac{2}{1}-\dfrac{1}{x}}{\dfrac{2x}{1}-\dfrac{3x}{x+1}}$$

$\{\text{LCD: } x(x+1)\}$

$$=\frac{2x^2+2x-x-1}{2x^3+2x^2-3x^2}$$

$$=\frac{2x^2+x-1}{2x^3-x^2}$$

$$=\frac{(2x-1)(x+1)}{x^2(2x-1)}$$

$$=\frac{x+1}{x^2}$$

37.

$$\cfrac{\dfrac{1}{1}-\dfrac{2}{x-2}}{\dfrac{x-6}{1}+\dfrac{10}{x+1}}$$

$\{\text{LCD: } (x-2)(x+1)\}$

$$=\frac{x^2-x-2-2x-2}{(x-6)(x+1)(x-2)+10(x-2)}$$

$$=\frac{x^2-3x-4}{(x-2)(x^2-5x+4)}$$

$$=\frac{(x-4)(x+1)}{(x-2)(x-4)(x-1)}$$

$$=\frac{x+1}{(x-2)(x-1)}$$

38.

$$\require{enclose}\begin{array}{r}x^4-x^3+2x^2-2x+2\\ x+1\enclose{longdiv}{x^5+0x^4+x^3+0x^2+0x-1}\\ \underline{x^5+x^4}\\ -x^4+x^3\\ \underline{-x^4-x^3}\\ 2x^3+0x^2\\ \underline{2x^3+2x^2}\\ -2x^2+0x\\ \underline{-2x^2-2x}\\ 2x-1\\ \underline{2x+2}\\ -3\end{array}$$

Thus, the quotient is:

$x^4-x^3+2x^2-2x+2$ with remainder -3.

or

$$x^4-x^3+2x^2-2x+2-\frac{3}{x+1}$$

39.

$$\require{enclose}\begin{array}{r}a^3+a^2\\ a+2\enclose{longdiv}{a^4+3a^3+2a^2}\\ \underline{a^4+2a^3}\\ a^3+2a^2\\ \underline{a^3+2a^2}\\ 0\end{array}$$

Thus, the quotient is:

a^3+a^2 with remainder 0.

40.

$$x^2 - 2\overline{\smash{\big)}\ 3x^5 - x^4 + 0x^3 + 0x^2 + 5x - 1}$$

quotient: $3x^3 - x^2 + 6x - 2$

$$\underline{3x^5 \qquad - 6x^3}$$
$$-x^4 + 6x^3 + 0x^2$$
$$\underline{-x^4 \qquad + 2x^2}$$
$$6x^3 - 2x^2 + 5x$$
$$\underline{6x^3 \qquad -12x}$$
$$-2x^2 + 17x - 1$$
$$\underline{-2x^2 \qquad + 4}$$
$$17x - 5$$

Thus, the quotient is:

$\left(3x^3 - x^2 + 6x - 2\right)$ with rem $\left(17x - 5\right)$

or

$$3x^3 - x^2 + 6x - 2 + \frac{17x - 5}{x^2 - 2}$$

41.

$$3x^2 - 1\overline{\smash{\big)}\ 3x^4 + 0x^3 + 2x^2 + 0x + 1}$$

quotient: $x^2 + 1$

$$\underline{3x^4 \qquad - x^2}$$
$$3x^2 + 0x + 1$$
$$\underline{3x^2 \qquad -1}$$
$$2$$

Thus, the quotient is:

$x^2 + 1$ with remainder 2

or

$$x^2 + 1 + \frac{2}{3x^2 - 1}$$

Section 6.1 Skills Check

1. a. $h(x) = 3x^3 + 5x^2 - x - 10$

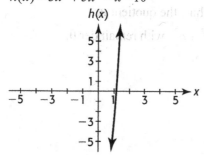

b.

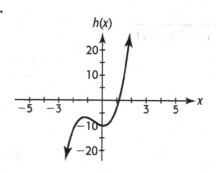

Window b) gives a complete graph.

3. a. $g(x) = 3x^4 - 12x^2$

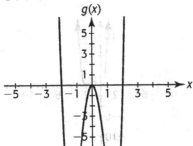

b.

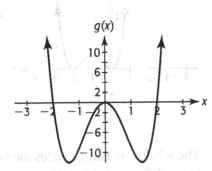

Window b) gives a complete graph.

5. a. The x-intercepts appear to be -2, 1, and 2.

b. The leading coefficient is positive since the graph rises to the right.

b. The function is cubic since the end behavior is "one end up and one end down".

7. a. The x-intercepts appear to be -1, 1, and 5.

b. The leading coefficient is negative since the graph falls to the right.

c. The function is cubic since the end behavior is "one end up and one end down".

9. a. The x-intercepts appear to be -1.5 and 1.5.

b. The leading coefficient is positive since the graph rises to the right.

c. The function is quartic since the end behavior is "both ends opening up".

11. matches with graph C since it is cubic with a positive leading coefficient.

13. matches with graph E since it is cubic with a negative leading coefficient and y-intercept $(0, -6)$.

15. matches with graph F since it is quartic with a positive leading coefficient and y-intercept $(0, 3)$.

17. a. The polynomial is 3rd degree, and the leading coefficient is 2.

b. The graph rises right and falls left because the leading coefficient is positive and the function is cubic.

c. $f(x) = 2x^3 - x$

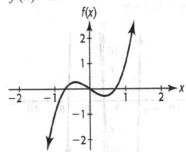

19. a. The polynomial is 3rd degree, and the leading coefficient is -2.

b. The graph falls right and rises left because the leading coefficient is negative and the function is cubic.

c. $f(x) = -2(x-1)(x^2-4)$

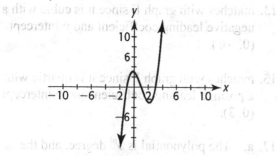

21. a. $y = x^3 - 3x^2 - x + 3$

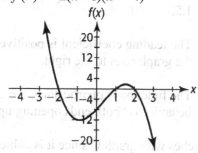

b.   Yes, the graph is complete. As suggested by the degree of the cubic function, three x-intercepts show, along with the y-intercept.

23. a. $y = 25x - x^3$

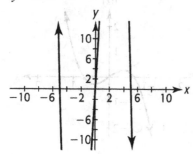

b.

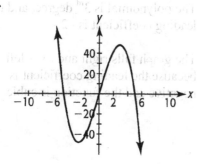

25. a. $y = x^4 - 4x^3 + 4x^2$

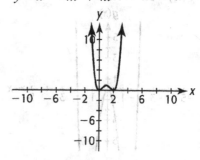

b.

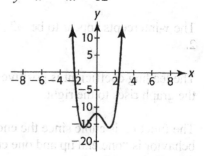

c.   The window in part b) yields the best view of the turning points.

27. a. $y = x^4 - 4x^2 - 12$

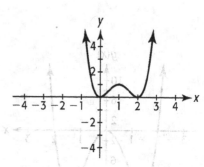

b.   The graph has three turning points.

c.   No, since the polynomial is degree 4, it has at most three turning points.

29.

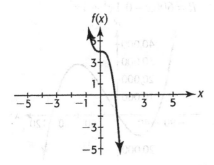

Answers will vary. One such graph is for the function, $f(x) = -4x^3 + 4$, as shown.

31.

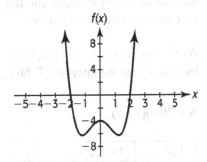

Answers will vary. One such graph is for the function, $f(x) = x^4 - 3x^2 - 4$, as shown.

33. a. $y = x^3 + 4x^2 + 5$

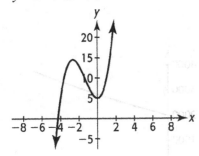

b.

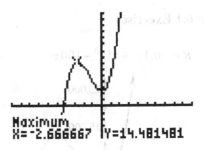

[−10, 10] by [−10, 30]

The local maximum is approximately $(-2.67, 14.48)$.

c.

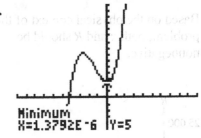

[−10, 10] by [−10, 30]

The local minimum is $(0, 5)$.

35. $y = x^4 - 4x^3 + 4x^2$

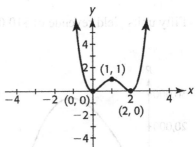

The local maximum is $(1, 1)$. The local minima are $(0, 0)$ and $(2, 0)$.

Section 6.1 Exercises

37. a. $R = -0.1x^3 + 11x^2 - 100x$

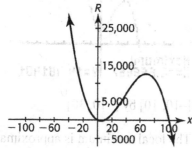

There are two turning points.

b. Based on the physical context of the problem, both x and R should be nonnegative.

c.

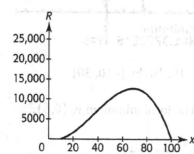

d. Fifty units yield revenue of $10,000.

39. a.

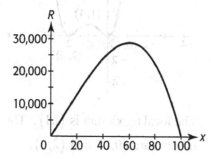

b. Selling 60 units yield a maximum daily revenue of $28,800.

c. $R = 600x - 0.1x^3 + 4x^2$

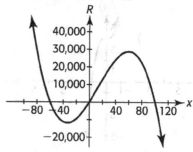

Answers will vary for the window.

d. The graph in part a) represents the physical situation better since both the number of units produced and the revenue must be nonnegative.

e. As shown in part a), the graph is increasing on the interval $(0, 60)$.

41. a. $S = 2000(1 + r)^3$

Rate, r	Future Value, $S(\$)$
0.00	2,000.00
0.05	2,315.25
0.10	2,662.00
0.15	3,041.75
0.20	3,456.00

b.

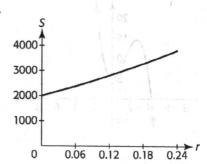

c. At the 20% rate, the investment yields $3456. At the 10% rate, the investment yields $2662. Therefore, the 20% rate yields $794 more.

d. The 10% rate is more realistic.

43. a. $y = 10.3x^3 - 400x^2 + 5590x - 8694$

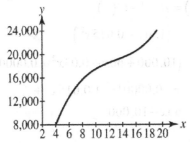

b.

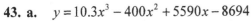

In 2020, when $x = 20$, the number of Starbucks stores is projected to be 25,506 stores.

c.   No; according to the model, the number of stores will not decrease.

45. a. $y = 0.00728x^3 + 0.0414x^2$
$$- 0.296x + 0.340$$

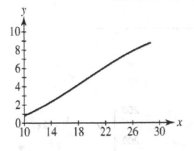

b.

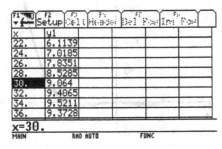

In 2020, when $x = 30$, the number of subscriberships is estimated to be approximately 9.064 billion.

c.

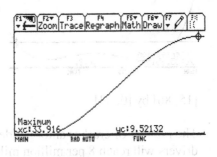

[0, 35] by [0, 10]

The maximum number of subscriberships is projected in the year 2024 when $x = 34$.

47. a. $y = 0.00001828x^4 - 0.003925x^3$
$$+ 0.3031x^2 - 9.907x + 118.2$$

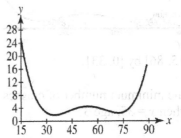

b.

When a driver is 22, when $x = 22$, the number of crashes is estimated to be approximately 9.4 per million miles.

c.

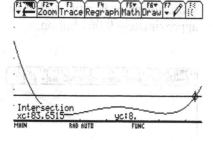

[15, 86] by [0, 33]

The number of crashes among older drivers will reach 8 per million miles around the age of 83, when $x = 83.7$.

d.

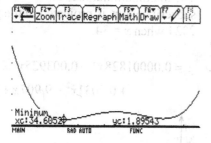

[15, 86] by [0, 33]

The minimum number of crashes occur when $x = 35$, age 35.

49. a.

$$P(x) = R(x) - C(x)$$
$$= (120x - 0.015x^2) -$$
$$(10,000 + 60x - 0.03x^2 + 0.00001x^3)$$
$$= -0.00001x^3 + 0.015x^2 +$$
$$60x - 10,000$$

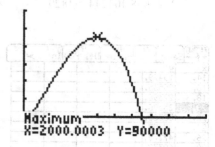

[0, 5000] by [−20,000, 120,000]

2000 units produced and sold yields a maximum profit.

b. The maximum profit is $90,000.

Section 6.2 Skills Check

1.

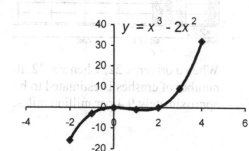

3.

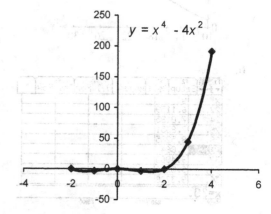

5. a.

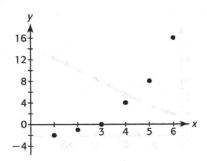

b. It appears that a cubic model will fit the data better.

7. a.

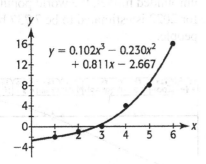

$$y = 0.102x^3 - 0.230x^2 + 0.811x - 2.667$$

b.

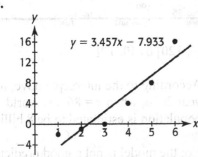

$$y = 3.457x - 7.933$$

c. It appears that a cubic model will fit the data better.

9. a.

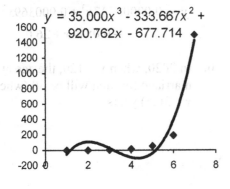

$$y = 35.000x^3 - 333.667x^2 + 920.762x - 677.714$$

b.

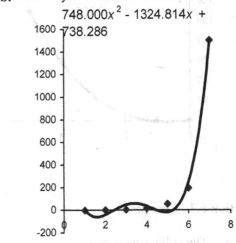

$$y = 12.515x^4 - 165.242x^3 + 748.000x^2 - 1324.814x + 738.286$$

11.

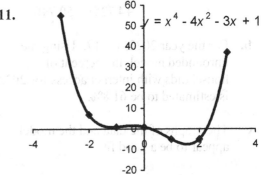

$$y = x^4 - 4x^2 - 3x + 1$$

13.

x	$f(x)$	First Difference	Second Difference	Third Difference
0	0			
1	1	1		
2	5	4	3	
3	24	19	15	12
4	60	36	17	2
5	110	50	14	−3

The function $f(x)$ is not exactly cubic.

15.

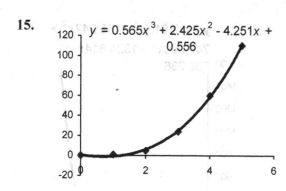

$$y = 0.565x^3 + 2.425x^2 - 4.251x + 0.556$$

Section 6.2 Exercises

17. a. The cubic equation is
$$y = -0.00556x^3 - 0.145x^2$$
$$+4.725x + 50.740$$

b. For the year 2022, $x = 17$. Using the unrounded model, the percent of households with internet access for 2022 is estimated to be 61.8%.

c. The graphs of the data and the model appear to be a good fit.

19. a.

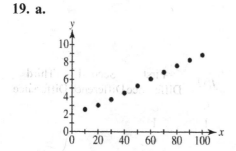

b. The equation is
$$y = -0.00000486x^3 + 0.000825x^2$$
$$+0.0336x + 2.11$$

c.

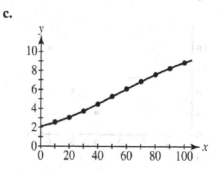

It is a good fit to the data.

d. For the year 2022, $x = 82$. Using the unrounded model, the world population for 2022 is estimated to be 7.737 billion people.

e.

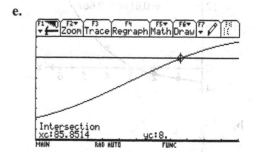

[0, 120] by [0, 10]

According to the intercept above, in the year 2026, when $x = 86$, the world population is estimated to be 8 billion.

f. No; the model is not a good predictor because it predicts that the world population will decrease.

21. a. The cubic function is:
$$y = 0.00000537x^3 + 0.000369x^2$$
$$-0.0844x + 26.0$$

b. In 2020, when $x = 120$, the age at first marriage for men will be 30 (when $x = 30.5$) years.

23. a. The equation is
$$y = -0.000487x^3 - 0.162x^2$$
$$+19.8x + 388$$

b.

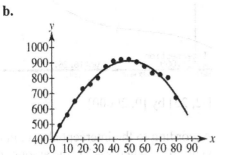

It is a good fit to the data.

c. For the year 2022, $x = 52$. Using the unrounded model, China's labor pool for 2022 is estimated to be 913 million people.

d.

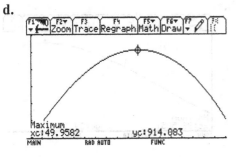

[-2, 90] by [400, 1000]

Using the maximum function, the maximum point will be (49.95, 914.08). This means that in the year 2020 (1970 + 50), the labor pool will have a maximum of approximately 914 million people.

25. a. The cubic function is:
$$y = 0.000078x^3 - 0.01069x^2 - 0.1818x$$
$$+64.6848$$

b.

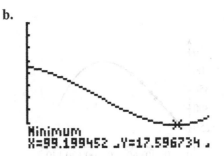

[0, 120] by [0, 100]

In 2000, when $x = 99.2$, the per cent of elderly men in the work force reached its minimum.

27. a.

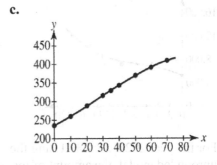

[0, 80] by [200, 450]

b. The equation is
$$y = -0.000233x^3 + 0.0186x^2$$
$$+2.33x + 235$$

c.

It is an excellent fit to the data.

d.

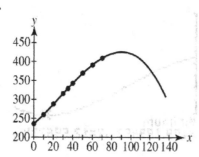

The model does not seem appropriate for the increased time period, because it predicts that the population of the U.S. will begin to decrease, and that does not seem likely.

29. a.

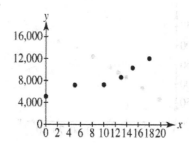

b. The equation is
$$y = 2.481x^3 - 52.251x^2 + 528.682x + 5192.580$$

c.

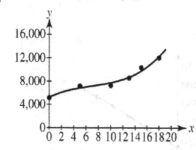

d. For the year 2023, $x = 23$. Using the unrounded model, the amount of federal tax per capita for 2023 is estimated to be $19,894 per person.

e.

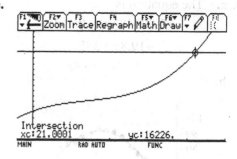

[-2, 25] by [0, 20000]

According to the intercept above, in the year 2021, when $x = 21$, the amount of federal tax per capita is estimated to be $16,226 per person.

31. a. The cubic function is:
$$y = -0.00007x^3 + 0.00567x^2 + 0.863x + 16.009$$

b.

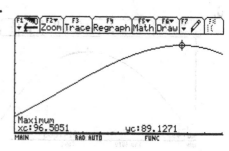

[0, 120] by [0, 100]

In 2046, when $x = 96.5$, the number of women in the work force will reach its maximum.

33. a.

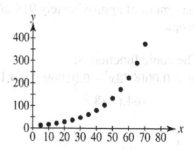

b. The quadratic function is:
$$y = 0.117x^2 - 3.79x + 45.3$$

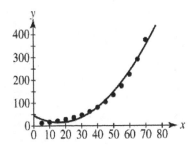

c. The cubic function is:
$$y = 0.00188x^3 - 0.0954x^2 + 2.79x$$
$$-2.70$$

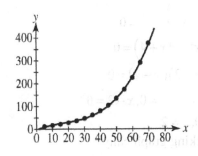

d. The cubic model is a better fit for the data.

35. a.

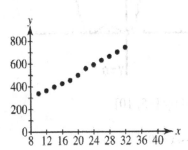

b. The equation is
$$y = -0.02189x^3 + 1.511x^2$$
$$-13.67x + 347.4$$

c.

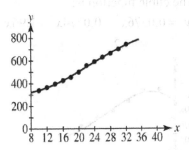

The model is a good fit to the data.

d. For the year 2026, $x = 26$. Using the unrounded model, the millions of metric tons of emission for 2026 is estimated to be 628.4 million metric tons.

e.

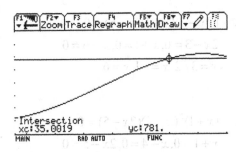

[8, 45] by [200, 1000]

According to the intercept above, in the year 2035, when $x = 35$, the number of carbon emissions is estimated to be 781 million metric tons.

37. a. The quadratic function is:
$$y = -0.169x^2 + 0.381x + 99.9$$

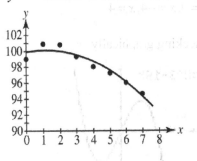

b. The cubic function is:
$$y = 0.0576x^3 - 0.0774x^2 + 1.96x + 99.2$$

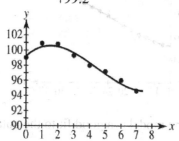

c. The cubic model is a better fit for the data.

Section 6.3 Skills Check

1. $(2x-3)(x+1)(x-6) = 0$

$2x-3=0, x+1=0, x-6=0$

$x = 3/2, x = -1, x = 6$

3. $(x+1)^2(x-4)(2x-5) = 0$

$x+1=0, x-4=0, 2x-5=0$

$x = -1, x = 4, x = 5/2$

5. $x^3 - 16x = 0$

$x(x^2 - 16) = 0$

$x(x+4)(x-4) = 0$

$x = 0, x+4=0, x-4=0$

$x = 0, x = -4, x = 4$

Checking graphically

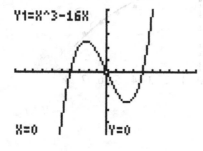

[−10, 10] by [−50, 50]

7. $x^4 - 4x^3 + 4x^2 = 0$

$x^2(x^2 - 4x + 4) = 0$

$x^2(x-2)(x-2) = 0$

$x^2 = 0 \Rightarrow x = 0, x - 2 = 0$

$x = 0, x = 2$

Checking graphically

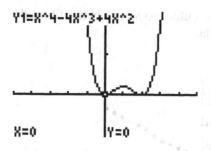

[−5, 5] by [−5, 10]

9. $4x^3 - 4x = 0$

$4x(x^2 - 1) = 0$

$4x(x+1)(x-1) = 0$

$4x = 0, x+1=0, x-1=0$

$x = 0, x = -1, x = 1$

Checking graphically

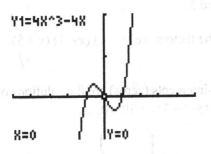

Y1=4X^3-4X

X=0   Y=0

[−5, 5] by [−5, 10]

11. $x^3 - 4x^2 - 9x + 36 = 0$

$\left(x^3 - 4x^2\right) + (-9x + 36) = 0$

$x^2(x-4) + (-9)(x-4) = 0$

$(x-4)\left(x^2 - 9\right) = 0$

$(x-4)(x+3)(x-3) = 0$

$x - 4 = 0, x + 3 = 0, x - 3 = 0$

$x = 4, x = -3, x = 3$

13. $3x^3 - 4x^2 - 12x + 16 = 0$

$\left(3x^3 - 4x^2\right) + (-12x + 16) = 0$

$x^2(3x-4) + (-4)(3x-4) = 0$

$(3x-4)\left(x^2 - 4\right) = 0$

$(3x-4)(x+2)(x-2) = 0$

$x = \dfrac{4}{3}, x = -2, x = 2$

15. $2x^3 - 16 = 0$

$2x^3 = 16$

$x^3 = 8$

$\sqrt[3]{x^3} = \sqrt[3]{8}$

$x = 2$

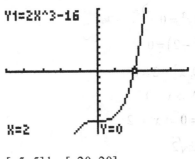

Y1=2X^3-16

X=2   Y=0

[−5, 5] by [−20, 20]

17. $\dfrac{1}{2}x^4 - 8 = 0$

$\dfrac{1}{2}x^4 = 8$

$2\left(\dfrac{1}{2}x^4\right) = 2(8)$

$x^4 = 16$

$\sqrt[4]{x^4} = \pm\sqrt[4]{16}$

$x = \pm 2$

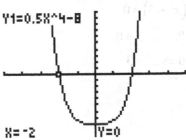

Y1=0.5X^4-8

X=−2   Y=0

[−5, 5] by [−10, 10]

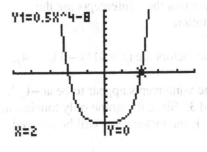

Y1=0.5X^4-8

X=2   Y=0

[−5, 5] by [−10, 10]

19. $4x^4 - 8x^2 = 0$

$4x^2\left(x^2 - 2\right) = 0$

$4x^2 = 0,\ x^2 - 2 = 0$

$4x^2 = 0 \Rightarrow x = 0$

$x^2 - 2 = 0 \Rightarrow x^2 = 2$

$\sqrt{x^2} = \pm\sqrt{2}$

$x = \pm\sqrt{2}$

$x = \pm\sqrt{2},\ x = 0$

21. $0.5x^3 - 12.5x = 0$

$0.5x\left(x^2 - 25\right) = 0$

$0.5x\left(x + 5\right)\left(x - 5\right) = 0$

$x = 0, x = -5, x = 5$

23. $x^4 - 6x^2 + 9 = 0$

$\left(x^2 - 3\right)\left(x^2 - 3\right) = 0$

$x^2 - 3 = 0,\ x^2 - 3 = 0$

$x^2 - 3 = 0 \Rightarrow x^2 = 3$

$\sqrt{x^2} = \pm\sqrt{3}$

$x = \pm\sqrt{3}$

25. a. $f(x) = 0$ implies $x = -3, x = 1, x = 4$. Note that the x-intercepts are the solutions.

b. The factors are $(x + 3)\,(x - 1)(x - 4)$.

27. a. The x-intercepts appear to be at −1, 2, and 3. Since the graph only touches at $x = -1$, the factor $x + 1$ will be squared.

b. The factors are $(x + 1)^2\,(x - 2)(x - 3)$.

29. a. The x-intercepts appear to be at −1, 1, and 5.

b. The factors are $(x + 1)\,(x - 1)\,(x - 5)$.

31. The x-intercepts (zeros) are the solutions of $4x^3 - 15x^2 - 31x + 30 = 0$.

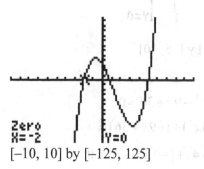

Zero
X=-2   Y=0

[−10, 10] by [−125, 125]

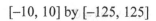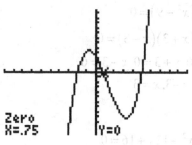

Zero
X=.75   Y=0

[−10, 10] by [−125, 125]

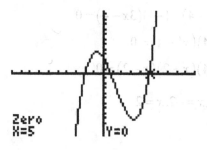

Zero
X=5   Y=0

[−10, 10] by [−125, 125]

$x = -2, x = 0.75 \text{ or } \dfrac{3}{4}, x = 5$

Section 6.3 Exercises

33. a. $R = 400x - x^3$

$400x - x^3 = 0$

$x(400 - x^2) = 0$

$x(20 - x)(20 + x) = 0$

$x = 0, \ 20 - x = 0, \ 20 + x = 0$

$x = 0, \ -x = -20, \ x = -20$

$x = 0, \ x = 20, \ x = -20$

In the physical context of the problem, selling zero units or selling 20 units will yield revenue of zero dollars. −20 units is not possible and so is eliminated.

b.   Yes.

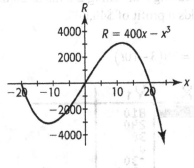

35. a. $R = (100,000 - 0.1x^2)x$

$(100,000 - 0.1x^2)x = 0$

$x = 0, \ 100,000 - 0.1x^2 = 0$

$-0.1x^2 = -100,000$

$x^2 = 1,000,000$

$x = \pm\sqrt{1,000,000}$

$x = 0, \ x = 1000, \ x = -1000$

In the physical context of the problem, selling zero units or selling 1000 units will yield revenue of zero dollars. −1000 units is not possible and so is eliminated.

b.   Yes.

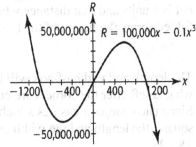

37. a.   Complete the table:

Rate	Future Value
4%	$2249.73
5%	$2315.25
7.25%	$2467.30
10.5%	$2698.47

b.

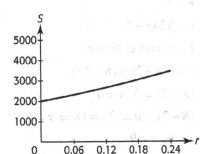

c. $2662 = 2000(1+r)^3$

$(1+r)^3 = \dfrac{2662}{2000} = 1.331$

$\sqrt[3]{(1+r)^3} = \sqrt[3]{1.331} = 1.10$

$1 + r = 1.10$

$r = 1.10 - 1$

$r = 0.10 = 10\%$

d. $3456 = 2000(1+r)^3$

$(1+r)^3 = \dfrac{3456}{2000} = 1.728$

$\sqrt[3]{(1+r)^3} = \sqrt[3]{1.728} = 1.20$

$1 + r = 1.20$

$r = 1.20 - 1$

$r = 0.20 = 20\%$

39. a. The height is x inches, since the distance cut is x units and that distance when folded forms the height of the box.

b. The length and width of box will be what is left after the corners are cut. Since each corner measures x inches square, the length and the width are $18 - 2x$.

c. $V = lwh$

$V = (18 - 2x)(18 - 2x)x$

$V = (324 - 36x - 36x + 4x^2)x$

$V = 324x - 72x^2 + 4x^3$

d. $V = 0$

$0 = 324x - 72x^2 + 4x^3$

From part c) above:

$0 = (18 - 2x)(18 - 2x)x$

$18 - 2x = 0, \ x = 0$

$18 - 2x = 0 \Rightarrow 2x = 18 \Rightarrow x = 9$

$x = 0, \ x = 9$

e. A box will not exist for either of the values calculated in part d) above. For both values of x, no tab will exist to fold up to form the box.

41. Since the profit is given in hundreds of dollars, $40,000 should be represented as 400 hundreds. Thus,

$400 = -x^3 + 2x^2 + 400x - 400$

$0 = -x^3 + 2x^2 + 400x - 800$

$x^3 - 2x^2 - 400x + 800 = 0$

$(x^3 - 2x^2) + (-400x + 800) = 0$

$x^2(x - 2) + (-400)(x - 2) = 0$

$(x - 2)(x^2 - 400) = 0$

$(x - 2)(x + 20)(x - 20) = 0$

$x = 2, x = -20, x = 20$

The negative answer does not make sense in the physical context of the problem. Producing and selling 2 units or 20 units yields a profit of $40,000.

43. a. $s = 30(3 - 10t)^3$

X	Y1
0	810
.1	240
.2	30
.3	0
.4	-30
.5	-240
.6	-810

X=0

t	0	0.1	0.2	0.3
s (cm/sec)	810	240	30	0

b.
$$0 = 30(3 - 10t)^3$$

$$(3 - 10t)^3 = \frac{0}{30} = 0$$

$$\sqrt[3]{(3 - 10t)^3} = \sqrt[3]{0}$$

$$3 - 10t = 0$$

$$t = \frac{-3}{-10}$$

$$t = 0.3$$

The solution in the table is the same as the solution found by the root method.

45. a. $y = -0.0001x^3 + 0.0088x^2 + 1.43x + 57.9$

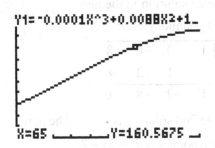

[0, 100] by [0, 220]

When $x = 65$, in the year 2015, the civilian labor force is projected to be 160.6 million people.

b.

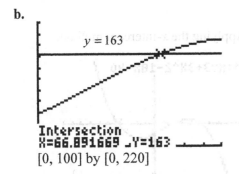

[0, 100] by [0, 220]

When $x = 66.9$, in the year 2017, the projected civilian work force will be 163 million.

47. Applying the intersection of graphs method for

$$y = -0.00007x^3 + 0.00567x^2 + 0.863x + 16.009$$

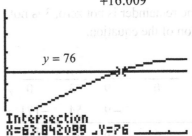

[0, 100] by [0, 150]

When $x = 63.8$, in the year 2014, the number of women in the work force is estimated to be 76 million.

49. Applying the intersection of graphs method for

$$y = 0.00188x^3 - 0.0954x^2 + 2.79x - 2.70$$

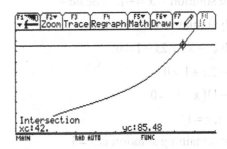

[0, 50] by [0, 100]

In 2042 ($x = 42$) the U.S. GDP is estimated to be $85.48 trillion.

Section 6.4 Skills Check

1.

$$3\overline{)\begin{array}{ccccc} 1 & -4 & 0 & 3 & 10 \\ & 3 & -3 & -9 & -18 \\ \hline 1 & -1 & -3 & -6 & -8 \end{array}}$$

$$x^3 - x^2 - 3x - 6 - \dfrac{8}{x-3}$$

3.

$$1\overline{)\begin{array}{ccccc} 2 & -3 & 0 & 1 & -7 \\ & 2 & -1 & -1 & 0 \\ \hline 2 & -1 & -1 & 0 & -7 \end{array}}$$

$$2x^3 - x^2 - x - \dfrac{7}{x-1}$$

5.

$$3 \overline{)\begin{array}{ccccc} 2 & -4 & 0 & 3 & 18 \\ & 6 & 6 & 18 & 63 \\ \hline 2 & 2 & 6 & 21 & 81 \end{array}}$$

Since the remainder is not zero, 3 is not a solution of the equation.

7.

$$-3 \overline{)\begin{array}{ccccc} -1 & 0 & -9 & 3 & 0 \\ & 3 & -9 & 54 & -171 \\ \hline -1 & 3 & -18 & 57 & -171 \end{array}}$$

Since the remainder is not zero, $x+3$ is not a factor.

9.

$$-1 \overline{)\begin{array}{cccc} -1 & 1 & 1 & -1 \\ & 1 & -2 & 1 \\ \hline -1 & 2 & -1 & 0 \end{array}}$$

One solution is $x = -1$. The new polynomial is $-x^2 + 2x - 1$.

Solve $-x^2 + 2x - 1 = 0$.

$x^2 - 2x + 1 = 0$

$(x-1)(x-1) = 0$

$x = 1, x = 1$

The remaining solution is $x = 1$ (a double solution).

11.

$$-5 \overline{)\begin{array}{ccccc} 1 & 2 & -21 & -22 & 40 \\ & -5 & 15 & 30 & -40 \\ \hline 1 & -3 & -6 & 8 & 0 \end{array}}$$

One solution is $x = -5$. The new polynomial is $x^3 - 3x^2 - 6x + 8$.

Synthetically dividing by the 2nd given solution in the new polynomial yields:

$$1 \overline{)\begin{array}{cccc} 1 & -3 & -6 & 8 \\ & 1 & -2 & -8 \\ \hline 1 & -2 & -8 & 0 \end{array}}$$

The 2nd solution is $x = 1$. The new polynomial is $x^2 - 2x - 8$.

Solve $x^2 - 2x - 8 = 0$.

$(x-4)(x+2) = 0$

$x = 4, x = -2$

The remaining solutions are $x = 4$, $x = -2$.

13. Applying the x-intercept method:

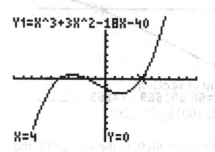

[–10, 10] by [–250, 250]

One solution appears to be $x = 4$.

$$4 \overline{)\begin{array}{cccc} 1 & 3 & -18 & -40 \\ & 4 & 28 & 40 \\ \hline 1 & 7 & 10 & 0 \end{array}}$$

The new polynomial is $x^2 + 7x + 10$.

Solve $x^2 + 7x + 10 = 0$.

$(x+2)(x+5) = 0$

$x = -2, x = -5$

The remaining solutions are $x = -5, x = -2$.

15. Applying the *x*-intercept method:

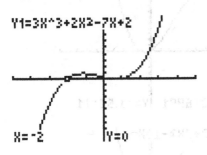

[–5, 5] by [–100, 100]

One solution appears to be $x = -2$.

$$-2 \overline{)\begin{array}{rrrr} 3 & 2 & -7 & 2 \\ & -6 & 8 & -2 \\ \hline 3 & -4 & 1 & 0 \end{array}}$$

The new polynomial is $3x^2 - 4x + 1$.

Solve $3x^2 - 4x + 1 = 0$.

$(3x - 1)(x - 1) = 0$

$x = \dfrac{1}{3}, x = 1$

The remaining solutions are $x = 1, x = \dfrac{1}{3}$.

17. $x^3 - 6x^2 + 5x + 12 = 0$

$\dfrac{p}{q} = \pm\left(\dfrac{1,2,3,4,6,12}{1}\right) = \pm(1,2,3,4,6,12)$

19. $9x^3 + 18x^2 + 5x - 4 = 0$

$\dfrac{p}{q} = \pm\left(\dfrac{1,2,4}{1,3,9}\right)$

$= \pm\left(1,2,4,\dfrac{1}{3},\dfrac{2}{3},\dfrac{4}{3},\dfrac{1}{9},\dfrac{2}{9},\dfrac{4}{9}\right)$

21. Applying the *x*-intercept method:

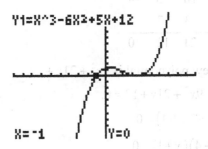

[–10, 10] by [–100, 100]

One solution appears to be $x = -1$.

$$-1 \overline{)\begin{array}{rrrr} 1 & -6 & 5 & 12 \\ & -1 & 7 & -12 \\ \hline 1 & -7 & 12 & 0 \end{array}}$$

The new polynomial is $x^2 - 7x + 12$.

Solve $x^2 - 7x + 12 = 0$.

$(x - 3)(x - 4) = 0$

$x = 3, x = 4$

The remaining solutions are $x = 3, x = 4$.

23. Applying the *x*-intercept method:

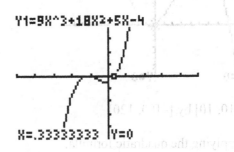

[–5, 5] by [–50, 50]

One solution appears to be $x = \dfrac{1}{3}$.

$$\frac{1}{3}\overline{)\begin{array}{rrrr} 9 & 18 & 5 & -4 \\ & 3 & 7 & 4 \\ \hline 9 & 21 & 12 & 0 \end{array}}$$

The new polynomial is $9x^2 + 21x + 12$.

Solve $9x^2 + 21x + 12 = 0$.

$3(3x^2 + 7x + 4) = 0$

$3(3x + 4)(x + 1) = 0$

$x = -\dfrac{4}{3}, x = -1$

The remaining solutions are $x = -\dfrac{4}{3}, x = -1$.

25. Factoring the common x term:

$x^3 = 10x - 7x^2$

$x^3 + 7x^2 - 10x = 0$

$x(x^2 + 7x - 10) = 0$

$x = 0, x^2 + 7x - 10 = 0$

One solution is $x = 0$.

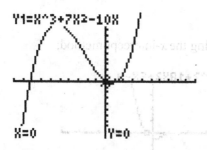

$[-10, 10]$ by $[-100, 120]$

Applying the quadratic formula:

$x = \dfrac{-7 \pm \sqrt{7^2 - 4(1)(-10)}}{2(1)}$

$x = \dfrac{-7 \pm \sqrt{89}}{2} \approx -8.217, \ 1.217$

The remaining solutions are $x = \dfrac{-7 \pm \sqrt{89}}{2}$,

both real numbers, which show on the graph.

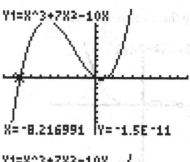

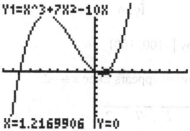

$[-10, 10]$ by $[-100, 120]$

27. Applying the x-intercept method:

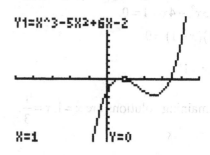

$[-5, 5]$ by $[-10, 10]$

It appears that $w = 1$ is a zero.

$$1\overline{)\begin{array}{rrrr} 1 & -5 & 6 & -2 \\ & 1 & -4 & 2 \\ \hline 1 & -4 & 2 & 0 \end{array}}$$

The new polynomial is $w^2 - 4w + 2$.

Applying the quadratic formula:

$$w = \frac{-(-4) \pm \sqrt{(-4)^2 - 4(1)(2)}}{2(1)}$$

$$w = \frac{4 \pm \sqrt{8}}{2}$$

$$w = \frac{4 \pm 2\sqrt{2}}{2} = \frac{2(2 \pm \sqrt{2})}{2}$$

$$w = 2 \pm \sqrt{2} \approx 0.586,\ 3.414$$

The remaining solutions are $w = 2 \pm \sqrt{2}$ both real numbers, which show on the graph.

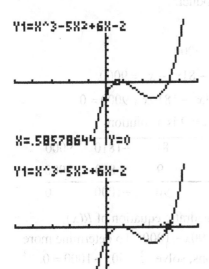

Y1=X^3-5X²+6X-2

X=.58578644   Y=0

Y1=X^3-5X²+6X-2

X=3.4142136   Y=1E⁻12

[−5, 5] by [−10, 10]

29. Applying the x-intercept method:

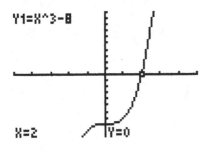

Y1=X^3-8

X=2   Y=0

[−5, 5] by [−10, 10]

It appears that $z = 2$ is a zero.

$$2 \overline{\smash{\big)}\ 1 \quad\ \ 0 \quad\ \ 0 \quad -8}$$
$$2 \quad\ \ 4 \quad\ \ 8$$
$$\overline{\ 1 \quad\ \ 2 \quad\ \ 4 \quad\ \ 0}$$

The new polynomial is $z^2 + 2z + 4$.

Applying the quadratic formula:

$$z = \frac{-2 \pm \sqrt{(2)^2 - 4(1)(4)}}{2(1)}$$

$$z = \frac{-2 \pm \sqrt{-12}}{2}$$

$$z = \frac{-2 \pm 2i\sqrt{3}}{2} = \frac{2(-1 \pm i\sqrt{3})}{2}$$

$$z = -1 \pm i\sqrt{3}$$

The remaining solutions are $z = -1 \pm i\sqrt{3}$ both imaginary, which do not show on the graph..

Section 6.4 Exercises

31. a.

$$50 \overline{\smash{\big)}\ -0.2 \quad\ \ 66 \quad -1600 \quad -60{,}000}$$
$$-10 \quad\ \ 2800 \quad\ 60{,}000$$
$$\overline{\ -0.2 \quad\ \ 56 \quad\ \ 1200 \quad\quad\quad\ 0}$$

The quadratic factor of $P(x)$ is $-0.2x^2 + 56x + 1200$.

b.

$$-0.2x^2 + 56x + 1200 = 0$$

$$-0.2(x^2 - 280x - 6000) = 0$$

$$-0.2(x + 20)(x - 300) = 0$$

$$x = -20,\ x = 300$$

In the context of the problem, only the positive solution is reasonable. Producing and selling 300 units results in break-even for the product.

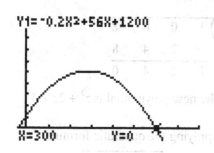

Y1= -0.2X²+56X+1200

X=300   Y=0

[−20, 400] by [−1000, 10,000]

33. a.

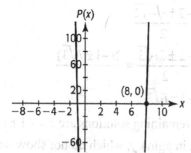

b. Based on the graph in part a), one *x*-intercept appears to be $x = 8$.

c.

$$8{\overline{\smash{\big)}\,\begin{array}{rrrr} -0.1 & 50.7 & -349.2 & -400 \\ & -0.8 & 399.2 & 400 \\ \hline -0.1 & 49.9 & 50 & 0 \end{array}}}$$

The quadratic factor of $P(x)$ is $-0.1x^2 + 49.9x + 50$.

d. Based on parts b) and c), one zero is $x = 8$. To find more zeros, solve $-0.1x^2 + 49.9x + 50 = 0$.

$$-0.1\left(x^2 - 499x - 500\right) = 0$$

$$-0.1\left(x - 500\right)\left(x + 1\right) = 0$$

$$x = 500, \; x = -1$$

The zeros are $x = 500$, $x = -1$, $x = 8$.

e. Based on the context of the problem, producing and selling 8 units or 500 units results in break-even for the product.

35.

$$R(x) = 9000$$

$$1810x - 81x^2 - x^3 = 9000$$

$$x^3 + 81x^2 - 1810x + 9000 = 0$$

Since $x = 9$ is a solution,

$$9{\overline{\smash{\big)}\,\begin{array}{rrrr} 1 & 81 & -1810 & 9000 \\ & 9 & 810 & -9000 \\ \hline 1 & 90 & -1000 & 0 \end{array}}}$$

The quadratic equation of $R(x)$ is $x^2 + 90x - 1000$. To determine more solutions, solve $x^2 + 90x - 1000 = 0$.

$$\left(x + 100\right)\left(x - 10\right) = 0$$

$$x = -100, x = 10$$

Revenue of $9000 is also achieved by selling 10 units.

37. a. $y = 244$

$$0.4566x^3 - 14.3085x^2 + 117.2978x + 107.8456 = 244$$

$$0.4566x^3 - 14.3085x^2 + 117.2978x - 136.1544 = 0$$

b.

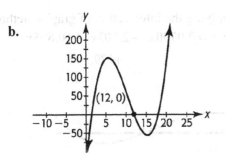

It appears that $x = 12$ is a solution.

c.

$$
\begin{array}{r|rrrr}
12) & 0.4566 & -14.3085 & 117.2978 & -136.1544 \\
    &   & 5.4792   & -105.9516 & 136.1544 \\
\hline
    & 0.4566 & -8.8293 & 11.3462 & 0 \\
\end{array}
$$

The quadratic equation of $P(x)$ is $0.4566x^2 - 8.8293x + 11.3462$.

d. $0.4566x^2 - 8.8293x + 11.3462 = 0$

$$x = \frac{-b \pm \sqrt{b^2 - 4ac}}{2a}$$

$$x = \frac{-(-8.8293) \pm \sqrt{(-8.8293)^2 - 4(0.4566)(11.3462)}}{2(0.4566)}$$

$$x = \frac{8.8293 \pm \sqrt{57.23383881}}{0.9132}$$

$x = 17.953,\ x = 1.384$

e. Based on the solutions in previous parts, the number of fatalities was 244 in 1982, 1992, and 1998.

39. a.

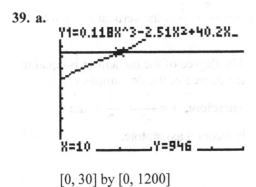

Y1=0.118X^3-2.51X²+40.2X_

X=10 _____Y=946 ___

[0, 30] by [0, 1200]

Ten years after 2000, in 2010, the U.S. per capita out-of-pocket cost for healthcare was $946.

b.

$$y = 946$$

$$0.118x^3 - 2.51x^2 + 40.2x + 677 = 946$$

$$0.118x^3 - 2.51x^2 + 40.2x - 269 = 0$$

Since $x = 10$ is a solution,

```
10)0.118   -2.51    40.2   -269
            1.18   -13.3    269
    ─────────────────────────────
     0.118  -1.33    26.9     0
```

The new polynomial is

$0.118x^2 - 1.33x + 26.9$.

To determine more solutions,

solve $0.118x^2 - 1.33x + 26.9 = 0$.

Applying the quadratic formula:

$$t = \frac{-(-1.33) \pm \sqrt{(-1.33)^2 - 4(0.118)(26.9)}}{2(0.118)}$$

$$t = \frac{1.33 \pm \sqrt{-10.9279}}{0.236}$$

$t = $ a non-real solution

Thus, there are no other years when the U.S. per capita out-of-pocket cost is $946.

41. Applying the intersection of graphs method for $y = 0.0790x^3 - 2.103x^2 + 10.695x$
$$+179.504$$

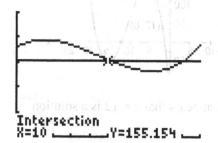

Intersection
X=10 _____ Y=155.154 ___

[0, 20] by [0, 250]

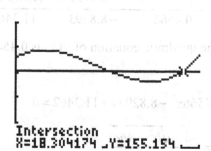

Intersection
X=18.304174 _ Y=155.154 ___

[0, 20] by [0, 250]

Based on the graphs, 155,154 births occurred in 2009, in addition to 2000.

Section 6.5 Skills Check

1. a. To find the vertical asymptote let
$x - 5 = 0$

$x = 5$ is the vertical asymptote.

b. The degree of the numerator is less than the degree of the denominator. Therefore, $y = 0$ is the horizontal asymptote.

3. a. To find the vertical asymptote let
$5 - 2x = 0$

$-2x = -5$

$x = \dfrac{-5}{-2} = \dfrac{5}{2}$ is the vertical asymptote.

b. The degree of the numerator is equal to the degree of the denominator.

Therefore, $y = \dfrac{1}{-2} = -\dfrac{1}{2}$ is the horizontal asymptote.

5. a. To find the vertical asymptote let

$x^2 - 1 = 0$

$(x+1)(x-1) = 0$

$x = -1, x = 1$ are the vertical asymptotes.

b. The degree of the numerator is greater than the degree of the denominator. Therefore, there is not a horizontal asymptote.

7. The function in part c) does not have a vertical asymptote. Its denominator cannot be zero. Parts a), b), and d) all have denominators that can equal zero for specific x-values.

9. matches with graph E that has a vertical asymptote at $x = 1$ and a horizontal asymptote at $y = 0$.

11. matches with graph F that has two vertical asymptotes at $x = 4$ and $x = -2$ (a result of factoring the function's denominator) , and a horizontal asymptote at $y = 0$.

13. matches with graph C that has a vertical asymptote at $x = 3$, but there is no horizontal asymptote since the degree of the numerator is greater than the degree of the denominator.

15. a. The degree of the numerator is equal to the degree of the denominator.

Therefore, $y = \dfrac{1}{1} = 1$ is the horizontal asymptote.

b. To find the vertical asymptote let

$x - 2 = 0$

$x = 2$ is the vertical asymptote.

c.

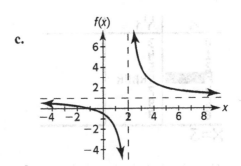

17. a. The degree of the numerator is less than the degree of the denominator. Therefore, $y = 0$ is the horizontal asymptote.

b. To find the vertical asymptote let

$1 - x^2 = 0$

$x^2 = 1$

$\sqrt{x^2} = \pm\sqrt{1}$

$x = \pm 1$

$x = 1, x = -1$ are the vertical asymptotes.

c.

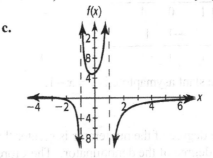

19. $y = \dfrac{x^2 - 9}{x - 3}$

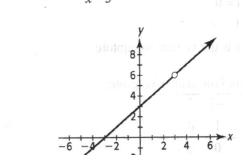

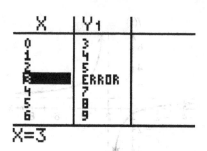

There is a hole in the graph at $x = 3$.

21. The degree of the numerator is greater than the degree of the denominator. Therefore, there is not a horizontal asymptote.

To find the vertical asymptote let
$$x + 1 = 0$$
$$x = -1$$
$x = -1$ is the vertical asymptote.

To find the slant asymptote:

$$-1 \overline{)\begin{array}{ccc} 1 & 0 & -4 \\ & -1 & 1 \\ \hline 1 & -1 & -3 \end{array}}$$

The slant asymaptote is $y = x - 1$.

23. The degree of the numerator is greater than the degree of the denominator. Therefore, there is not a horizontal asymptote.

To find the vertical asymptote let
$$x - 1 = 0$$
$$x = 1$$
$x = 1$ is the vertical asymptote.

To find the slant asymptote:

$$1 \overline{)\begin{array}{ccc} 1 & -1 & 2 \\ & 1 & 0 \\ \hline 1 & 0 & 2 \end{array}}$$

The slant asymaptote is $y = x$.

25. $f(x) = \dfrac{x^2}{x - 1}$

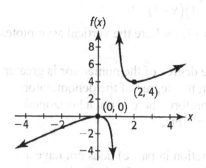

Based on the model, the turning points appear to be $(0, 0)$ and $(2, 4)$.

27. $g(x) = \dfrac{x^2 + 4}{x}$

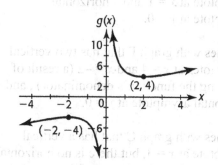

Based on the model, the turning points appear to be $(-2, -4)$ and $(2, 4)$.

29. a. $y = \dfrac{1 - x^2}{x - 2}$

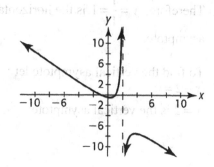

b.

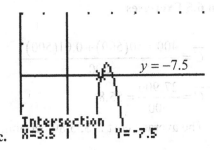

Based on the model and the table, when $x \approx 1$, $y = 0$ and when $x = 3$, $y = -8$.

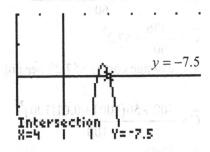

$y = -7.5$

Intersection X=3.5 Y=-7.5

c.

$[0, 8]$ by $[-7.7, -7.3]$

$y = -7.5$

Intersection X=4 Y=-7.5

$[0, 8]$ by $[-7.7, -7.3]$

Based on the models, it appears that when $y = -7.5$, then $x = 3.5$ or $x = 4$.

d.

$$-7.5 = \frac{1 - x^2}{x - 2} \qquad\qquad \text{LCD}: x - 2$$

$$-7.5(x - 2) = \left(\frac{1 - x^2}{x - 2}\right)(x - 2)$$

$$-7.5x + 15 = 1 - x^2$$

$$x^2 - 7.5x + 14 = 0$$

$$10(x^2 - 7.5x + 14) = 10(0)$$

$$10x^2 - 75x + 140 = 0$$

$$(x - 4)(10x - 35) = 0$$

$$x - 4 = 0, \quad 10x - 35 = 0$$

$$x = 4, x = 3.5$$

Both solutions check.

31. a. $y = \dfrac{3 - 2x}{x}$

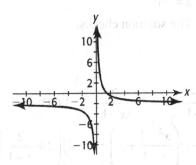

b.

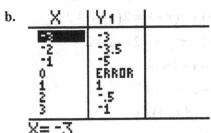

Based on the model and the table, when $x = -3$, $y = -3$ and when $x = 3$, $y = -1$.

c.

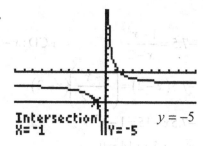

$[-10, 10]$ by $[-10, 10]$

Based on the graph, it appears that when $y = -5$, then $x = -1$.

d. $-5 = \dfrac{3 - 2x}{x}$ LCD: x

$-5x = \left(\dfrac{3 - 2x}{x}\right)x$

$-5x = 3 - 2x$

$-3x = 3$

$x = -1$

The solution checks.

33.

$\dfrac{x^2 + 1}{x - 1} + x = 2 + \dfrac{2}{x - 1}$ LCD: $x - 1$

$(x - 1)\left(\dfrac{x^2 + 1}{x - 1} + x\right) = (x - 1)\left(2 + \dfrac{2}{x - 1}\right)$

$x^2 + 1 + x(x - 1) = 2(x - 1) + 2$

$x^2 + 1 + x^2 - x = 2x - 2 + 2$

$2x^2 - x + 1 = 2x$

$2x^2 - 3x + 1 = 0$

$(2x - 1)(x - 1) = 0$

$x = \dfrac{1}{2}, \quad x = 1$

$x = 1$ does not check since the denominator $x - 1 = 0$.

The only solution is $x = \dfrac{1}{2}$.

35. $y = \dfrac{k}{x^4}$ is the inverse function format.

$5 = \dfrac{k}{(-1)^4}$

$k = 5$

$y = \dfrac{5}{(0.5)^4} = 80$

Section 6.5 Exercises

37. a. $\overline{C} = \dfrac{400 + 50(500) + 0.01(500)^2}{500}$

$\overline{C} = \dfrac{27,900}{500} = 55.8$

The average cost is $55.80 per unit.

b. $\overline{C} = \dfrac{400 + 50(60) + 0.01(60)^2}{60}$

$\overline{C} = \dfrac{3436}{60} = 57.2\overline{6}$

The average cost is $57.27 per unit.

c. $\overline{C} = \dfrac{400 + 50(100) + 0.01(100)^2}{100}$

$\overline{C} = \dfrac{5500}{100} = 55$

The average cost is $55 per unit.

d. No. Consider the graph of the function where x = 600 units. The average cost per unit is then $56.67.

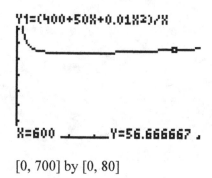

$[0, 700]$ by $[0, 80]$

39. a. $y = \dfrac{400(5)}{5+20} = \dfrac{2000}{25} = 80$

$5000 in monthly advertising expenditures results in a monthly sales volume of $80,000.

b. When $x = -20$, the denominator is zero and the function is undefined. Since advertising expenditures cannot be negative, x cannot be -20 in the context of the problem.

41. a. $\overline{C}(x) = \dfrac{1000 + 30x + 0.1x^2}{x}$

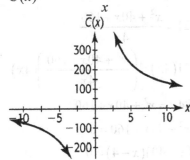

b.

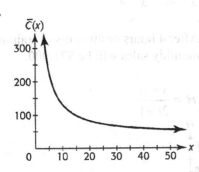

c. The window in part b) fits the context of the problem where both x and $\overline{C}(x)$ are greater than zero.

d.

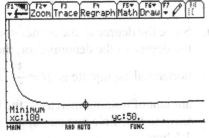

[0, 250] by [0, 300]

The minimum average cost of $50 occurs when $x = 100$. Since x is measured in hundreds of units produced, 10,000 units are produced.

43. a. $f(t) = \dfrac{30 + 40t}{5 + 2t}$

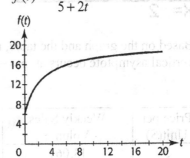

b. $f(0) = 6$. The initial number of employees for the startup company is 6.

c. $f(12) \approx 17.586$. After 12 months, the number of employees for the startup company is approximately 18.

45. a. To find the vertical asymptote let
$100 - p = 0$.
$-p = -100$
$p = 100$ is the vertical asymptote.

b. Since $p \neq 100$, 100% of the impurities can not be removed from the waste water.

47. a. $V = \dfrac{640}{(p+2)^2}$

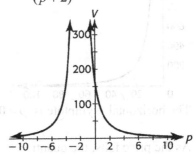

X	Y1
-5	71.111
-4	160
-3	640
-2	ERROR
-1	640
0	160
1	71.111

X= -2

Based on the graph and the table, the vertical asymptote occurs at $p = -2$.

b.

Price per Unit($)	Weekly Sales Volume
5	13,061
20	1322
50	237
100	62
200	16
500	3

c. The domain of the function in the context of the problem is $p \geq 0$. There is no vertical asymptote on the restricted domain.

d. The horizontal asymptote is $V = 0$. As the price grows without bound, the sales volume approaches zero units.

49. a. $p = \dfrac{100,000}{(q+1)^2}$

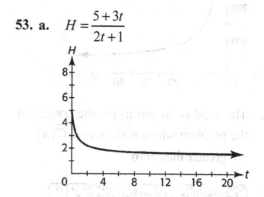

b. The horizontal asymptote is $p = 0$.

c. As the price falls, the quantity demanded increases.

51. a.

$$S = \frac{40}{x} + \frac{x}{4} + 10 \qquad \text{LCD: } 4x$$

$$\left(\frac{40}{x}\right)\left(\frac{4}{4}\right) + \frac{x}{4}\left(\frac{x}{x}\right) + \left(\frac{10}{1}\right)\left(\frac{4x}{4x}\right)$$

$$\frac{160}{4x} + \frac{x^2}{4x} + \frac{40x}{4x}$$

$$\frac{x^2 + 40x + 160}{4x}$$

$$S = \frac{x^2 + 40x + 160}{4x}$$

b. $21 = \dfrac{x^2 + 40x + 160}{4x}$

$$21(4x) = \left(\frac{x^2 + 40x + 160}{4x}\right)(4x)$$

$$84x = x^2 + 40x + 160$$

$$x^2 - 44x + 160 = 0$$

$$(x - 40)(x - 4) = 0$$

$$x = 40, \quad x = 4$$

After 4 hours or 40 hours of training, the monthly sales will be $21,000.

53. a. $H = \dfrac{5 + 3t}{2t + 1}$

b. Since the degree of the numerator equals the degree of the denominator, the horizontal asymptote is $H = \dfrac{3}{2}$. As the amount of training increases, the time it takes to assemble one unit approaches 1.5 hours.

c.

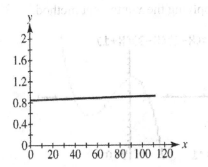

X=17

It takes 17 days of training to reduce the time it takes to assemble one unit to 1.6 hours.

55. a.

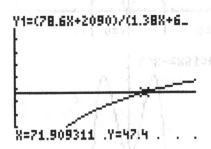

b. According to the model, in the year 2020, the predicted ratio of women's to men's age at first marriage will be 0.927.

c. Let $x = 500$. The function value will be:

$$y = \frac{0.00117(500)^2 - 0.0948(500) + 22.4}{0.00132(500)^2 - 0.127(500) + 26.3}$$

$$y \approx 0.91$$

Let $x = 1000$. The function value will be:

$$y = \frac{0.00117(1000)^2 - 0.0948(1000) + 22.4}{0.00132(1000)^2 - 0.127(1000) + 26.3}$$

$$y \approx 0.90$$

It does not appear to increase indefinitely.

57. a. $p(t) = \dfrac{78.6t + 2090}{1.38t + 64.1}$ with t equal to the number of years after 1950.

For $x = 80$ (2030 – 1950), $p(80) =$ 48.0%. Yes, this agrees with the given data for the year 2030.

b. The maximum possible percent of women in the workforce, according to the model, would occur at the horizontal asymptote. Since the degree of the numerator is equal to the degree of the denominator, the horizontal asymptote is at $y = \dfrac{78.6}{1.38} = 56.957 \approx 57\%$.

c. The percent will reach 47.4% when $x = 71.9$, in the year 2022 (1950 + 72).

Y1=(78.6X+2090)/(1.38X+6_

X=71.909311 Y=47.4 . . .

[0, 100] by [40, 60]

Section 6.6 Skills Check

1. $16x^2 - x^4 \geq 0$
    Applying the x-intercept method:

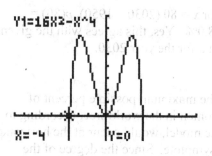

[–10, 10] by [–20, 80]

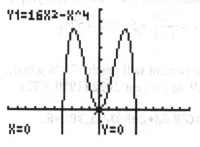

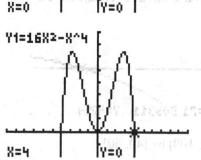

The function is greater than or equal to zero on the interval $[-4, 4]$ or when $-4 \leq x \leq 4$.

3. $2x^3 - x^4 < 0$
    Applying the x-intercept method:

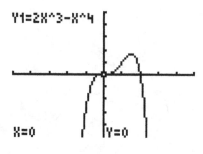

[–5, 5] by [–5, 5]

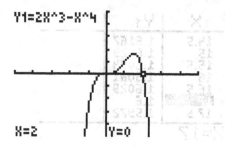

The function is less than zero on the interval $(-\infty, 0) \cup (2, \infty)$ or when $x < 0$ or $x > 2$.

5. $(x-1)(x-3)(x+1) \geq 0$
    Applying the x-intercept method:

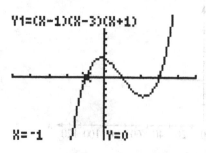

[–5, 5] by [–10, 10]

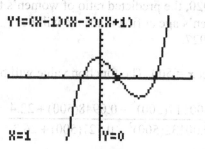

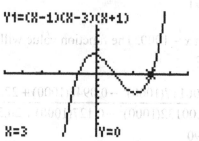

The function is greater than or equal to zero on the interval $[-1, 1] \cup [3, \infty)$ or when $-1 \leq x \leq 1$ or $x \geq 3$.

7. $\dfrac{4-2x}{x} > 2$

Applying the intersection of graphs method:

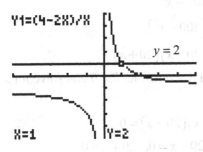

Y1=(4-2X)/X $y = 2$

X=1   Y=2

[–5, 5] by [–10, 10]

Note that the graphs intersect when $x = 1$. Also note that a vertical asymptote occurs at $x = 0$. Therefore, $\dfrac{4-2x}{x} > 2$ on the interval $(0,1)$ or when $0 < x < 1$.

9. $\dfrac{x}{2} + \dfrac{x-2}{x+1} \le 1$

Applying the intersection of graphs method:

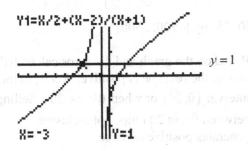

Y1=X/2+(X-2)/(X+1) $y = 1$

X=-3   Y=1

[–10, 10] by [–5, 5]

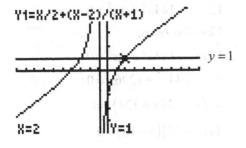

Y1=X/2+(X-2)/(X+1) $y = 1$

X=2   Y=1

Note that the graphs intersect when $x = -3$ and $x = 2$. Also note that a vertical asymptote occurs at $x + 1 = 0$ or $x = -1$. Therefore, $\dfrac{x}{2} + \dfrac{x-2}{x+1} \le 1$ on the interval $(-\infty, -3] \cup (-1, 2]$ or when $x \le -3$ or $-1 < x \le 2$.

11. $(x-1)^3 > 27$

Applying the intersection of graphs method:

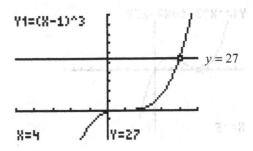

Y1=(X-1)^3 $y = 27$

X=4   Y=27

[–5, 5] by [–15, 50]

Note that the graphs intersect when $x = 4$. Therefore $(x-1)^3 > 27$ on the interval $(4, \infty)$ or when $x > 4$.

13. $(x-1)^3 < 64$

Applying the intersection of graphs method:

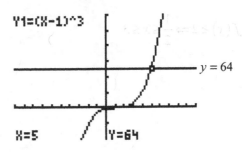

Y1=(X-1)^3 $y = 64$

X=5   Y=64

[–10, 10] by [–50, 150]

Note that the graphs intersect when $x = 5$. Therefore, $(x-1)^3 < 64$ on the interval $(-\infty, 5)$ or when $x < 5$.

15. $-x^3 - 10x^2 - 25x \leq 0$
Applying the x-intercept method:

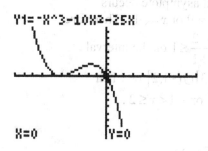

[–10, 10] by [–100, 100]

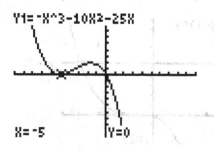

Note that the x-intercepts are $x = -5$ and $x = 0$. Therefore, $-x^3 - 10x^2 - 25x \leq 0$ on the interval $\{-5\} \cup [0, \infty)$ or when $x = -5$ or $x \geq 0$.

17. a. $f(x) < 0 \Rightarrow x < -3$ or $0 < x < 2$

b. $f(x) \geq 0 \Rightarrow -3 \leq x \leq 0$ or $x \geq 2$

19. $f(x) \geq 2 \Rightarrow \dfrac{1}{2} \leq x \leq 3$

Section 6.6 Exercises

21.
$$R = 400x - x^3$$
$$= x\left(400 - x^2\right)$$
$$= x(20 - x)(20 + x)$$

To find the zeros, let $R = 0$ and solve for x.

$R = 0$

$x(20 - x)(20 + x) = 0$

$x = 0, \ 20 - x = 0, \ 20 + x = 0$

$x = 0, \ x = 20, \ x = -20$

Note that since x represents product sales, only positive values of x make sense in the context of the problem.

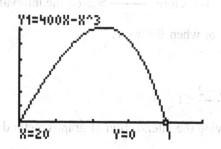

[0, 25] by [–500, 3500]

Based on the graph and the zeros calculated above, the revenue is positive, $R > 0$, in the interval $(0, 20)$ or when $0 < x < 20$. Selling between 0 and 20 units, not inclusive, generates positive revenue.

23. a. $V > 0$
$$1296x - 144x^2 + 4x^3 > 0$$
Find the zeros:
$$1296x - 144x^2 + 4x^3 = 0$$
$$4x^3 - 144x^2 + 1296x = 0$$
$$4x\left(x^2 - 36x + 324\right) = 0$$
$$4x(x - 18)(x - 18) = 0$$
$$4x = 0, \ x - 18 = 0, \ x - 18 = 0$$
$$x = 0, \ x = 18, \ x = 18$$

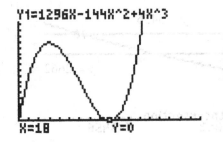

[0, 36] by [–500, 5000]

Based on the graph and the zeros calculated above, the volume is positive in the interval $(0,18)\cup(18,\infty)$ or when $0 < x < 18$ or $x > 18$.

b. In the context of the problem, the largest possible cut is 18 centimeters. Therefore, to generate a positive volume, the size of the cut, x, must be in the interval $(0,18)$ or $0 < x < 18$.

25. $C(x) \geq 1200$

$3x^3 - 6x^2 - 300x + 1800 \geq 1200$

$3x^3 - 6x^2 - 300x + 600 \geq 0$

Find the zeros:

$3x^3 - 6x^2 - 300x + 600 = 0$

$3\left(x^3 - 2x^2 - 100x + 200\right) = 0$

$3\left[\left(x^3 - 2x^2\right) + \left(-100x + 200\right)\right] = 0$

$3\left[x^2\left(x - 2\right) + \left(-100\right)\left(x - 2\right)\right] = 0$

$3\left(x - 2\right)\left(x^2 - 100\right) = 0$

$3\left(x - 2\right)\left(x + 10\right)\left(x - 10\right) = 0$

$x - 2 = 0, \quad x + 10 = 0, \quad x - 10 = 0$

$x = 2, \quad x = -10, \quad x = 10$

Sign chart:

Function	---	+++	---	+++
3	+++	+++	+++	+++
$(x-10)$	---	---	---	+++
$(x+10)$	---	+++	+++	+++
$(x-2)$	---	---	+++	+++
	-10	2	10	

Based on the sign chart, the function is greater than zero on the intervals $(-10,2)$ and $(10,\infty)$. Considering the context of the problem, the number of units cannot be negative. The endpoints of the interval would be part of the solution because the question uses the phrase "at least." Therefore, total cost is at least \$120,000 if $0 \leq x \leq 2$ or if $x \geq 10$. In interval notation the solution is $\left[0,2\right]\cup\left[10,\infty\right]$.

Applying the x-intercept method:

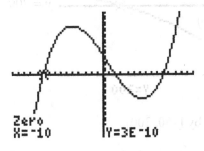

[–15, 15] by [–2000, 2000]

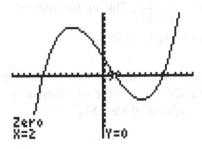

[–15, 15] by [–2000, 2000]

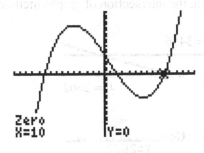

[–15, 15] by [–2000, 2000]

In context of the problem, the number of units must be greater than or equal to zero. Therefore, based on the graphs, the total cost is greater than or equal to $120,000 on the intervals $[0,2] \cup [10,\infty]$ or when $0 \leq x \leq 2$ or $x \geq 10$.

27. $y = \dfrac{400x}{x+20}$

Applying the intersection of graphs method:

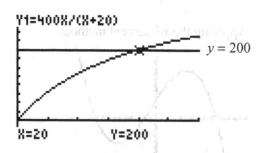

[0, 30] by [−50, 300]

Note that the graphs intersect when $x = 20$.

Therefore, $\dfrac{400x}{x+20} \geq 200$ on the interval $[20,\infty)$ or when $x \geq 20$.

Therefore, since x is in thousands of dollars, spending $20,000 or more on advertising results in sales of at least $200,000.

29. $S = 2000(1+r)^3$

Applying the intersection of graphs method

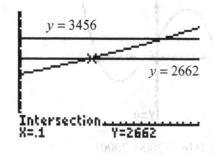

[0, 0.25] by [−500, 4500]

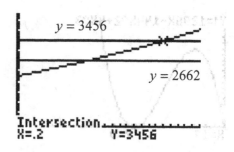

Note that the graphs intersect when $r = 0.10$ or $r = 0.20$. Therefore,

$2662 \leq 2000(1+r)^3 \leq 3456$ on the interval $[0.10, 0.20]$ or when $0.10 \leq r \leq 0.20$.

Therefore, interest rates between 10% and 20%, inclusive, generate future values between $2662 and $3456, inclusive.

31. $R = 4000x - 0.1x^3$

Applying the intersection of graphs method:

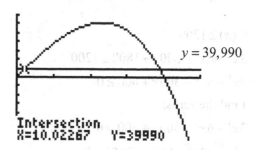

[0, 250] by [−350,000, 350,000]

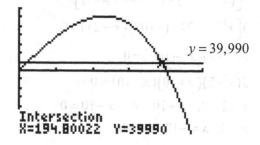

Note that the graphs intersect when $x \approx 10.02$ and when $x \approx 194.80$. Therefore, $R = 4000x - 0.1x^3 \geq 39,990$ on the interval $[10.02, 194.80]$ or when $10.02 \leq x \leq 194.80$.

Therefore, producing and selling between 10 units and 194 units, inclusive, generates a revenue of at least $39,990.

33.

Considering the supply function and solving for q:

$6p - q = 180$

$-q = 180 - 6p$

$q = 6p - 180$

Considering the demand function and solving for q:

$(p + 20)q = 30,000$

$q = \dfrac{30,000}{p + 20}$

Supply > Demand

$6p - 180 > \dfrac{30,000}{p + 20}$

Applying the intersection of graphs method:

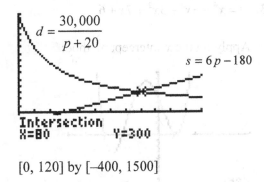

[0, 120] by [−400, 1500]

When the price is above $80, supply exceeds demand.

35. a. $f(t) = \dfrac{30 + 40t}{5 + 2t}$

Applying the intersection of graphs method:

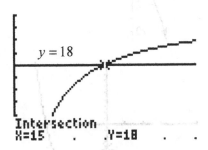

[0, 30] by [15, 20]

Note that the graphs intersect when $t = 15$. Therefore, $f(t) = \dfrac{30 + 40t}{5 + 2t} < 18$ on the interval $[0,15)$ or when $0 \le t < 15$.

b.   For the first 15 months of the operation, the number of employees is below 18.

Chapter 6 Skills Check

1.   The degree of the polynomial is the highest exponent. In this case, the degree of the polynomial is 4.

3. $y = -4x^3 + 4x^2 + 1$

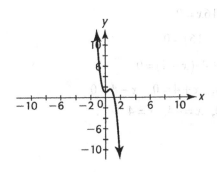

Yes. The graph is complete on the given viewing window.

5. a. $y = x^3 - 3x^2 - 4$

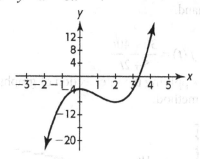

b.

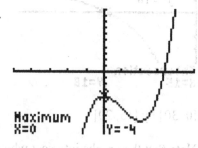

Maximum
X=0   Y=-4

$[-5, 5]$ by $[-10, 10]$

The local maximum is $(0, -4)$.

c.

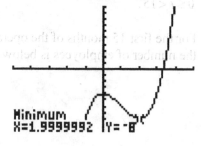

Minimum
X=1.9999992 Y=-8

$[-5, 5]$ by $[-10, 10]$

The local minimum is $(2, -8)$.

7. $x^3 - 16x = 0$

$x(x^2 - 16) = 0$

$x(x+4)(x-4) = 0$

$x = 0, \quad x+4 = 0, \quad x-4 = 0$

$x = 0, \quad x = -4, \quad x = 4$

9. $x^4 - x^3 - 20x^2 = 0$

$x^2(x^2 - x - 20) = 0$

$x^2(x-5)(x+4) = 0$

$x^2 = 0, \quad x-5 = 0, \quad x+4 = 0$

$x = 0, \quad x = 5, \quad x = -4$

11. $4x^3 - 20x^2 - 4x + 20 = 0$

$4(x^3 - 5x^2 - x + 5) = 0$

$4[(x^3 - 5x^2) + (-x+5)] = 0$

$4[x^2(x-5) + -1(x-5)] = 0$

$4(x-5)(x^2 - 1) = 0$

$4(x-5)(x+1)(x-1) = 0$

$x-5 = 0, \quad x+1 = 0, \quad x-1 = 0$

$x = 5, \quad x = -1, \quad x = 1$

13. $y = x^4 - 3x^3 - 3x^2 + 7x + 6$

Applying the x-intercept method:

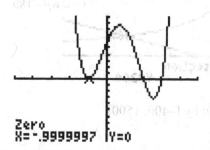

Zero
X=-.9999997 Y=0

$[-5, 5]$ by $[-10, 10]$

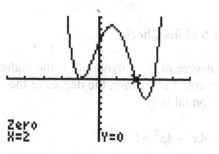

Zero
X=2   Y=0

$[-5, 5]$ by $[-10, 10]$

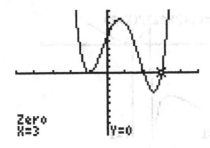

[–5, 5] by [–10, 10]

Based on the graphs, the solutions are $x = -1, x = 2, x = 3$.

15. $(x-4)^3 = 8$

$$\sqrt[3]{(x-4)^3} = \sqrt[3]{8}$$

$$x - 4 = 2$$

$$x = 6$$

17.
```
2) 4  -3   0   2  -8
        8  10  20  44
   ──────────────────
   4   5  10  22  36
```

$$4x^3 + 5x^2 + 10x + 22 + \frac{36}{x-2}$$

19. $y = 3x^3 - x^2 - 12x + 4$

Applying the x-intercept method:

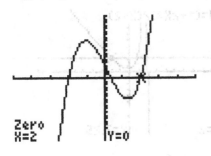

[–5, 5] by [–20, 20]

It appears that $x = 2$ is a zero.

```
2) 3  -1  -12   4
        6   10  -4
   ───────────────
   3   5   -2   0
```

The new polynomial is $3x^2 + 5x - 2$.

Applying the quadratic formula:

$$x = \frac{-(5) \pm \sqrt{(5)^2 - 4(3)(-2)}}{2(3)}$$

$$x = \frac{-5 \pm \sqrt{49}}{6}$$

$$x = \frac{-5 \pm 7}{6}$$

$$x = \frac{-5 + 7}{6} = \frac{2}{6} = \frac{1}{3}$$

or

$$x = \frac{-5 - 7}{6} = \frac{-12}{6} = -2$$

The solutions are $x = 2, x = -2, x = \dfrac{1}{3}$

21. a. $y = \dfrac{1 - x^2}{x + 2}$

To find the y-intercept, let $x = 0$ and solve for y.

$$y = \frac{1 - (0)^2}{0 + 2} = \frac{1}{2}$$

$$\left(0, \frac{1}{2}\right)$$

To find x-intercepts, let the numerator equal zero and solve for x.

$$1 - x^2 = 0$$

$$x^2 = 1$$

$$\sqrt{x^2} = \pm\sqrt{1}$$

$$x = \pm 1$$

$$(-1, 0), (1, 0)$$

b. To find the vertical asymptote let

$$x + 2 = 0$$

$$x = -2$$

$x = -2$ is the vertical asymptote.

The degree of the numerator is greater than the degree of the denominator. Therefore, there is not a horizontal asymptote.

c. To find the slant asymptote:

$$-2\overline{)\begin{array}{ccc} -1 & 0 & 1 \\ & 2 & -4 \\ \hline -1 & 2 & -3 \end{array}}$$

The slant asymaptote is $y = 2 - x$.

d.

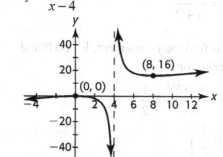

23. $y = \dfrac{x^2}{x-4}$

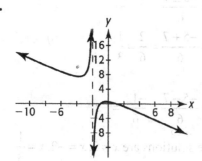

The local maximum is $(0,0)$, while the

local minimum is $(8,16)$.

25. a. $y = \dfrac{1+2x^2}{x+2}$

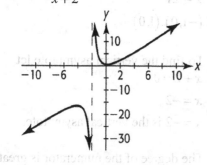

b. Y1=(1+2X^2)/(X+2)

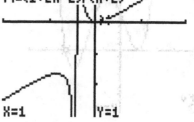

[−10, 10] by [−30, 10]

Y1=(1+2X^2)/(X+2)

X=3   Y=3.8

When $x = 1$, $y = 1$. When $x = 3$, $y = 3.8$.

c. Y1=(1+2X2)/(X+2)

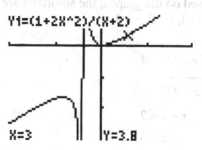

X=-.875   Y=2.25

[−10, 10] by [−10, 10]

Y1=(1+2X2)/(X+2)

X=2   Y=2.25

If $y = 2.25$, then $x = -0.875$ or $x = 2$.

d.

$$\frac{9}{4} = \frac{1 + 2x^2}{x + 2} \qquad \text{LCD: } 4(x + 2)$$

$$4(x+2)\left(\frac{9}{4}\right) = 4(x+2)\left(\frac{1+2x^2}{x+2}\right)$$

$$9(x+2) = 4(1+2x^2)$$

$$9x + 18 = 4 + 8x^2$$

$$8x^2 - 9x - 14 = 0$$

$$(8x + 7)(x - 2) = 0$$

$$8x + 7 = 0 \Rightarrow 8x = -7 \Rightarrow x = -\frac{7}{8}$$

$$x - 2 = 0 \Rightarrow x = 2$$

The solutions are $x = 2, x = -\dfrac{7}{8}$.

27. $y = x^3 + x^2 + 2x - 4$

Applying the x-intercept method:

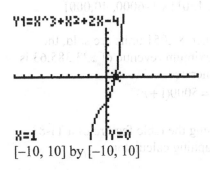

Y1=X^3+X2+2X-4

X=1   Y=0
[−10, 10] by [−10, 10]

It appears that $x = 1$ is a zero.

$$\begin{array}{r|rrrr} 1) & 1 & 1 & 2 & -4 \\ & & 1 & 2 & 4 \\ \hline & 1 & 2 & 4 & 0 \end{array}$$

The new polynomial is $x^2 + 2x + 4$.

Set the new polynomial equal to zero and solve.

$$x^2 + 2x + 4 = 0$$

$$x = \frac{-b \pm \sqrt{b^2 - 4ac}}{2a}$$

$$x = \frac{-(2) \pm \sqrt{(2)^2 - 4(1)(4)}}{2(1)}$$

$$x = \frac{-2 \pm \sqrt{-12}}{2}$$

$$x = \frac{-2 \pm 2i\sqrt{3}}{2} = \frac{2(-1 \pm 1i\sqrt{3})}{2}$$

$$x = -1 \pm i\sqrt{3}$$

The solutions are $x = 1, x = -1 \pm i\sqrt{3}$.

29. $x^3 - 5x^2 \geq 0$

$$x^2(x - 5) \geq 0$$

$$x^2(x - 5) = 0$$

$$x - 5 = 0, \quad x^2 = 0$$

$$x = 5, \quad x = 0$$

Sign chart:

Function	---	---	+++
x^2	+++	+++	+++
$(x - 5)$	---	---	+++
	0	5	

Based on the sign chart, the function is greater than or equal to zero on the interval $[5, \infty)$ or when $x \geq 5$. In addition, the function is equal to zero when $x = 0$.

31. $2 < \dfrac{4x - 6}{x}$

$\dfrac{4x - 6}{x} > 2$

Applying the intersection of graphs method:

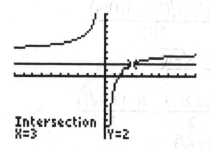

Intersection
X=3   Y=2

$[-10, 10]$ by $[-10, 10]$

Note that the graphs intersect when $x = 3$. Also note that a vertical asymptote occurs at $x = 0$. Therefore, $\dfrac{4x - 6}{x} > 2$ on the interval $(-\infty, 0) \cup (3, \infty)$ or when $x < 0$ or $x > 3$.

Chapter 6 Review Exercises

33. a. $R = -0.1x^3 + 15x^2 - 25x$

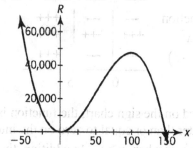

b.

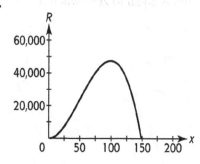

c.

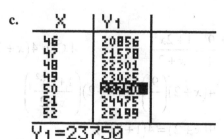

Y1=23750

When 50,000 units are produced and sold, the revenue is $23,750.

34. $R = -0.1x^3 + 13.5x^2 - 150x$

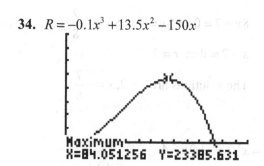

Maximum
X=84.051256 Y=23385.631

$[0, 150]$ by $[-6000, 40,000]$

When 84,051 units are sold, the maximum revenue of $23,385.63 is generated.

35. a. $S = 5000(1 + r)^6$

Using the table feature of a TI-83 graphing calculator:

Rate, r	Future Value, $S(\$)$
0.01	5307.60
0.05	6700.48
0.10	8857.81
0.15	11,565.30

b.

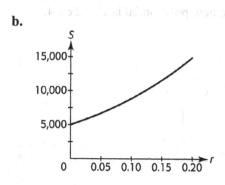

c.

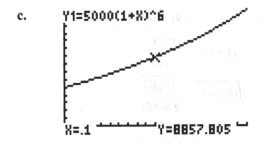

[0, 0.2] by [−1000, 15,000]

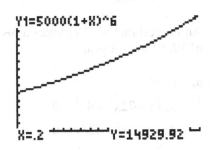

[0, 0.2] by [−1000, 15,000]

The difference in the future values is $14{,}929.92 - 8857.81 = \$6072.11$.

36. $y = 120x^2 - 20x^3$

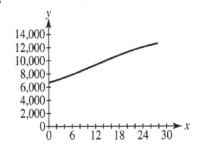

[0, 8] by [0, 1000]

An intensity level of 4 allows the maximum amount of photosynthesis to take place.

37. a.

b. Using this model, the yearly income poverty threshold for a single person in 2017 (when $x = 27$) will be \$12,513.

c. Using the maximum value function on the calculator:

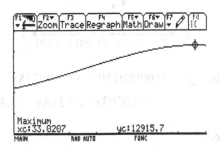

[0, 35] by [0, 15000]

Based on the graph, in the year 2023 (when $x = 33$), the poverty threshold for a single person will be a maximum.

38. a.

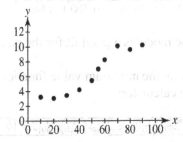

b. $y = -0.0000596x^3 + 0.00933x^2$
$\qquad -0.308x + 5.63$

c.

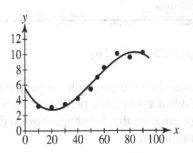

d. Using the unrounded model, the number of elderly Americans in the workforce in 2060 will be 8.578 million.

39. a.

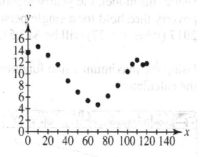

b. $y = -0.00000110x^4 + 0.000278x^3$

$-0.0203x^2 + 0.318x + 13.4$

c.

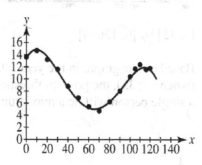

The model is a good fit for the data.

d. Using the minimum value function on the calculator:

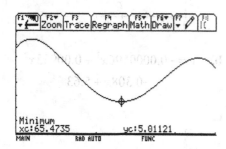

$[0, 125]$ by $[0, 16]$

Based on the graph, in the year 1966 (when $x = 66$), the percent of the U.S. population that is foreign-born will be a minimum.

40. $9261 = 8000(1+r)^3$

$(1+r)^3 = \dfrac{9261}{8000}$

$\sqrt[3]{(1+r)^3} = \sqrt[3]{\dfrac{9261}{8000}}$

$1+r = 1.05$

$r = 1.05 - 1$

$r = 0.05$

An interest rate of 5% creates a future value of $9261 after 3 years.

41. a. $V = 0$

$324x - 72x^2 + 4x^3 = 0$

$4x^3 - 72x^2 + 324x = 0$

$4x(x^2 - 18x + 81) = 0$

$4x(x-9)(x-9) = 0$

$4x = 0, \quad x - 9 = 0, \quad x - 9 = 0$

$x = 0, \quad x = 9, \quad x = 9$

b. If the values of x from part a) are used to cut squares from corners of a piece of tin, no box can be created. Either no square is cut or the squares encompass all the tin. Therefore, the volume of the box is zero.

c. Reasonable values of x would allow for a box to be created. An x-value larger than zero and less than half the length of the edge of the piece of tin would allow for a box to be created. Therefore, reasonable values are

$0 < x < \dfrac{18}{2}$ or $0 < x < 9$.

42. a. $P(x) = -0.2x^3 + 20.5x^2 - 48.8x - 120$

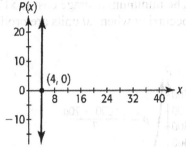

The x-intercept is $(4, 0)$.

b.
$$4\overline{)\,-0.2 \quad 20.5 \quad -48.8 \quad -120}$$
$$ \quad\quad -0.8 \quad\quad 78.8 \quad\quad 120$$
$$\overline{\; -0.2 \quad\; 19.7 \quad\quad\; 30 \quad\quad\quad 0}$$

The remaining quadratic factor
is $-0.2x^2 + 19.7x + 30$.

c. Set the remaining polynomial equal to
zero and solve.

$$-0.2x^2 + 19.7x + 30 = 0$$

$$x = \frac{-b \pm \sqrt{b^2 - 4ac}}{2a}$$

$$x = \frac{-(19.7) \pm \sqrt{(19.7)^2 - 4(-0.2)(30)}}{2(-0.2)}$$

$$x = \frac{-19.7 \pm \sqrt{412.09}}{-0.4}$$

$$x = \frac{-19.7 \pm 20.3}{-0.4}$$

$$x = \frac{-19.7 + 20.3}{-0.4}, \; x = \frac{-19.7 - 20.3}{-0.4}$$

$$x = -1.5, \; x = 100$$

The solutions are $x = 4$, $x = -1.5$,
and $x = 100$.

d. Since negative solutions do not make
sense in the context of the question,
break-even occurs when 400 units are
produced or when 10,000 units are
produced.

43. a.

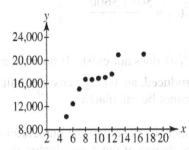

b. $y = 10.3x^3 - 400x^2 + 5590x - 8690$

c.

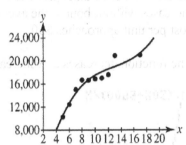

d. Using the unrounded model, the number
of worldwide Starbucks stores in 2019 is
predicted to be 23,574.

44. $C = \dfrac{0.3t}{t^2 + 1}$

a. Since the degree of the numerator is less
than the degree of the denominator, the
horizontal asymptote is $C = 0$.

b. As the time increases, the concentration
of the drug approaches zero percent.

c. Y1=0.3X/(X²+1)

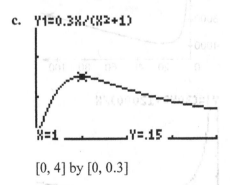

[0, 4] by [0, 0.3]

The maximum drug concentration is
15%, occurring after one hour.

45. $\overline{C}(x) = \dfrac{50x + 5600}{x}$

a. $\overline{C}(0)$ does not exist. If no units are produced, an average cost per unit cannot be calculated.

b.   Since the degree of the numerator equals the degree of the denominator, the horizontal asymptote is $\overline{C}(x) = \dfrac{50}{1} = 50$.

As the number of units produced increases without bound, the average cost per unit approaches $50.

c.   The function decreases as x increases.

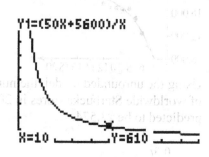

[0, 20] by [0, 5000]

46. a. $\overline{C} = \dfrac{30x^2 + 12,000}{x}$

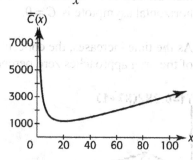

b.

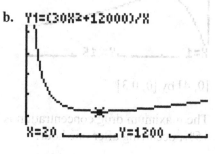

[0, 50] by [0, 6000]

The minimum average cost is $1200, occurring when 20 units are produced.

47.

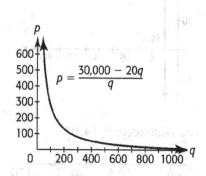

48. a. $C(x) = 200 + 20x + \dfrac{180}{x}$

$C(x) = \dfrac{200x}{x} + \dfrac{20x^2}{x} + \dfrac{180}{x}$

$C(x) = \dfrac{20x^2 + 200x + 180}{x}$

b.

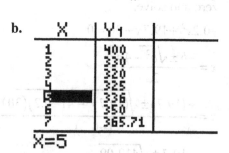

Using 5 plates creates a cost of $336.

c.

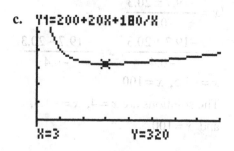

[0, 8] by [−100, 600]

Using 3 plates creates a minimum cost of $320.

49. a.

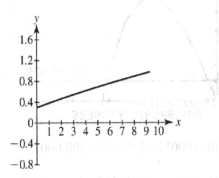

b. For 2015, $x = 2015 - 2010 = 5$.

$$f(5) = \frac{21.35(5) + 69.59}{4.590(5) + 233.1}$$

$$= \frac{176.34}{256.05} = 0.6886$$

Therefore, in the year 2015, approximately 69% of cell phones in the U.S. will be smart phones.

c. $f(10) = \dfrac{21.35(10) + 69.59}{4.590(10) + 233.1}$

$$= \frac{283.09}{279} = 1.015$$

This indicates that there are more smart phone users than cell phone users, which does not make sense because a smart phone is a special type of cell phone. Thus, the model is not valid after 2019.

50. $S = \dfrac{40}{x} + \dfrac{x}{4} + 10$

Applying the intersection of graphs method:

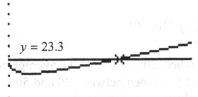

[4, 80] by [−10, 50]

Note that $S \geq 23.3$ occurs on the interval $[50, \infty)$ or when $x \geq 50$. Fifty or more hours of training results in sales greater than $23,300.

51. a.

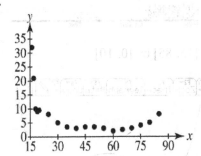

b. $y = 0.00001828x^4 - 0.003925x^3$
$$+ 0.3031x^2 - 9.907x + 118.2$$

c.

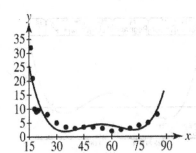

d. Applying the intersection of graphs method:

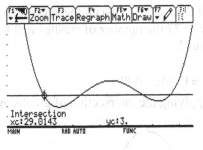

[15, 85] by [0, 10]

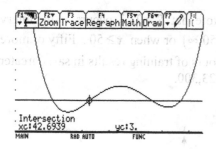

[15, 85] by [0, 10]

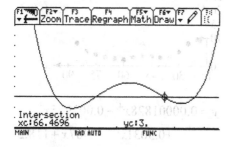

[15, 85] by [0, 10]

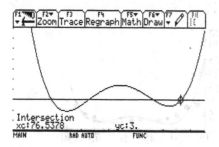

[15, 85] by [0, 10]

Note that $y < 3$ when $29 \leq x \leq 43$ and $67 \leq x \leq 77$. Between the ages 29 to 43 and 67 to 77 the number of crashes will be less than 3.

52. $y = 1200x - 0.003x^3$

Applying the intersection of graphs method:

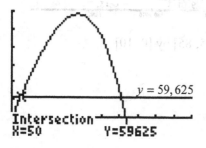

[0, 1000] by [−50,000, 300,000]

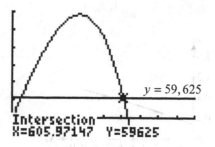

[0, 1000] by [−50,000, 300,000]

Note that $R \geq 59,625$ when $x \geq 50$ and $x \leq 605.97$. Selling 50 or more units but no more than 605 units creates a revenue stream of at least $59,625.

53. $\overline{C} = \dfrac{100 + 30x + 0.1x^2}{x}$

Applying the intersection of graphs method

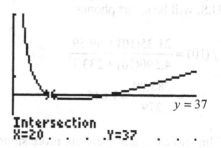

[0, 100] by [30, 50]

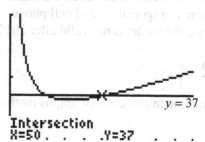

[0, 100] by [30, 50]

$\overline{C} \leq 37$ when $20 \leq x \leq 50$. The average cost is at most $37 when between 20 and 50 units, inclusive, are produced.

54. $p = \dfrac{100C}{9600 + C}$

Applying the intersection of graphs method:

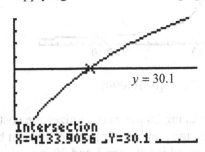

$y = 30.1$

Intersection
X=4133.9056 _Y=30.1

[0, 10,000] by [0, 50]

Note that $p \geq 30.1$ when $x \geq 4133.91$. To remove at least 30.1% of the particulate pollution will cost at least $4133.91.

Group Activity/Extended Applications

1. Global Climate Change

1. a. $p \neq 100$

  b. D: $[0, \infty)$
      R: $[0, 242{,}000)$

2. $C(60) = \$3630$ hundred $= \$363{,}000$.
    $C(80) = \$9680$ hundred $= \$968{,}000$. This result means the annual cost in dollars of removing 80% of the particulate pollution from the smokestack of a power plant is $968,000.

3. The vertical asymptote of this function is $p = 100$, and means it is impossible to remove 100% of the particulate pollution. Algebraically, the horizontal asymptote is $C = 0$, but in the context of the problem (within the domain), there is no horizontal asymptote that makes any sense.

4. The y-intercept is $(0, 0)$ and means that it would cost $0 if no particulate pollution amount is removed.

5. The CFO of this company should recommend to pay the fine since the cost to remove 80% of the pollution ($968,000) is higher than the fine of $700,000.

6. If the company has already paid $363,000 to remove 60% of the pollution, the difference to remove 20% more is only $605,000 which is less than the $700,000 fine. Advise the company to remove the 20% difference rather than pay the fine.

7. Answer may vary, but at least $1,000,000 would be higher than the cost to remove 80% of the pollution.

8. The cost to remove 90% is $2,178,000 which is significantly higher than paying the current fine. It would seem it is not worth the cost.

2. Printing

A. 1. Assuming the printer uses 10 plates, then $1000 \cdot 10 = 10,000$ impressions can be made per hour. If 10,000 impressions are made per hour, it will take $\dfrac{100,000}{10,000} = 10$ hours to complete all the invitations.

2. Since it costs $128 per hour to run the press, the cost of using 10 plates is $10 \cdot 128 = \$1280$.

3. The 10 plates cost $10 \cdot 8 = \$80$.

4. The total cost of finishing the job is $1280 + 80 = \$1360$

B. 1. Let x = number of plates. Then, the cost of the plates is $8x$.

2. Using x plates implies x invitations can be made per impression.

3. $1000x$ invitations per hour Creating all 100,000 invitations would require $\dfrac{100,000}{1000x} = \dfrac{100}{x}$ hours.

4. $C(x) = 8x + 128\left(\dfrac{100}{x}\right)$

$C(x) = 8x + \left(\dfrac{12,800}{x}\right)$, where

x represents the number of plates and $C(x)$ represents the cost of the 100,000 invitations in dollars.

5.

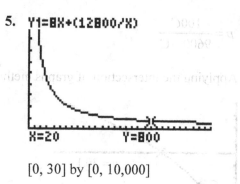

[0, 30] by [0, 10,000]

Producing 20 plates minimizes the cost, since the number of plates, x, that can be produced is between 1 and 20 inclusive.

6. Considering a table of values yields

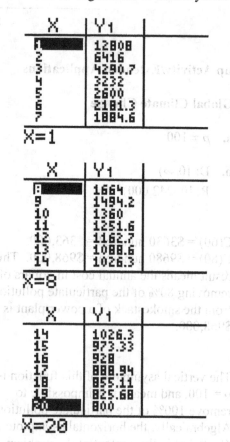

Producing 20 plates creates a minimum cost of $800 for printing the 100,000 invitations.

CHAPTER 7
Systems of Equations and Matrices

Chapter 7 Algebra Toolbox

1. The constant of proportionality is 12.

2. $y = 10x$

$x = \dfrac{y}{10}$

$x = \dfrac{1}{10}y$

The constant of proportionality is $\dfrac{1}{10}$.

3. Is $\dfrac{5}{6} = \dfrac{10}{18}$?

Is $5 \times 18 = 6 \times 10$?

$90 \neq 60$

No. The ratios are not proportional.

4. Is $\dfrac{8}{3} = \dfrac{24}{9}$?

Is $8 \times 9 = 3 \times 24$?

$72 = 72$

Yes. The ratios are proportional.

5. Does $\dfrac{4}{6} = \dfrac{2}{3}$?

$4 \times 3 = 6 \times 2$

$12 = 12$

Yes. The pairs are proportional.

6. Does $\dfrac{5.1}{2.3} = \dfrac{51}{23}$?

$5.1 \times 23 = 2.3 \times 51$

$117.3 = 117.3$

Yes. The pairs are proportional.

7. $y = kx$

$18 = k(2)$

$2k = 18$

$k = 9$

$y = 9x$

Let $x = 11$.

$y = 9(11)$

$y = 99$

8. $x = ky$

$3 = k(18)$

$18k = 3$

$k = \dfrac{3}{18} = \dfrac{1}{6}$

$x = \dfrac{1}{6}y$

Let $x = 13$.

$13 = \dfrac{1}{6}y$

$6(13) = 6\left(\dfrac{1}{6}y\right)$

$y = 78$

9. If the pairs of triples are proportional, then there exists a number k such that $y = kx$ for all pairs (x, y).

$y = kx$

$20 = k(8)$

$k = \dfrac{20}{8} = 2.5$

$y = 2.5x$

Checking the other pairs:

$12.5 = 2.5(6)$

$12.5 \neq 15$

The pairs of triples are not proportional.

10. If the pairs of triples are proportional, then there exists a number k such that $y = kx$ for all pairs (x, y).

$$y = kx$$
$$8.2 = k(4.1)$$
$$k = \frac{8.2}{4.1} = 2$$
$$y = 2x$$

Checking the other pairs:
$$13.6 = 6.8(2)$$
$$13.6 = 13.6$$
and
$$18.6 = 2(9.3)$$
$$18.6 = 18.6$$

The pairs of triples are proportional.

11. If the pairs of triples are proportional, then there exists a number k such that $y = kx$ for all pairs (x, y).

$$y = kx$$
$$-3.4 = k(-1)$$
$$k = \frac{-3.4}{-1} = 3.4$$
$$y = 3.4x$$

Checking the other pairs yields
$$10.2 = 3.4(6)$$
$$10.2 \neq 20.4$$

The pairs of triples are not proportional.

12. If the pairs of triples are proportional, then there exists a number k such that $y = kx$ for all pairs (x, y).
$$y = kx$$
$$2 = k(5)$$
$$k = \frac{2}{5} = 0.4$$
$$y = 0.4x$$

Checking the other pairs:
$$-3 = 0.4(-12)$$
$$-3 \neq -4.8$$

The pairs of triples are not proportional.

13. Substituting yields:

$$x - 2y + z = 8$$
$$1 - 2(-2) + 3 = 8$$
$$8 = 8$$
and
$$2x - y + 2z = 10$$
$$2(1) - (-2) + 2(3) = 10$$
$$2 + 2 + 6 = 10$$
$$10 = 10$$
and
$$3x - 2y + z = 5$$
$$3(1) - 2(-2) + 3 = 5$$
$$10 \neq 5$$

The given values do not satisfy the system of equations.

14. Substituting yields:

$$x - 2y + 2z = -9$$
$$-1 - 2(4) + 2(0) = -9$$
$$-9 = -9$$

and

$$2x - 3y + z = -14$$
$$2(-1) - 3(4) + 0 = -14$$
$$-14 = -14$$

and

$$x + 2y - 3z = 7$$
$$-1 + 2(4) - 3(0) = 7$$
$$7 = 7$$

The given values satisfy the system of equations.

15. Substituting yields:

$$x + y - z = 4$$
$$5 + (-2) - (-1) = 4$$
$$4 = 4$$

and

$$2x + 3y - z = 5$$
$$2(5) + 3(-2) - (-1) = 5$$
$$5 = 5$$

and

$$3x - 2y + 5z = 14$$
$$3(5) - 2(-2) + 5(-1) = 14$$
$$15 + 4 - 5 = 14$$
$$14 = 14$$

The given values do satisfy the system of equations.

16. Substituting yields:

$$x - 2y - 3z = 2$$
$$2 - 2(3) - 3(-2) = 2$$
$$2 = 2$$

and

$$2x + 3y + 3z = 19$$
$$2(2) + 3(3) + 3(-2) = 19$$
$$4 + 9 - 6 = 19$$
$$7 \neq 19$$

The given values do not satisfy the system of equations.

17. Since the coefficients of the variables are proportional, but the constants are not, the planes are parallel.

18. Since the coefficients of the variables are proportional and the constants are in the same proportion, the planes are the same.

19. Since the coefficients of the variables are not proportional, the planes are different. They intersect along a line. The planes are neither the same nor parallel.

20. Since the coefficients of the variables are proportional and the constants are in the same proportion, the planes are the same.

Section 7.1 Skills Check

1. Since z is isolated, back substitution yields:
$$y + 3(3) = 11$$
$$y + 9 = 11$$
$$y = 2$$
and
$$x + 2y - z = 3$$
$$x + 2(2) - 3 = 3$$
$$x + 1 = 3$$
$$x = 2$$
The solutions are $x = 2$, $y = 2$, $z = 3$.

3. Since z is isolated, back substitution yields:
$$y + 3(-2) = 3$$
$$y - 6 = 3$$
$$y = 9$$
and
$$x + 2y - z = 6$$
$$x + 2(9) - (-2) = 6$$
$$x + 18 + 2 = 6$$
$$x + 20 = 6$$
$$x = -14$$
The solutions are $x = -14$, $y = 9$, $z = -2$.

5. $\begin{cases} x - y - 4z = 0 \\ y + 2z = 4 \\ 3y + 7z = 22 \end{cases} \xrightarrow{-3Eq2+Eq3\to Eq3}$

$\begin{cases} x - y - 4z = 0 \\ y + 2z = 4 \\ z = 10 \end{cases}$

Since z is isolated, back substitution yields:

$$y + 2(10) = 4$$
$$y + 20 = 4$$
$$y = -16$$
and
$$x - y - 4z = 0$$
$$x - (-16) - 4(10) = 0$$
$$x + 16 - 40 = 0$$
$$x - 24 = 0$$
$$x = 24$$
The solutions are $x = 24$, $y = -16$, $z = 10$.

7. $\begin{cases} x - 2y + 3z = 0 \\ y - 2z = -1 \\ y + 5z = 6 \end{cases} \xrightarrow{-1Eq2+Eq3\to Eq3}$

$\begin{cases} x - 2y + 3z = 0 \\ y - 2z = -1 \\ 7z = 7 \end{cases} \xrightarrow{\frac{1}{7}Eq3\to Eq3}$

$\begin{cases} x - 2y + 3z = 0 \\ y - 2z = -1 \\ z = 1 \end{cases}$

Since z is isolated, back substitution yields:
$$y - 2(1) = -1$$
$$y - 2 = -1$$
$$y = 1$$
and
$$x - 2y + 3z = 0$$
$$x - 2(1) + 3(1) = 0$$
$$x - 2 + 3 = 0$$
$$x + 1 = 0$$
$$x = -1$$
The solutions are $x = -1$, $y = 1$, $z = 1$.

9.
$$\begin{cases} x+2y-2z=0 \\ x-y+4z=3 \\ x+2y+2z=3 \end{cases} \xrightarrow[\substack{-1\,Eq1+Eq2\to Eq2 \\ -1\,Eq1+Eq3\to Eq3}]{}$$

$$\begin{cases} x+2y-2z=0 \\ -3y+6z=3 \\ 4z=3 \end{cases} \xrightarrow[]{\frac{1}{4}Eq3\to Eq3}$$

$$\begin{cases} x+2y-2z=0 \\ -3y+6z=3 \\ z=\dfrac{3}{4} \end{cases}$$

Since z is isolated, back substitution yields:
$$-3y+6\left(\frac{3}{4}\right)=3$$
$$-3y+\frac{9}{2}=3$$
$$-3y=-\frac{3}{2}$$
$$y=\frac{1}{2}$$
and
$$x+2y-2z=0$$
$$x+2\left(\frac{1}{2}\right)-2\left(\frac{3}{4}\right)=0$$
$$x+1-\frac{3}{2}=0$$
$$x-\frac{1}{2}=0$$
$$x=\frac{1}{2}$$

The solutions are $x=\dfrac{1}{2}, \ y=\dfrac{1}{2}, z=\dfrac{3}{4}$.

11.
$$\begin{cases} x+3y-z=0 \\ x-2y+z=8 \\ x-6y+2z=6 \end{cases} \xrightarrow[\substack{-1\,Eq1+Eq2\to Eq2 \\ -1\,Eq1+Eq3\to Eq3}]{}$$

$$\begin{cases} x+3y-z=0 \\ -5y+2z=8 \\ -9y+3z=6 \end{cases} \xrightarrow[]{9\,Eq2-5\,Eq3\to Eq3}$$

$$\begin{cases} x+3y-z=0 \\ -5y+2z=8 \\ 3z=42 \end{cases} \xrightarrow[]{\frac{1}{3}Eq3\to Eq3}$$

$$\begin{cases} x+3y-z=0 \\ -5y+2z=8 \\ z=14 \end{cases}$$

Since z is isolated, back substitution yields:
$$-5y+2(14)=8$$
$$-5y+28=8$$
$$-5y=-20$$
$$y=4$$
and
$$x+3y-z=0$$
$$x+3(4)-14=0$$
$$x+12-14=0$$
$$x=2$$

The solutions are $x=2, \ y=4, \ z=14.$

13.
$$\begin{cases} 2x+4y-14z=0 \\ 3x+5y+z=19 \\ x+4y-z=12 \end{cases} \xrightarrow[\substack{-3\,Eq1+2\,Eq2\to Eq2 \\ Eq1-2\,Eq3\to Eq3}]{}$$

$$\begin{cases} 2x+4y-14z=0 \\ -2y+44z=38 \\ -4y-12z=-24 \end{cases} \xrightarrow[]{-2\,Eq2+Eq3\to Eq3}$$

$$\begin{cases} 2x+4y-14z=0 \\ -2y+44z=38 \\ -100z=-100 \end{cases} \xrightarrow[]{-\frac{1}{100}Eq3\to Eq3}$$

$$\begin{cases} 2x+4y-14z=0 \\ -2y+44z=38 \\ z=1 \end{cases}$$

Since z is isolated, back substitution yields:
$$-2y + 44(1) = 38$$
$$-2y + 44 = 38$$
$$-2y = -6$$
$$y = 3$$

and
$$2x + 4y - 14z = 0$$
$$2x + 4(3) - 14(1) = 0$$
$$2x + 12 - 14 = 0$$
$$2x = 2$$
$$x = 1$$

The solutions are $x = 1$, $y = 3$, $z = 1$.

15. $\begin{cases} 3x - 4y + 6z = 10 \\ 2x - 4y - 5z = -14 \\ x + 2y - 3z = 0 \end{cases} \xrightarrow[\substack{-2Eq1 + 3Eq2 \leftrightarrow Eq2 \\ Eq1 - 3Eq3 \leftrightarrow Eq3}]{}$

$\begin{cases} 3x - 4y + 6z = 10 \\ -4y - 27z = -62 \\ -10y + 15z = 10 \end{cases} \xrightarrow[5Eq2 - 2Eq3 \leftrightarrow Eq3]{}$

$\begin{cases} 3x - 4y + 6z = 10 \\ -4y - 27z = -62 \\ -165z = -330 \end{cases} \xrightarrow[-\frac{1}{165}Eq3 \leftrightarrow Eq3]{}$

$\begin{cases} 3x - 4y + 6z = 10 \\ -4y - 27z = -62 \\ z = 2 \end{cases}$

Since z is isolated, back substitution yields:
$$-4y - 27(2) = -62$$
$$-4y - 54 = -62$$
$$-4y = -8$$
$$y = 2$$

and
$$3x - 4y + 6z = 10$$
$$3x - 4(2) + 6(2) = 10$$
$$3x - 8 + 12 = 10$$
$$3x = 6$$
$$x = 2$$

The solutions are $x = 2$, $y = 2$, $z = 2$.

17. $\begin{cases} x - 3y + z = -2 \\ y - 2z = -4 \end{cases}$

Let z be any real number.
$$y = 2z - 4$$
and
$$x = 3y - z - 2$$
$$= 3(2z - 4) - z - 2$$
$$= 6z - 12 - z - 2$$
$$= 5z - 14$$

There are infinitely many solutions to the system all fitting the form
$x = 5z - 14$, $y = 2z - 4$, z.

19. $\begin{cases} x - y + 2z = 4 \\ y + 8z = 16 \end{cases}$

Let z be any real number.
$$y = 16 - 8z$$
and
$$x = y - 2z + 4$$
$$= (16 - 8z) - 2z + 4$$
$$= 20 - 10z$$

There are infinitely many solutions to the system all fitting the form
$x = 20 - 10z$, $y = 16 - 8z$, z.

21. $\begin{cases} x - y + 2z = -1 \\ 2x + 2y + 3z = -3 \\ x + 3y + z = -2 \end{cases} \xrightarrow[-1Eq1+Eq3 \to Eq3]{-2Eq1+Eq2 \to Eq2}$

$\begin{cases} x - y + 2z = -1 \\ \quad 4y - z = -1 \\ \quad 4y - z = -1 \end{cases} \xrightarrow{-1Eq2+Eq3 \to Eq3}$

$\begin{cases} x - y + 2z = -1 \\ \quad 4y - z = -1 \\ \quad\quad 0 = 0 \end{cases}$

Let z be any real number.

$4y - z = -1$

$4y = z - 1$

$y = \dfrac{1}{4}z - \dfrac{1}{4}$

and

$x = y - 2z - 1$

$\quad = \left(\dfrac{1}{4}z - \dfrac{1}{4}\right) - 2z - 1$

$\quad = -\dfrac{7}{4}z - \dfrac{5}{4}$

There are infinitely many solutions
to the system all fitting the form

$x = -\dfrac{7}{4}z - \dfrac{5}{4}, \quad y = \dfrac{1}{4}z - \dfrac{1}{4}, \quad z.$

23. $\begin{cases} 3x - 4y - 6z = 10 \\ 2x - 4y - 5z = -14 \\ x \quad\quad - z = 0 \end{cases} \xrightarrow[Eq1-3Eq3 \to Eq3]{Eq2-2Eq3 \to Eq2}$

$\begin{cases} 3x - 4y - 6z = 10 \\ \quad -4y - 3z = -14 \\ \quad -4y - 3z = 10 \end{cases} \xrightarrow{-1Eq2+Eq3 \to Eq3}$

$\begin{cases} 3x - 4y - 6z = 10 \\ \quad -4y - 3z = -14 \\ \quad\quad 0 = 24 \end{cases}$

Since $0 \neq 24$, the system is inconsistent and
has no solution.

25. $\begin{cases} 2x + 3y - 14z = 16 \\ 4x + 5y - 30z = 34 \\ x + y - 8z = 9 \end{cases} \xrightarrow[Eq1-2Eq3 \to Eq3]{Eq2-4Eq3 \to Eq2}$

$\begin{cases} 2x + 3y - 14z = 16 \\ \quad y + 2z = -2 \\ \quad y + 2z = -2 \end{cases} \xrightarrow{-1Eq2+Eq3 \to Eq3}$

$\begin{cases} 2x + 3y - 14z = 16 \\ \quad y + 2z = -2 \\ \quad\quad 0 = 0 \end{cases}$

Let z be any real number.

$y = -2z - 2$

and

$x = -y + 8z + 9$

$\quad = -(-2z - 2) + 8z + 9$

$\quad = 10z + 11$

There are infinitely many solutions
to the system all fitting the form

$x = 10z + 11, \quad y = -2z - 2, \quad z.$

Section 7.1 Exercises

27. $\begin{cases} x + y + z = 60 \\ 15{,}000x + 25{,}000y + 45{,}000z = 1{,}400{,}000 \\ 30x + 40y + 50z = 2200 \end{cases}$ $\xrightarrow[\ -30\,Eq1 + Eq3 \to Eq3\ ]{-15{,}000\,Eq1 + Eq2 \to Eq2}$

$\begin{cases} x + y + z = 60 \\ 10{,}000y + 30{,}000z = 500{,}000 \\ 10y + 20z = 400 \end{cases}$ $\xrightarrow{\ Eq2 - 1000\,Eq3 \to Eq3\ }$

$\begin{cases} x + y + z = 60 \\ 10{,}000y + 30{,}000z = 500{,}000 \\ 10{,}000z = 100{,}000 \end{cases}$ $\xrightarrow{\ \frac{1}{10{,}000}Eq3 \to Eq3\ }$

$\begin{cases} x + y + z = 60 \\ 10{,}000y + 30{,}000z = 500{,}000 \\ z = 10 \end{cases}$

Since z is isolated, back substitution yields:

$$10{,}000y + 30{,}000(10) = 500{,}000$$
$$10{,}000y + 300{,}000 = 500{,}000$$
$$10{,}000y = 200{,}000$$
$$y = 20$$

and

$$x + y + z = 60$$
$$x + 20 + 10 = 60$$
$$x = 30$$

The solution to the system is $x = 30$, $y = 20$, $z = 10$. The agency should purchase 30 compact cars, 20 midsize cars, and 10 luxury cars.

29. a. $x + y = 2600$

b. $40x$

c. $60y$

d. $40x + 60y = 120{,}000$

e. $\begin{cases} x+y=2600 \\ 40x+60y=120,000 \end{cases} \xrightarrow{\;-40\,Eq1+Eq2\rightarrow Eq2\;}$

$\begin{cases} x+y=2600 \\ 20y=16,000 \end{cases} \xrightarrow{\;\frac{1}{20}Eq2\rightarrow Eq2\;}$

$\begin{cases} x+y=2600 \\ y=800 \end{cases}$

Since y is isolated, back substitution yields:

$x+y=2600$

$x+800=2600$

$x=1800$

The solution to the system is $x=1800$, $y=800$. The concert promoter must sell 1800 \$40 tickets and 800 \$60 tickets.

31. a. $x+y+z=400,000$

  b. $7.5\%x+8\%y+9\%z=33,700$, or rewriting
$0.075x+0.08y+0.09z=33,700$

  c. $z=x+y$, or rewriting
$x+y-z=0$

  d. $\begin{cases} x+y+z=400,000 \\ 0.075x+0.08y+0.09z=33,700 \\ x+y-z=0 \end{cases} \xrightarrow[\;-1\,Eq1+Eq3\rightarrow Eq3\;]{\;-0.075\,Eq1+Eq2\rightarrow Eq2\;}$

$\begin{cases} x+y+z=400,000 \\ 0.005y+0.015z=3700 \\ -2z=-400,000 \end{cases} \xrightarrow{\;-\frac{1}{2}Eq3\rightarrow Eq3\;}$

$\begin{cases} x+y+z=400,000 \\ 0.005y+0.015z=3700 \\ z=200,000 \end{cases}$

Since z is isolated, back substitution yields:

$$0.005y + 0.015(200,000) = 3700$$
$$0.005y + 3000 = 3700$$
$$0.005y = 700$$
$$y = 140,000$$

and

$$x + y + z = 400,000$$
$$x + 140,000 + 200,000 = 400,000$$
$$x = 60,000$$

The solution to the system is $x = 60,000$, $y = 140,000$, $z = 200,000$. $60,000 is invested in the property returning 7.5%, $140,000 is invested in the property returning 8%, and $200,000 is invested in the property returning 9%.

33. Let A = the number of units of product A, B = the number of units of product B, and C = the number of units of product C.

$$\begin{cases} 24A + 20B + 40C = 8000 \\ 40A + 30B + 60C = 12,400 \\ 150A + 180B + 200C = 52,600 \end{cases} \begin{array}{l} {}_{-5\,Eq1+3\,Eq2\rightarrow Eq2} \\ \xrightarrow{-25\,Eq1+4\,Eq3\rightarrow Eq3} \end{array}$$

$$\begin{cases} 24A + 20B + 40C = 8000 \\ -10B - 20C = -2800 \\ 220B - 200C = 10,400 \end{cases} \xrightarrow{22\,Eq2+Eq3\rightarrow Eq3}$$

$$\begin{cases} 24A + 20B + 40C = 8000 \\ -10B - 20C = -2800 \\ -640C = -51,200 \end{cases} \xrightarrow{-\frac{1}{640}Eq3\rightarrow Eq3}$$

$$\begin{cases} 24A + 20B + 40C = 8000 \\ -10B - 20C = -2800 \\ C = 80 \end{cases}$$

Since C is isolated, back substitution yields:

$$-10B - 20(80) = -2800$$
$$-10B - 1600 = -2800$$
$$-10B = -1200$$
$$B = 120$$

and

$$24A + 20B + 40C = 8000$$
$$24A + 20(120) + 40(80) = 8000$$
$$24A + 2400 + 3200 = 8000$$
$$24A = 2400$$
$$A = 100$$

The solution to the system is $A = 100$, $B = 120$, $C = 80$. To meet the restrictions imposed on volume, weight, and value, 100 units of product A, 120 units of product B, and 80 of product C can be carried.

35. a. $x + y + z = 500,000$ means that the sum of the money invested in the three accounts is \$500,000. $0.08x + 0.10y + 0.14z = 49,000$ means that the sum of the interest earned on the three accounts in one year is \$49,000.

b. $\begin{cases} x + y + z = 500,000 \\ 0.08x + 0.10y + 0.14z = 49,000 \end{cases} \xrightarrow{-0.8Eq1 + Eq2 \to Eq2}$

$\begin{cases} x + y + z = 500,000 \\ 0.02y + 0.06z = 9,000 \end{cases}$

Let $z = z$.

$$0.02y + 0.06z = 9000$$
$$0.02y = 9000 - 0.06z$$
$$y = 450,000 - 3z$$

and

$$x + (450,000 - 3z) + z = 500,000$$
$$x - 2z + 450,000 = 500,000$$
$$x = 2z + 50,000$$

The solution is $x = 2z + 50,000$, $y = 450,000 - 3z$, $z = z$.

c. $z \geq 0$ implies that $x = 2z + 50{,}000 \geq 0$.

For $y \geq 0$, $450{,}000 - 3z \geq 0$.

Solving for z,

$-3z \geq -450{,}000$

$z \leq \dfrac{-450{,}000}{-3}$

$z \leq 150{,}000$

For all investments to be non-negative, $0 \leq z \leq 150{,}000$.

---

37. Let $x =$ the number of Acclaim units produced, $y =$ the number of Bestfrig units produced, and $z =$ the number of Cool King units produced.

$$\begin{cases} 5x + 4y + 4.5z = 300 \\ 2x + 1.4y + 1.7z = 120 \\ 1.4x + 1.2y + 1.3z = 210 \end{cases} \xrightarrow[\;-0.7\,Eq2 + Eq3 \to Eq3\;]{-2\,Eq1 + 5\,Eq2 \to Eq2}$$

$$\begin{cases} 5x + 4y + 4.5z = 300 \\ \quad -1y - 0.5z = 0 \\ \quad 0.22y + 0.11z = 126 \end{cases} \xrightarrow[\;\;]{0.22\,Eq2 + Eq3 \to Eq3}$$

$$\begin{cases} 5x + 4y + 4.5z = 300 \\ \quad -1y - 0.5z = 0 \\ \qquad\qquad 0 = 126 \end{cases}$$

Since $0 \neq 126$, the system is inconsistent and has no solution. It is not possible to satisfy the given manufacturing conditions.

---

39. Let $x =$ grams of food I, $y =$ grams of food II, and $z =$ grams of food III.

$$\begin{cases} 12\%x + 14\%y + 8\%z = 6.88 \\ 8\%x + 6\%y + 16\%z = 6.72 \\ 10\%x + 10\%y + 12\%z = 6.80 \end{cases}$$

or

$$\begin{cases} 0.12x + 0.14y + 0.08z = 6.88 \\ 0.08x + 0.06y + 0.16z = 6.72 \\ 0.10x + 0.10y + 0.12z = 6.80 \end{cases}$$

$$\begin{cases} 0.12x + 0.14y + 0.08z = 6.88 \\ 0.08x + 0.06y + 0.16z = 6.72 \\ 0.10x + 0.10y + 0.12z = 6.80 \end{cases} \xrightarrow[]{\substack{-2\,Eq1+3\,Eq2\rightarrow Eq2 \\ -5\,Eq1+6\,Eq3\rightarrow Eq3}}$$

$$\begin{cases} 0.12x + 0.14y + 0.08z = 6.88 \\ -0.10y + 0.32z = 6.40 \\ -0.10y + 0.32z = 6.40 \end{cases} \xrightarrow[]{-1\,Eq2+Eq3\rightarrow Eq3}$$

$$\begin{cases} 0.12x + 0.14y + 0.08z = 6.88 \\ -0.10y + 0.32z = 6.40 \\ 0 = 0 \end{cases}$$

Let $z = z$.

$-0.10y + 0.32z = 6.40$

$-0.10y = 6.40 - 0.32z$

$y = 3.2z - 64$

and

$0.12x + 0.14y + 0.08z = 6.88$

$0.12x + 0.14(3.2z - 64) + 0.08z = 6.88$

$0.12x + 0.448z - 8.96 + 0.08z = 6.88$

$0.12x + 0.528z - 8.96 = 6.88$

$0.12x = 15.84 - 0.528z$

$x = \dfrac{15.84 - 0.528z}{0.12}$

$x = 132 - 4.4z$

The solution is $x = 132 - 4.4z,\ y = 3.2z - 64,\ z = z$.

Note that $z \geq 0$. $y \geq 0$ implies that $3.2z - 64 \geq 0$.

$\qquad 3.2z \geq 64$

$\qquad\quad z \geq \dfrac{64}{3.2}$

$\qquad\quad z \geq 20$

$x \geq 0$ implies that $132 - 4.4z \geq 0$.

$\qquad -4.4z \geq -132$

$\qquad\quad z \leq \dfrac{-132}{-4.4}$

$\qquad\quad z \leq 30$

Therefore, $20 \leq z \leq 30$.

Section 7.2 Skills Check

1. $\begin{bmatrix} 1 & 1 & -1 & | & 4 \\ 1 & -2 & -1 & | & -2 \\ 2 & 2 & 1 & | & 11 \end{bmatrix}$

3. $\begin{bmatrix} 5 & -3 & 2 & | & 12 \\ 3 & 6 & -9 & | & 4 \\ 2 & 3 & -4 & | & 9 \end{bmatrix}$

5. Since the matrix is in reduced row-echelon form, the solution is $x = -1$, $y = 4$, $z = -2$.

7. $\begin{bmatrix} 1 & 1 & -1 & | & 4 \\ 1 & -2 & -1 & | & -2 \\ 2 & 2 & 1 & | & 11 \end{bmatrix} \xrightarrow{\begin{array}{c} -1R_1 + R_2 \to R_2 \\ -2R_1 + R_3 \to R_3 \end{array}}$

$\begin{bmatrix} 1 & 1 & -1 & | & 4 \\ 0 & -3 & 0 & | & -6 \\ 0 & 0 & 3 & | & 3 \end{bmatrix} \xrightarrow{\begin{array}{c} \left(-\frac{1}{3}\right)R_2 \to R_2 \\ \left(\frac{1}{3}\right)R_3 \to R_3 \end{array}}$

$\begin{bmatrix} 1 & 1 & -1 & | & 4 \\ 0 & 1 & 0 & | & 2 \\ 0 & 0 & 1 & | & 1 \end{bmatrix}$

Back substituting to find the solutions

$\begin{cases} x + y - z = 4 \\ \quad\quad y = 2 \\ \quad\quad\quad z = 1 \end{cases}$

$x + 2 - 1 = 4$

$x + 1 = 4$

$x = 3$

The solutions are $x = 3$, $y = 2$, $z = 1$.

9. $\begin{bmatrix} 2 & -3 & 4 & | & 13 \\ 1 & -2 & 1 & | & 3 \\ 2 & -3 & 1 & | & 4 \end{bmatrix} \xrightarrow{R_2 \leftrightarrow R_1}$

$\begin{bmatrix} 1 & -2 & 1 & | & 3 \\ 2 & -3 & 4 & | & 13 \\ 2 & -3 & 1 & | & 4 \end{bmatrix} \xrightarrow{\begin{array}{c} -2R_1 + R_2 \to R_2 \\ -2R_1 + R_3 \to R_3 \end{array}}$

$\begin{bmatrix} 1 & -2 & 1 & | & 3 \\ 0 & 1 & 2 & | & 7 \\ 0 & 1 & -1 & | & -2 \end{bmatrix} \xrightarrow{-1R_2 + R_3 \to R_3}$

$\begin{bmatrix} 1 & -2 & 1 & | & 3 \\ 0 & 1 & 2 & | & 7 \\ 0 & 0 & -3 & | & -9 \end{bmatrix} \xrightarrow{\left(-\frac{1}{3}\right)R_3 \to R_3}$

$\begin{bmatrix} 1 & -2 & 1 & | & 3 \\ 0 & 1 & 2 & | & 7 \\ 0 & 0 & 1 & | & 3 \end{bmatrix}$

Back substituting to find the solutions

$\begin{cases} x - 2y + z = 3 \\ \quad\quad y + 2z = 7 \\ \quad\quad\quad z = 3 \end{cases}$

$y + 2(3) = 7$

$y + 6 = 7$

$y = 1$

and

$x - 2(1) + (3) = 3$

$x + 1 = 3$

$x = 2$

The solutions are $x = 2$, $y = 1$, $z = 3$.

11. Since the third row of the given augmented matrix implies $0 = 1$, the system is inconsistent and has no solution.

13. Since the third row of the augmented matrix states $0 = 0$, the system has infinitely many solutions. Let z be any real number. Then,

$y - 5z = 5$

$y = 5z + 5$

and

$x + 3z = 2$

$x = 2 - 3z$

There are infinitely many solutions to the system of the form

$x = 2 - 3z, \ y = 5z + 5, \ z.$

15. $\begin{bmatrix} 1 & 1 & -1 & | & 0 \\ 1 & -2 & -1 & | & 6 \\ 2 & 2 & 1 & | & 3 \end{bmatrix} \xrightarrow[\ -2R_1+R_3 \to R_3\ ]{-1R_1+R_2 \to R_2}$

$\begin{bmatrix} 1 & 1 & -1 & | & 0 \\ 0 & -3 & 0 & | & 6 \\ 0 & 0 & 3 & | & 3 \end{bmatrix} \xrightarrow[\ \left(\frac{1}{3}\right)R_3 \to R_3\ ]{\left(-\frac{1}{3}\right)R_2 \to R_2}$

$\begin{bmatrix} 1 & 1 & -1 & | & 0 \\ 0 & 1 & 0 & | & -2 \\ 0 & 0 & 1 & | & 1 \end{bmatrix}$

Back substituting to find the solutions

$\begin{cases} x + y - z = 0 \\ \quad\quad y = -2 \\ \quad\quad\quad z = 1 \end{cases}$

$x - 2 - 1 = 0$

$x - 3 = 0$

$x = 3$

The solutions are $x = 3, \ y = -2, \ z = 1.$

17. $\begin{bmatrix} 3 & -2 & 5 & | & 15 \\ 1 & -2 & -2 & | & -1 \\ 2 & -2 & 0 & | & 0 \end{bmatrix} \xrightarrow{R_2 \leftrightarrow R_1}$

$\begin{bmatrix} 1 & -2 & -2 & | & -1 \\ 3 & -2 & 5 & | & 15 \\ 2 & -2 & 0 & | & 0 \end{bmatrix} \xrightarrow[\ -2R_1+R_3 \to R_3\ ]{-3R_1+R_2 \to R_2}$

$\begin{bmatrix} 1 & -2 & -2 & | & -1 \\ 0 & 4 & 11 & | & 18 \\ 0 & 2 & 4 & | & 2 \end{bmatrix} \xrightarrow[\ \left(\frac{1}{2}\right)R_3 \to R_3\ ]{\left(\frac{1}{4}\right)R_2 \to R_2}$

$\begin{bmatrix} 1 & -2 & -2 & | & -1 \\ 0 & 1 & \frac{11}{4} & | & \frac{9}{2} \\ 0 & 1 & 2 & | & 1 \end{bmatrix} \xrightarrow{-1R_2+R_3 \to R_3}$

$\begin{bmatrix} 1 & -2 & -2 & | & -1 \\ 0 & 1 & \frac{11}{4} & | & \frac{9}{2} \\ 0 & 0 & \frac{-3}{4} & | & -\frac{7}{2} \end{bmatrix} \xrightarrow{\left(-\frac{4}{3}\right)R_3 \to R_3}$

$\begin{bmatrix} 1 & -2 & -2 & | & -1 \\ 0 & 1 & \frac{11}{4} & | & \frac{9}{2} \\ 0 & 0 & 1 & | & \frac{14}{3} \end{bmatrix}$

Back substituting to find the solutions

$$\begin{cases} x - 2y - 2z = -1 \\ \quad y + \dfrac{11}{4}z = \dfrac{9}{2} \\ \qquad\qquad z = \dfrac{14}{3} \end{cases}$$

$$y + \frac{11}{4}\left(\frac{14}{3}\right) = \frac{9}{2}$$

$$y + \frac{77}{6} = \frac{27}{6}$$

$$y = -\frac{50}{6}$$

$$y = -\frac{25}{3}$$

and

$$x - 2\left(-\frac{25}{3}\right) - 2\left(\frac{14}{3}\right) = -1$$

$$x + \frac{50}{3} - \frac{28}{3} = -1$$

$$x + \frac{22}{3} = -\frac{3}{3}$$

$$x = -\frac{25}{3}$$

The solutions are $x = -\dfrac{25}{3}$, $y = -\dfrac{25}{3}$, $z = \dfrac{14}{3}$.

19. $\begin{bmatrix} 2 & 3 & 4 & | & 5 \\ 6 & 7 & 8 & | & 9 \\ 2 & 1 & 1 & | & 1 \end{bmatrix} \xrightarrow[\substack{-3R_1+R_2\to R_2 \\ -1R_1+R_3\to R_3}]{}$

$\begin{bmatrix} 2 & 3 & 4 & | & 5 \\ 0 & -2 & -4 & | & -6 \\ 0 & -2 & -3 & | & -4 \end{bmatrix} \xrightarrow[\substack{-\frac{1}{2}R_2\to R_2 \\ -\frac{1}{2}R_3\to R_3}]{}$

$\begin{bmatrix} 2 & 3 & 4 & | & 5 \\ 0 & 1 & 2 & | & 3 \\ 0 & 1 & \frac{3}{2} & | & 2 \end{bmatrix} \xrightarrow[\substack{-1R_2+R_3\to R_3}]{}$

$\begin{bmatrix} 2 & 3 & 4 & | & 5 \\ 0 & 1 & 2 & | & 3 \\ 0 & 0 & -\frac{1}{2} & | & -1 \end{bmatrix} \xrightarrow[\substack{\left(\frac{1}{2}\right)R_1\to R_1 \\ -2R_3\to R_3}]{}$

$\begin{bmatrix} 1 & \frac{3}{2} & 2 & | & \frac{5}{2} \\ 0 & 1 & 2 & | & 3 \\ 0 & 0 & 1 & | & 2 \end{bmatrix}$

Back substituting to find the solutions

$$\begin{cases} x + \dfrac{3}{2}y + 2z = \dfrac{5}{2} \\ \quad\;\; y + 2z = 3 \\ \qquad\qquad z = 2 \end{cases}$$

$y + 2(2) = 3$

$y + 4 = 3$

$y = -1$

and

$x + \dfrac{3}{2}(-1) + 2(2) = \dfrac{5}{2}$

$x - \dfrac{3}{2} + 4 = \dfrac{5}{2}$

$x + \dfrac{5}{2} = \dfrac{5}{2}$

$x = 0$

The solutions are $x = 0$, $y = -1$, $z = 2$.

21. $\begin{bmatrix} 1 & -1 & 1 & -1 & | & -2 \\ 2 & 0 & 4 & 1 & | & 5 \\ 2 & -3 & 1 & 0 & | & -5 \\ 0 & 1 & 2 & 20 & | & 4 \end{bmatrix} \xrightarrow[\;-2R_1+R_3\to R_3\;]{-2R_1+R_2\to R_2}$

$\begin{bmatrix} 1 & -1 & 1 & -1 & | & -2 \\ 0 & 2 & 2 & 3 & | & 9 \\ 0 & -1 & -1 & 2 & | & -1 \\ 0 & 1 & 2 & 20 & | & 4 \end{bmatrix} \xrightarrow[\;R_3+R_4\to R_4\;]{2R_3+R_2\to R_2}$

$\begin{bmatrix} 1 & -1 & 1 & -1 & | & -2 \\ 0 & 0 & 0 & 7 & | & 7 \\ 0 & -1 & -1 & 2 & | & -1 \\ 0 & 0 & 1 & 22 & | & 3 \end{bmatrix} \xrightarrow[\;R_4\to R_3\;]{\substack{R_3\to R_2 \\ R_2\to R_4}}$

$\begin{bmatrix} 1 & -1 & 1 & -1 & | & -2 \\ 0 & -1 & -1 & 2 & | & -1 \\ 0 & 0 & 1 & 22 & | & 3 \\ 0 & 0 & 0 & 7 & | & 7 \end{bmatrix} \xrightarrow[\;\left(\frac{1}{7}\right)R_4\to R_4\;]{-1R_2\to R_2}$

$\begin{bmatrix} 1 & -1 & 1 & -1 & | & -2 \\ 0 & 1 & 1 & -2 & | & 1 \\ 0 & 0 & 1 & 22 & | & 3 \\ 0 & 0 & 0 & 1 & | & 1 \end{bmatrix}$

Back substituting to find the solutions

$$\begin{cases} x - y + z - w = -2 \\ y + z - 2w = 1 \\ z + 22w = 3 \\ w = 1 \end{cases}$$

$z + 22(1) = 3$

$z = -19$

and

$y + (-19) - 2(1) = 1$

$y - 19 - 2 = 1$

$y - 21 = 1$

$y = 22$

and

$x - (22) + (-19) - (1) = -2$

$x - 22 - 19 - 1 = -2$

$x - 42 = -2$

$x = 40$

The solutions are $x = 40$, $y = 22$,

$z = -19$, $w = 1$.

23. $\begin{bmatrix} -2 & 3 & 2 & | & 13 \\ -2 & -2 & 3 & | & 0 \\ 4 & 1 & 4 & | & 11 \end{bmatrix}$

Using the calculator to generate the reduced
row-echelon form of the matrix yields

```
rref([A])
        [[1 0 0 0]
         [0 1 0 3]
         [0 0 1 2]]
```

The solution to the system is
$x = 0, y = 3, z = 2$.

25. $\begin{bmatrix} 2 & 5 & 6 & | & 6 \\ 3 & -2 & 2 & | & 4 \\ 5 & 3 & 8 & | & 10 \end{bmatrix}$

Using the calculator to generate the reduced
row-echelon form of the matrix yields

$$\begin{bmatrix} 1 & 0 & \dfrac{22}{19} & | & \dfrac{32}{19} \\ 0 & 1 & \dfrac{14}{19} & | & \dfrac{10}{19} \\ 0 & 0 & 0 & | & 0 \end{bmatrix}$$

Since the third row of the augmented
matrix states $0 = 0$, the system has
infinitely many solutions. Let z be
any real number. Then,

$$y + \frac{14}{19}z = \frac{10}{19}$$

$$y = \frac{10}{19} - \frac{14}{19}z$$

and

$$x + \frac{22}{19}z = \frac{32}{19}$$

$$x = \frac{32}{19} - \frac{22}{19}z$$

There are infinitely many solutions
to the system of the form

$$x = \frac{32}{19} - \frac{22}{19}z, \quad y = \frac{10}{19} - \frac{14}{19}z, \ z.$$

27. $\begin{bmatrix} -1 & -5 & 3 & | & -2 \\ 3 & 7 & 2 & | & 5 \\ 4 & 12 & -1 & | & 7 \end{bmatrix}$

Using the calculator to generate the reduced
row-echelon form of the matrix yields

```
rref([A])▶Frac
...1 0 31/8   11/8...
...0 1 -11/8 1/8 ...
...0 0 0       0    ...
```

or

$$\begin{bmatrix} 1 & 0 & \dfrac{31}{8} & | & \dfrac{11}{8} \\ 0 & 1 & -\dfrac{11}{8} & | & \dfrac{1}{8} \\ 0 & 0 & 0 & | & 0 \end{bmatrix}$$

Since the third row of the augmented matrix states $0 = 0$, the system has infinitely many solutions. Let z be any real number. Then,

$$y - \frac{11}{8}z = \frac{1}{8}$$

$$y = \frac{11}{8}z + \frac{1}{8}$$

and

$$x + \frac{31}{8}z = \frac{11}{8}$$

$$x = \frac{11}{8} - \frac{31}{8}z$$

There are infinitely many solutions to the system of the form

$$x = \frac{11}{8} - \frac{31}{8}z, \ y = \frac{11}{8}z + \frac{1}{8}, \ z.$$

29. $\begin{bmatrix} 2 & -3 & 2 & | & 5 \\ 4 & 1 & -3 & | & 6 \end{bmatrix}$

Using the calculator to generate the reduced row-echelon form of the matrix yields

```
rref([B])▶Frac
…1 0 -1/2 23/14…
…0 1 -1    -4/7 …
```

or

$$\begin{bmatrix} 1 & 0 & -\dfrac{1}{2} & \Big| & \dfrac{23}{14} \\ 0 & 1 & -1 & \Big| & -\dfrac{4}{7} \end{bmatrix}$$

Since the augmented matrix contains two equations with three variables, the system has infinitely many solutions. Let z be any real number. Then,

$$y - z = -\frac{4}{7}$$

$$y = z - \frac{4}{7}$$

and

$$x - \frac{1}{2}z = \frac{23}{14}$$

$$x = \frac{1}{2}z + \frac{23}{14}$$

There are infinitely many solutions to the system of the form

$$x = \frac{1}{2}z + \frac{23}{14}, \ y = z - \frac{4}{7}, \ z.$$

31. $\begin{bmatrix} 1 & 0 & -3 & -3 & | & -2 \\ 1 & 1 & 1 & 3 & | & 2 \\ 2 & 1 & -2 & -2 & | & 0 \\ 3 & 2 & -1 & 1 & | & 2 \end{bmatrix}$

Using the calculator to generate the reduced row-echelon form of the matrix yields

```
rref([C])▶Frac
[[1 0 -3 0 -2]
 [0 1 4  0 4 ]
 [0 0 0  1 0 ]
 [0 0 0  0 0 ]]
```

Since the fourth row of the augmented matrix states $0 = 0$, the system has infinitely many solutions.

$w = 0$

Let z be any real number. Then,

$y + 4z = 4$

$y = 4 - 4z$

and

$x - 3z = -2$

$x = 3z - 2$

There are infinitely many solutions to the system of the form

$x = 3z - 2, \; y = 4 - 4z, \; z, \; w = 0.$

Section 7.2 Exercises

33. Let x, y, and z represent the number of tickets in each section.

Note that $x = 2(y + z)$, which can be rewritten as $x - 2y - 2z = 0$.

$$\begin{bmatrix} 1 & 1 & 1 & | & 3600 \\ 1 & -2 & -2 & | & 0 \\ 40 & 70 & 100 & | & 192{,}000 \end{bmatrix}$$

Using the calculator to generate the reduced row-echelon form of the matrix yields

```
rref([A])▶Frac
    [[1 0 0 2400]
     [0 1 0 800 ]
     [0 0 1 400 ]]
```

The solution to the system is $x = 2400, \; y = 800, \; z = 400$. The theater owner should sell 2400 $40 tickets, 800 $70 tickets, and 400 $100 tickets.

35. a. Let $x =$ the points per each true-false question, $y =$ the points per each multiple-choice question, and $z =$ the points per each essay question.

$15x + 10y + 5z = 100$

and

$2x = y$, or rewriting

$2x - y = 0$

and

$3x = z$, or rewriting

$3x - z = 0$

The system is

$$\begin{cases} 15x + 10y + 5z = 100 \\ 2x - y = 0 \\ 3x - z = 0 \end{cases}$$

b.
$$\begin{bmatrix} 15 & 10 & 5 & | & 100 \\ 2 & -1 & 0 & | & 0 \\ 3 & 0 & -1 & | & 0 \end{bmatrix}$$

Using the calculator to generate the reduced row-echelon form of the matrix yields

```
rref([A])
    [[1 0 0 2]
     [0 1 0 4]
     [0 0 1 6]]
```

The solution to the system is $x = 2, \; y = 4, \; z = 6$. On the exam, true-false questions are worth two points, multiple-choice questions are worth four points, and essay questions are worth six points.

37. Let x = the number of Plan I units,
y = the number of Plan II units, and
z = the number of Plan III units.

$$\begin{cases} 4x+8y+14z=42 \\ 2x+4y+6z=20 \\ 6y+6z=18 \end{cases}$$

$$\begin{bmatrix} 4 & 8 & 14 & 42 \\ 2 & 4 & 6 & 20 \\ 0 & 6 & 6 & 18 \end{bmatrix}$$

Using the calculator to generate the reduced row-echelon form of the matrix yields

```
rref([A])
        [[1 0 0 3]
         [0 1 0 2]
         [0 0 1 1]]
```

The solution to the system is $x=3$, $y=2$, $z=1$. The client needs to purchase 3 units of Plan I, 2 units of Plan II, and 1 unit of Plan III to achieve the investment objectives.

39. Let x = grams of Food I, y = grams of Food II, and z = grams of Food III.

$$\begin{cases} 12\%x+15\%y+28\%z=3.74 \\ 8\%x+6\%y+16\%z=2.04 \\ 15\%x+2\%y+6\%z=1.35 \end{cases}$$

or

$$\begin{cases} 0.12x+0.15y+0.28z=3.74 \\ 0.08x+0.06y+0.16z=2.04 \\ 0.15x+0.02y+0.06z=1.35 \end{cases}$$

$$\begin{bmatrix} 0.12 & 0.15 & 0.28 & 3.74 \\ 0.08 & 0.06 & 0.16 & 2.04 \\ 0.15 & 0.02 & 0.06 & 1.35 \end{bmatrix}$$

Using the calculator to generate the reduced row-echelon form of the matrix yields

```
rref([A])
        [[1 0 0 5]
         [0 1 0 6]
         [0 0 1 8]]
```

The solution to the system is $x=5, y=6, z=8$. The psychologist recommends 5 grams of Food I, 6 grams of Food II, and 8 grams of Food III each day.

41. Let x = the number of type A clients,
y = the number of type B clients,
and z = the number of type C clients.

$$\begin{cases} x+y+z=500 \\ 400x+1000y+600z=300,000 \\ 600x+400y+200z=200,000 \end{cases}$$

$$\begin{bmatrix} 1 & 1 & 1 & 500 \\ 400 & 1000 & 600 & 300,000 \\ 600 & 400 & 200 & 200,000 \end{bmatrix}$$

Using the calculator to generate the reduced row-echelon form of the matrix yields

```
rref([A])▶Frac
        [[1 0 0 200]
         [0 1 0 100]
         [0 0 1 200]]
```

200 type A clients, 100 type B clients, and 200 type C clients can be served under the given conditions.

43. a. Let x = the number of portfolio I units, y = the number of portfolio II units, and z = the number of portfolio III units.

$$\begin{cases} 10x+12y+10z=180 \\ 2x+8y+6z=140 \\ 3x+5y+4z=110 \end{cases}$$

b. $\begin{bmatrix} 10 & 12 & 10 & | & 180 \\ 2 & 8 & 6 & | & 140 \\ 3 & 5 & 4 & | & 110 \end{bmatrix}$

Using the calculator to generate the reduced row-echelon form of the matrix yields a last row containing all zeros along with an augmented 1.

```
rref([A])▸Frac
    [[1 0 1/7 0]
     [0 1 5/7 0]
     [0 0 0   1]]
```

Therefore, the system is inconsistent and has no solution. The client cannot achieve the desired investment results.

45. a. Let $x =$ the number of $40,000 cars, $y =$ the number of $30,000 cars, and $z =$ the number of $20,000.

$$\begin{cases} x + y + z = 4 \\ 40,000x + 30,000y + 20,000z = 100,000 \end{cases}$$

b. Since the system has more variables than equations, the system can not have a unique solution.

c. $\begin{bmatrix} 1 & 1 & 1 & | & 4 \\ 40,000 & 30,000 & 20,000 & | & 100,000 \end{bmatrix}$

Using the calculator to generate the reduced row-echelon form of the matrix yields

```
rref([A])
    [[1 0 -1 -2]
     [0 1 2  6]]
```

Let $z = z$.
$y + 2z = 6$
$y = 6 - 2z$
and
$x - z = -2$
$x = z - 2$

The solution to the system is
$x = z - 2, \; y = 6 - 2z, \; z = z.$

d. The only values of z that make sense in the context of the problem are $z = 2$ and $z = 3$. Other values of z create negative solutions for x and y. If $z = 2$, the young man purchases two $20,000 cars, two $30,000 cars, and zero $40,000 cars. If $z = 3$, the young man purchases three $20,000 cars, zero $30,000 cars, and one $40,000 car.

47. a. Let $x =$ the amount invested at 4%, $y =$ the amount invested at 8%, and $z =$ the amount invested at 10%.

$$\begin{cases} x + y + z = 500,000 \\ 4\%x + 8\%y + 10\%z = 35,000 \end{cases}$$

or

$$\begin{cases} x + y + z = 500,000 \\ 0.04x + 0.08y + 0.10z = 35,000 \end{cases}$$

b. $\begin{bmatrix} 1 & 1 & 1 & | & 500,000 \\ 0.04 & 0.08 & 0.10 & | & 35,000 \end{bmatrix}$

Using the calculator to generate the reduced row-echelon form of the matrix yields:

$\begin{bmatrix} 1 & 0 & -0.5 & | & 125,000 \\ 0 & 1 & 1.5 & | & 375,000 \end{bmatrix}$

Let $z = z$.

$y + 1.5z = 375,000$

$y = 375,000 - 1.5z$

and

$x - 0.5z = 125,000$

$x = 0.5z + 125,000$

The solution is

$x = 0.5z + 125,000,$

$y = 375,000 - 1.5z,$

$z = z$

with all nonnegative values.

c. Let $y = 225,000$ and solve for z.
$$y = 375,000 - 1.5z$$
$$225,000 = 375,000 - 1.5z$$
$$-150,000 = -1.5z$$
$$100,000 = z$$

Let $z = 100,000$ and solve for x.
$$x = 0.5z + 125,000$$
$$x = 0.5(100,000) + 125,000$$
$$x = 50,000 + 125,000$$
$$x = 175,000$$

Therefore, she should invest $175,000 at 4%, $225,000 at 8%, and $100,000 at 10%.

49. $\begin{cases} x_1 - x_2 = -550 \\ x_2 - x_3 = -1300 \\ x_3 - x_4 = 1200 \\ x_1 - x_4 = -650 \end{cases}$

$$\begin{bmatrix} 1 & -1 & 0 & 0 & | & -550 \\ 0 & 1 & -1 & 0 & | & -1300 \\ 0 & 0 & 1 & -1 & | & 1200 \\ 1 & 0 & 0 & -1 & | & -650 \end{bmatrix}$$

Using the calculator to generate the reduced row-echelon form of the matrix yields

```
rref([C])▶Frac
…1 0 0 -1 -650]
…0 1 0 -1 -100]
…0 0 1 -1 1200]
…0 0 0 0 0   ]]
```

Let $x_4 = $ any real number, then

$x_1 - x_4 = -650$

$x_1 = x_4 - 650$

and

$x_2 - x_4 = -100$

$x_2 = x_4 - 100$

and

$x_3 - x_4 = 1200$

$x_3 = x_4 + 1200$

The solution is

$x_1 = x_4 - 650, \quad x_2 = x_4 - 100,$

$x_3 = x_4 + 1200, \quad x_4.$

Since all the variables must be positive in the physical context of the problem,
$$x_4 \geq 650.$$

51. a. Water flow at A is $x_1 + x_2 = 400,000$.
Water flow at B is $x_1 = 100,000 + x_4$.
Water flow at D is $x_3 + x_4 = 100,000$.

Rewriting the equations yields the following system
$$\begin{cases} x_1 + x_2 = 400,000 \\ x_1 - x_4 = 100,000 \\ x_3 + x_4 = 100,000 \\ x_2 - x_3 = 200,000 \end{cases}$$
The matrix is
$$\begin{bmatrix} 1 & 1 & 0 & 0 & | & 400,000 \\ 1 & 0 & 0 & -1 & | & 100,000 \\ 0 & 0 & 1 & 1 & | & 100,000 \\ 0 & 1 & -1 & 0 & | & 200,000 \end{bmatrix}$$

b. Using the calculator to generate the reduced row-echelon form of the matrix yields

```
rref([A])
[[1 0 0 -1 1000…
 [0 1 0 1  3000…
 [0 0 1 1  1000…
 [0 0 0 0  0    …
```

```
rref([A])
…0 0 -1 100000]
…1 0 1  300000]
…0 1 1  100000]
…0 0 0  0      ]]
```

The system has infinitely many solutions. Let x_4 be any number. Then,

$$x_3 = 100,000 - x_4,$$

$$x_2 = 300,000 - x_4, \text{ and}$$

$$x_1 = 100,000 + x_4.$$

Water flow from A to B is 100,000 plus the water flow from B to D. Water flow from A to C is 300,000 minus the water flow from B to D. Water flow from C to D is 100,000 minus the water flow from B to D. Water flow is measured by the number of gallons of water moving from one intersection to another.

Section 7.3 Skills Check

1. Only matrices of the same dimensions can be added. Therefore, matrices A, D, and E can be added, since they all have 3 rows and 3 columns. Furthermore, matrices B and F can be added, since they both have 2 rows and 3 columns. Note that a matrix can always be added to itself. Therefore, the following sums can be calculated:

$$A + D, A + E, D + E$$

$$B + F$$

$$A + A, B + B, C + C, D + D, E + E, F + F$$

3. To add the matrices, add the corresponding entries.

$$\begin{bmatrix} 1 & 3 & -2 \\ 3 & 1 & 4 \\ -5 & 3 & 6 \end{bmatrix} + \begin{bmatrix} 2 & 3 & 1 \\ 3 & 4 & -1 \\ 2 & 5 & 1 \end{bmatrix}$$

$$\begin{bmatrix} 1+2 & 3+3 & (-2)+1 \\ 3+3 & 1+4 & 4+(-1) \\ -5+2 & 3+5 & 6+1 \end{bmatrix}$$

$$\begin{bmatrix} 3 & 6 & -1 \\ 6 & 5 & 3 \\ -3 & 8 & 7 \end{bmatrix}$$

5. $3A$

$$= \begin{bmatrix} 3(1) & 3(3) & 3(-2) \\ 3(3) & 3(1) & 3(4) \\ 3(-5) & 3(3) & 3(6) \end{bmatrix}$$

$$= \begin{bmatrix} 3 & 9 & -6 \\ 9 & 3 & 12 \\ -15 & 9 & 18 \end{bmatrix}$$

7. $2D - 4A$

$$= 2\begin{bmatrix} 2 & 3 & 1 \\ 3 & 4 & -1 \\ 2 & 5 & 1 \end{bmatrix} - 4\begin{bmatrix} 1 & 3 & -2 \\ 3 & 1 & 4 \\ -5 & 3 & 6 \end{bmatrix}$$

$$= \begin{bmatrix} 4 & 6 & 2 \\ 6 & 8 & -2 \\ 4 & 10 & 2 \end{bmatrix} + \begin{bmatrix} -4 & -12 & 8 \\ -12 & -4 & -16 \\ 20 & -12 & -24 \end{bmatrix}$$

$$= \begin{bmatrix} 4+(-4) & 6+(-12) & 2+8 \\ 6+(-12) & 8+(-4) & -2+(-16) \\ 4+20 & 10+(-12) & 2+(-24) \end{bmatrix}$$

$$= \begin{bmatrix} 0 & -6 & 10 \\ -6 & 4 & -18 \\ 24 & -2 & -22 \end{bmatrix}$$

9. a. AD

$$= \begin{bmatrix} 1 & 3 & -2 \\ 3 & 1 & 4 \\ -5 & 3 & 6 \end{bmatrix}\begin{bmatrix} 2 & 3 & 1 \\ 3 & 4 & -1 \\ 2 & 5 & 1 \end{bmatrix}$$

$$= \begin{bmatrix} 2+9-4 & 3+12-10 & 1-3-2 \\ 6+3+8 & 9+4+20 & 3-1+4 \\ -10+9+12 & -15+12+30 & -5-3+6 \end{bmatrix}$$

$$= \begin{bmatrix} 7 & 5 & -4 \\ 17 & 33 & 6 \\ 11 & 27 & -2 \end{bmatrix}$$

DA

$$= \begin{bmatrix} 2 & 3 & 1 \\ 3 & 4 & -1 \\ 2 & 5 & 1 \end{bmatrix}\begin{bmatrix} 1 & 3 & -2 \\ 3 & 1 & 4 \\ -5 & 3 & 6 \end{bmatrix}$$

$$= \begin{bmatrix} 2+9-5 & 6+3+3 & -4+12+6 \\ 3+12+5 & 9+4-3 & -6+16-6 \\ 2+15-5 & 6+5+3 & -4+20+6 \end{bmatrix}$$

$$= \begin{bmatrix} 6 & 12 & 14 \\ 20 & 10 & 4 \\ 12 & 14 & 22 \end{bmatrix}$$

b. The products are different.

c. The dimensions of the matrices are the same. Both matrices are 3×3.

11. a. $DE = \begin{bmatrix} 2 & 3 & 1 \\ 3 & 4 & -1 \\ 2 & 5 & 1 \end{bmatrix} \begin{bmatrix} 9 & 2 & -7 \\ -5 & 0 & 5 \\ 7 & -4 & -1 \end{bmatrix}$

$= \begin{bmatrix} 18-15+7 & 4+0-4 & -14+15-1 \\ 27-20-7 & 6+0+4 & -21+20+1 \\ 18-25+7 & 4+0+-4 & -14+25-1 \end{bmatrix}$

$= \begin{bmatrix} 10 & 0 & 0 \\ 0 & 10 & 0 \\ 0 & 0 & 10 \end{bmatrix}$

$ED = \begin{bmatrix} 9 & 2 & -7 \\ -5 & 0 & 5 \\ 7 & -4 & -1 \end{bmatrix} \begin{bmatrix} 2 & 3 & 1 \\ 3 & 4 & -1 \\ 2 & 5 & 1 \end{bmatrix}$

$= \begin{bmatrix} 18+6-14 & 27+8-35 & 9-2-7 \\ -10+0+10 & -15+0+25 & -5+0+5 \\ 14-12-2 & 21-16-5 & 7+4-1 \end{bmatrix}$

$= \begin{bmatrix} 10 & 0 & 0 \\ 0 & 10 & 0 \\ 0 & 0 & 10 \end{bmatrix}$

Note that $DE = ED$.

b. $\dfrac{1}{10}DE$

$= \dfrac{1}{10} \begin{bmatrix} 10 & 0 & 0 \\ 0 & 10 & 0 \\ 0 & 0 & 10 \end{bmatrix}$

$= \begin{bmatrix} \dfrac{1}{10}(10) & \dfrac{1}{10}(0) & \dfrac{1}{10}(0) \\ \dfrac{1}{10}(0) & \dfrac{1}{10}(10) & \dfrac{1}{10}(0) \\ \dfrac{1}{10}(0) & \dfrac{1}{10}(0) & \dfrac{1}{10}(10) \end{bmatrix}$

$= \begin{bmatrix} 1 & 0 & 0 \\ 0 & 1 & 0 \\ 0 & 0 & 1 \end{bmatrix}$

Note that the solution is the 3×3 identity matrix.

13. $A + B$

$= \begin{bmatrix} 1 & 5 \\ 3 & 2 \end{bmatrix} + \begin{bmatrix} 2a & 3b \\ -c & -2d \end{bmatrix}$

$= \begin{bmatrix} 1+2a & 5+3b \\ 3-c & 2-2d \end{bmatrix}$

15. $3A - 2B$

$= 3\begin{bmatrix} a & b \\ c & d \end{bmatrix} - 2\begin{bmatrix} 1 & 2 \\ 3 & 4 \end{bmatrix}$

$= \begin{bmatrix} 3a & 3b \\ 3c & 3d \end{bmatrix} + \begin{bmatrix} -2 & -4 \\ -6 & -8 \end{bmatrix}$

$= \begin{bmatrix} 3a-2 & 3b-4 \\ 3c-6 & 3d-8 \end{bmatrix}$

17. The new matrix will be $m \times k$.

19. a. The product of two matrices can be calculated if the number of columns of the first matrix is equal to the number of rows of the second matrix. Therefore, BA can be calculated since B has 2 columns and A has 2 rows.

b. The matrix formed by the product will by 4×3.

21. AB

$$= \begin{bmatrix} a & b & c \\ d & e & f \end{bmatrix} \begin{bmatrix} 1 & 2 \\ 3 & 4 \\ 5 & 6 \end{bmatrix}$$

$$= \begin{bmatrix} a+3b+5c & 2a+4b+6c \\ d+3e+5f & 2d+4e+6f \end{bmatrix}$$

BA

$$= \begin{bmatrix} 1 & 2 \\ 3 & 4 \\ 5 & 6 \end{bmatrix} \begin{bmatrix} a & b & c \\ d & e & f \end{bmatrix}$$

$$= \begin{bmatrix} a+2d & b+2e & c+2f \\ 3a+4d & 3b+4e & 3c+4f \\ 5a+6d & 5b+6e & 5c+6f \end{bmatrix}$$

23. AB

$$= \begin{bmatrix} 1 & 5 \\ 3 & 2 \end{bmatrix} \begin{bmatrix} 2 & 3 \\ -1 & -2 \end{bmatrix}$$

$$= \begin{bmatrix} 2-5 & 3-10 \\ 6-2 & 9-4 \end{bmatrix}$$

$$= \begin{bmatrix} -3 & -7 \\ 4 & 5 \end{bmatrix}$$

BA

$$= \begin{bmatrix} 2 & 3 \\ -1 & -2 \end{bmatrix} \begin{bmatrix} 1 & 5 \\ 3 & 2 \end{bmatrix}$$

$$= \begin{bmatrix} 2+9 & 10+6 \\ -1-6 & -5-4 \end{bmatrix}$$

$$= \begin{bmatrix} 11 & 16 \\ -7 & -9 \end{bmatrix}$$

25. AB

$$= \begin{bmatrix} 1 & -1 & 2 \\ 3 & 4 & 4 \end{bmatrix} \begin{bmatrix} 3 & 1 \\ 1 & 3 \\ -2 & 1 \end{bmatrix}$$

$$= \begin{bmatrix} 3-1-4 & 1-3+2 \\ 9+4-8 & 3+12+4 \end{bmatrix}$$

$$= \begin{bmatrix} -2 & 0 \\ 5 & 19 \end{bmatrix}$$

BA

$$= \begin{bmatrix} 3 & 1 \\ 1 & 3 \\ -2 & 1 \end{bmatrix} \begin{bmatrix} 1 & -1 & 2 \\ 3 & 4 & 4 \end{bmatrix}$$

$$= \begin{bmatrix} 3+3 & -3+4 & 6+4 \\ 1+9 & -1+12 & 2+12 \\ -2+3 & 2+4 & -4+4 \end{bmatrix}$$

$$= \begin{bmatrix} 6 & 1 & 10 \\ 10 & 11 & 14 \\ 1 & 6 & 0 \end{bmatrix}$$

Section 7.3 Exercises

27. a. $A = \begin{bmatrix} 1211.5 & 192.4 & 165.5 \\ 80.8 & 79.0 & 78.4 \\ 1.88 & 1.68 & 1.95 \\ 8.3 & 12.5 & 10.2 \end{bmatrix}$

$B = \begin{bmatrix} 1341.3 & 148.7 & 93.3 \\ 73.8 & 69.4 & 69.2 \\ 1.56 & 2.16 & 3.05 \\ 19.6 & 41.8 & 20.9 \end{bmatrix}$

b.

$A - B$

$$= \begin{bmatrix} 1211.5 & 192.4 & 165.5 \\ 80.8 & 79.0 & 78.4 \\ 1.88 & 1.68 & 1.95 \\ 8.3 & 12.5 & 10.2 \end{bmatrix} - \begin{bmatrix} 1341.3 & 148.7 & 93.3 \\ 73.8 & 69.4 & 69.2 \\ 1.56 & 2.16 & 3.05 \\ 19.6 & 41.8 & 20.9 \end{bmatrix}$$

$$= \begin{bmatrix} -129.8 & 43.7 & 72.2 \\ 7 & 9.6 & 9.2 \\ 0.32 & -0.48 & -1.10 \\ -11.3 & -29.3 & -10.7 \end{bmatrix}$$

c. The negative entries in row 4 indicate that the infant mortality rate of all three countries are projected to decline which is a positive change for the countries.

29. a. $D = M - F$

$$D = \begin{bmatrix} 475 & 796 & 388 & -686 & -5613 \\ 492 & 669 & 374 & 41 & -6029 \\ 500 & 909 & 379 & 842 & -6644 \\ 502 & 932 & 425 & 1671 & -7357 \end{bmatrix}$$

The matrix shows the projected difference between male and female populations for different ages and years.

b. Female population is projected to be greater than male population for ages 18 – 64 in 2015 and for those over 64 in all years 2015 – 2030.

31. a. The original matrix is

$$A = \begin{bmatrix} 0.94 & 107.91 \\ 0.90 & 119.74 \end{bmatrix}.$$

Reducing the currency returned by 5% implies that 95% (i.e., 100% – 5%) of the value will be returned. Therefore the new matrix would be 95%A.

$0.95A$

$$= \begin{bmatrix} 0.95(0.94) & 0.95(107.91) \\ 0.95(0.90) & 0.95(119.74) \end{bmatrix}$$

$$= \begin{bmatrix} 0.89 & 102.51 \\ 0.86 & 113.75 \end{bmatrix}$$

b. The yen has increased in value during this period.

33. To compute the required matrix, the costs of TV, radio, and newspaper advertisements need to be multiplied by the number of each type advertisement targeted to the various audiences. The matrix is represented by BA.

$$BA = \begin{bmatrix} 30 & 45 & 35 \\ 25 & 32 & 40 \\ 22 & 12 & 30 \end{bmatrix} \begin{bmatrix} 12 \\ 15 \\ 5 \end{bmatrix}$$

Using technology to calculate the product yields:

```
[B] [A]
    [[1210]
     [980 ]
     [594 ]]
```

Therefore, the cost of advertising to singles is \$1210, the cost of advertising to males 35–55 is \$980, and the cost of advertising to females 65+ is \$594.

35. a. To calculate the total cost of various products per department, for each product multiply the quantity needed by the unit cost and add the results. The matrix form of the multiplication corresponds to

$$\begin{bmatrix} 60 & 40 & 20 \\ 40 & 20 & 40 \end{bmatrix} \begin{bmatrix} 600 & 560 \\ 300 & 200 \\ 300 & 400 \end{bmatrix}.$$

Using technology to calculate the product yields

```
[A] [B]
 [[54000 49600]
  [42000 42400]]
```

	DeTuris	Marriott
Department A	54,000	49,600
Department B	42,000	42,400

39. a.
$$\begin{bmatrix} 449 & 706 & 917 & 955 & 1003 & 768 \\ 436 & 663 & 737 & 722 & 742 & 598 \end{bmatrix}$$

b. To minimize the cost of the purchase, Department A should purchase the products from Marriott, while Department B should purchase the products from DeTuris.

37. $\begin{bmatrix} R \\ D \end{bmatrix} = \begin{bmatrix} 0.90 & 0.20 \\ 0.10 & 0.80 \end{bmatrix} \begin{bmatrix} a \\ b \end{bmatrix}$

Let $a = 0.50$ and $b = 0.50$.

$$\begin{bmatrix} R \\ D \end{bmatrix} = \begin{bmatrix} 0.90 & 0.20 \\ 0.10 & 0.80 \end{bmatrix} \begin{bmatrix} 0.50 \\ 0.50 \end{bmatrix}$$

Using technology to calculate the product yields

```
[A] [B]
      [[.55]
       [.45]]
```

$$\begin{bmatrix} R \\ D \end{bmatrix} = \begin{bmatrix} 0.55 \\ 0.45 \end{bmatrix} = \begin{bmatrix} 55\% \\ 45\% \end{bmatrix}$$

Based on the model, in the next election Republicans will receive 55% of the vote, while Democrats will receive 45% of the vote.

b. To generate a new matrix representing the median weekly earnings for men of 10% and the median weekly earnings for women of 25%, the matrix from part a) should be multiplied by the matrix $\begin{bmatrix} 1.10 & 0 \\ 0 & 1.25 \end{bmatrix}$.

$$\begin{bmatrix} 1.10 & 0 \\ 0 & 1.25 \end{bmatrix} \begin{bmatrix} 449 & 706 & 917 & 955 & 1003 & 768 \\ 436 & 663 & 737 & 722 & 742 & 598 \end{bmatrix}$$

Using technology to calculate the product yields:

$$[A][B] = \begin{bmatrix} 493.90 & 776.60 & 1008.70 & 1050.50 & 1103.30 & 844.80 \\ 545.00 & 828.75 & 921.25 & 902.50 & 927.50 & 747.50 \end{bmatrix}$$

Section 7.4 Skills Check

1. **a.** AB

$$= \begin{bmatrix} 3 & 1 \\ 4 & 2 \end{bmatrix} \begin{bmatrix} 1 & -0.5 \\ -2 & 1.5 \end{bmatrix}$$

$$= \begin{bmatrix} 3-2 & -1.5+1.5 \\ 4-4 & -2+3 \end{bmatrix}$$

$$= \begin{bmatrix} 1 & 0 \\ 0 & 1 \end{bmatrix}$$

BA

$$= \begin{bmatrix} 1 & -0.5 \\ -2 & 1.5 \end{bmatrix} \begin{bmatrix} 3 & 1 \\ 4 & 2 \end{bmatrix}$$

$$= \begin{bmatrix} 3-2 & 1-1 \\ -6+6 & -2+3 \end{bmatrix}$$

$$= \begin{bmatrix} 1 & 0 \\ 0 & 1 \end{bmatrix}$$

  b. Since $AB = BA = I$, A and B are inverses of one another.

3. Using technology to calculate AB yields

[A] [B]
```
[[1 0 0]
 [0 1 0]
 [0 0 1]]
```

Likewise, using technology to calculate BA yields

[B] [A]
```
[[1 0 0]
 [0 1 0]
 [0 0 1]]
```

Since $AB = BA = I$, A and B are inverses of one another.

5. $[A \mid I]$

$$\begin{bmatrix} 1 & 3 & | & 1 & 0 \\ 2 & 7 & | & 0 & 1 \end{bmatrix} \xrightarrow{-2R_1+R_2 \to R_2}$$

$$\begin{bmatrix} 1 & 3 & | & 1 & 0 \\ 0 & 1 & | & -2 & 1 \end{bmatrix} \xrightarrow{-3R_2+R_1 \to R_1}$$

$$\begin{bmatrix} 1 & 0 & | & 7 & -3 \\ 0 & 1 & | & -2 & 1 \end{bmatrix}$$

$[I \mid A^{-1}]$

$$A^{-1} = \begin{bmatrix} 7 & -3 \\ -2 & 1 \end{bmatrix}$$

7. Using technology to calculate A^{-1} yields

[A]⁻¹►Frac
```
[[-1/6 -1/3 1 ]
 [-1/3 1/3  0 ]
 [1/3  2/3  -1]]
```

9. Using technology to calculate A^{-1} yields

[A]⁻¹►Frac
```
[[-1/3 -1 1/3]
 [1    1  0  ]
 [-1/3 0  1/3]]
```

11. Using technology to calculate A^{-1} yields

[A]⁻¹
```
[[.9  .2 -.7]
 [-.5 0   .5]
 [.7  -.4 -.1]]
```

13. Using technology to calculate C^{-1} yields

[C]⁻¹
```
[[1 0 1 1]
 [0 1 1 1]
 [0 0 1 0]
 [0 0 0 1]]
```

15. $AX = \begin{bmatrix} 2 \\ 4 \end{bmatrix}$

$A^{-1}(AX) = A^{-1}\left(\begin{bmatrix} 2 \\ 4 \end{bmatrix} \right)$

$IX = \begin{bmatrix} 1 & 2 \\ 4 & 3 \end{bmatrix}\begin{bmatrix} 2 \\ 4 \end{bmatrix}$

$X = \begin{bmatrix} 2+8 \\ 8+12 \end{bmatrix}$

$X = \begin{bmatrix} 10 \\ 20 \end{bmatrix}$

17. $\begin{bmatrix} -1 & 1 & 0 \\ -2 & 3 & -2 \\ 2 & -2 & 1 \end{bmatrix}\begin{bmatrix} x \\ y \\ z \end{bmatrix} = \begin{bmatrix} 3 \\ 5 \\ 8 \end{bmatrix}$

Let $A = \begin{bmatrix} -1 & 1 & 0 \\ -2 & 3 & -2 \\ 2 & -2 & 1 \end{bmatrix}$

Applying technology to calculate A^{-1}

$A^{-1} = \begin{bmatrix} 1 & 1 & 2 \\ 2 & 1 & 2 \\ 2 & 0 & 1 \end{bmatrix}$

Solving for X

$A^{-1}(AX) = A^{-1}\left(\begin{bmatrix} 3 \\ 5 \\ 8 \end{bmatrix} \right)$

$IX = \begin{bmatrix} 1 & 1 & 2 \\ 2 & 1 & 2 \\ 2 & 0 & 1 \end{bmatrix}\begin{bmatrix} 3 \\ 5 \\ 8 \end{bmatrix}$

$X = \begin{bmatrix} x \\ y \\ z \end{bmatrix} = \begin{bmatrix} 3+5+16 \\ 6+5+16 \\ 6+0+8 \end{bmatrix}$

$\begin{bmatrix} x \\ y \\ z \end{bmatrix} = \begin{bmatrix} 24 \\ 27 \\ 14 \end{bmatrix}$

$x = 24,\ y = 27,\ z = 14$

19. $\begin{bmatrix} 4 & -3 & 1 \\ -6 & 5 & -2 \\ 1 & -1 & 1 \end{bmatrix}\begin{bmatrix} x \\ y \\ z \end{bmatrix} = \begin{bmatrix} 2 \\ -3 \\ 1 \end{bmatrix}$

Let $A = \begin{bmatrix} 4 & -3 & 1 \\ -6 & 5 & -2 \\ 1 & -1 & 1 \end{bmatrix}$

Applying technology to calculate A^{-1}

$A^{-1} = \begin{bmatrix} 3 & 2 & 1 \\ 4 & 3 & 2 \\ 1 & 1 & 2 \end{bmatrix}$

Solving for X

$A^{-1}(AX) = A^{-1}\left(\begin{bmatrix} 2 \\ -3 \\ 1 \end{bmatrix} \right)$

$IX = \begin{bmatrix} 3 & 2 & 1 \\ 4 & 3 & 2 \\ 1 & 1 & 2 \end{bmatrix}\begin{bmatrix} 2 \\ -3 \\ 1 \end{bmatrix}$

$X = \begin{bmatrix} x \\ y \\ z \end{bmatrix} = \begin{bmatrix} 6-6+1 \\ 8-9+2 \\ 2-3+2 \end{bmatrix}$

$\begin{bmatrix} x \\ y \\ z \end{bmatrix} = \begin{bmatrix} 1 \\ 1 \\ 1 \end{bmatrix}$

$x = 1,\ y = 1,\ z = 1$

21. $\begin{bmatrix} 2 & 1 & 1 \\ 1 & 4 & 2 \\ 2 & 1 & 2 \end{bmatrix}\begin{bmatrix} x \\ y \\ z \end{bmatrix} = \begin{bmatrix} 4 \\ 4 \\ 3 \end{bmatrix}$

Let $A = \begin{bmatrix} 2 & 1 & 1 \\ 1 & 4 & 2 \\ 2 & 1 & 2 \end{bmatrix}$

Applying technology to calculate A^{-1}

$A^{-1} = \begin{bmatrix} \dfrac{6}{7} & -\dfrac{1}{7} & -\dfrac{2}{7} \\ \dfrac{2}{7} & \dfrac{2}{7} & -\dfrac{3}{7} \\ -1 & 0 & 1 \end{bmatrix}$

$A^{-1}(AX) = A^{-1}\left(\begin{bmatrix} 4 \\ 4 \\ 3 \end{bmatrix} \right)$

$IX = \begin{bmatrix} \dfrac{6}{7} & -\dfrac{1}{7} & -\dfrac{2}{7} \\ \dfrac{2}{7} & \dfrac{2}{7} & -\dfrac{3}{7} \\ -1 & 0 & 1 \end{bmatrix}\begin{bmatrix} 4 \\ 4 \\ 3 \end{bmatrix}$

Solving for X

$X = \begin{bmatrix} x \\ y \\ z \end{bmatrix} = \begin{bmatrix} \dfrac{24}{7} - \dfrac{4}{7} - \dfrac{6}{7} \\ \dfrac{8}{7} + \dfrac{8}{7} - \dfrac{9}{7} \\ -4 + 0 + 3 \end{bmatrix}$

$\begin{bmatrix} x \\ y \\ z \end{bmatrix} = \begin{bmatrix} \dfrac{14}{7} \\ \dfrac{7}{7} \\ -1 \end{bmatrix}$

$\begin{bmatrix} x \\ y \\ z \end{bmatrix} = \begin{bmatrix} 2 \\ 1 \\ -1 \end{bmatrix}$

$x = 2, \ y = 1, \ z = -1$

23. $\begin{bmatrix} 1 & 0 & 1 & 1 \\ 0 & 1 & 1 & 1 \\ 2 & 5 & 1 & 0 \\ 0 & 3 & 0 & 1 \end{bmatrix}\begin{bmatrix} x_1 \\ x_2 \\ x_3 \\ x_4 \end{bmatrix} = \begin{bmatrix} 90 \\ 72 \\ 108 \\ 144 \end{bmatrix}$

Let $A = \begin{bmatrix} 1 & 0 & 1 & 1 \\ 0 & 1 & 1 & 1 \\ 2 & 5 & 1 & 0 \\ 0 & 3 & 0 & 1 \end{bmatrix}$.

Applying technology to calculate A^{-1}

$A^{-1} = \begin{bmatrix} \dfrac{7}{9} & -\dfrac{8}{9} & \dfrac{1}{9} & \dfrac{1}{9} \\ -\dfrac{2}{9} & \dfrac{1}{9} & \dfrac{1}{9} & \dfrac{1}{9} \\ -\dfrac{4}{9} & \dfrac{11}{9} & \dfrac{2}{9} & -\dfrac{7}{9} \\ \dfrac{2}{3} & -\dfrac{1}{3} & -\dfrac{1}{3} & \dfrac{2}{3} \end{bmatrix}$

Solving for X

$A^{-1}(AX) = A^{-1}\left(\begin{bmatrix} 90 \\ 72 \\ 108 \\ 144 \end{bmatrix} \right)$

$IX = \begin{bmatrix} \dfrac{7}{9} & -\dfrac{8}{9} & \dfrac{1}{9} & \dfrac{1}{9} \\ -\dfrac{2}{9} & \dfrac{1}{9} & \dfrac{1}{9} & \dfrac{1}{9} \\ -\dfrac{4}{9} & \dfrac{11}{9} & \dfrac{2}{9} & -\dfrac{7}{9} \\ \dfrac{2}{3} & -\dfrac{1}{3} & -\dfrac{1}{3} & \dfrac{2}{3} \end{bmatrix}\begin{bmatrix} 90 \\ 72 \\ 108 \\ 144 \end{bmatrix}$

$X = \begin{bmatrix} x_1 \\ x_2 \\ x_3 \\ x_4 \end{bmatrix} = \begin{bmatrix} 34 \\ 16 \\ -40 \\ 96 \end{bmatrix}$

$x_1 = 34, \ x_2 = 16, \ x_3 = -40, \ x_4 = 96$

Section 7.4 Exercises

25.

$$\begin{bmatrix} A \\ B \end{bmatrix} = \begin{bmatrix} \dfrac{3}{5} & \dfrac{1}{3} \\ \dfrac{2}{5} & \dfrac{2}{3} \end{bmatrix} \begin{bmatrix} a \\ b \end{bmatrix}$$

Let $A = 150,000$ and $B = 120,000$.

$$\begin{bmatrix} 150,000 \\ 120,000 \end{bmatrix} = \begin{bmatrix} \dfrac{3}{5} & \dfrac{1}{3} \\ \dfrac{2}{5} & \dfrac{2}{3} \end{bmatrix} \begin{bmatrix} a \\ b \end{bmatrix}$$

Let $D = \begin{bmatrix} \dfrac{3}{5} & \dfrac{1}{3} \\ \dfrac{2}{5} & \dfrac{2}{3} \end{bmatrix}$. Using technology to calculate D^{-1} yields $D^{-1} = \begin{bmatrix} \dfrac{5}{2} & -\dfrac{5}{4} \\ -\dfrac{3}{2} & \dfrac{9}{4} \end{bmatrix}$.

Multiplying both sides by D^{-1} yields

$$\begin{bmatrix} \dfrac{5}{2} & -\dfrac{5}{4} \\ -\dfrac{3}{2} & \dfrac{9}{4} \end{bmatrix} \begin{bmatrix} 150,000 \\ 120,000 \end{bmatrix} = \begin{bmatrix} \dfrac{5}{2} & -\dfrac{5}{4} \\ -\dfrac{3}{2} & \dfrac{9}{4} \end{bmatrix} \begin{bmatrix} \dfrac{3}{5} & \dfrac{1}{3} \\ \dfrac{2}{5} & \dfrac{2}{3} \end{bmatrix} \begin{bmatrix} a \\ b \end{bmatrix}.$$

Using technology to carry out the multiplication:

$$\begin{bmatrix} 225,000 \\ 45,000 \end{bmatrix} = \begin{bmatrix} 1 & 0 \\ 0 & 1 \end{bmatrix} \begin{bmatrix} a \\ b \end{bmatrix}$$

$$\begin{bmatrix} a \\ b \end{bmatrix} = \begin{bmatrix} 225,000 \\ 45,000 \end{bmatrix}$$

Last year Company X "a" had 225,000 customers, and Company Y "b" had 45,000 customers.

27. $\begin{bmatrix} R \\ D \end{bmatrix} = \begin{bmatrix} 0.90 & 0.20 \\ 0.10 & 0.80 \end{bmatrix} \begin{bmatrix} r \\ d \end{bmatrix}$

Let $R = 0.55$ and $D = 0.45$.

$$\begin{bmatrix} 0.55 \\ 0.45 \end{bmatrix} = \begin{bmatrix} 0.90 & 0.20 \\ 0.10 & 0.80 \end{bmatrix} \begin{bmatrix} r \\ d \end{bmatrix}$$

Let $A = \begin{bmatrix} 0.90 & 0.20 \\ 0.10 & 0.80 \end{bmatrix}$.

Using technology to calculate A^{-1} yields

```
[A]⁻¹▶Frac
   [[8/7   -2/7]
    [-1/7  9/7 ]]
```

$$A^{-1} = \begin{bmatrix} \dfrac{8}{7} & -\dfrac{2}{7} \\ -\dfrac{1}{7} & \dfrac{9}{7} \end{bmatrix}$$

Multiplying both sides by A^{-1} yields $\begin{bmatrix} \dfrac{8}{7} & -\dfrac{2}{7} \\ -\dfrac{1}{7} & \dfrac{9}{7} \end{bmatrix}\begin{bmatrix} 0.55 \\ 0.45 \end{bmatrix} = \begin{bmatrix} \dfrac{8}{7} & -\dfrac{2}{7} \\ -\dfrac{1}{7} & \dfrac{9}{7} \end{bmatrix}\begin{bmatrix} 0.90 & 0.20 \\ 0.10 & 0.80 \end{bmatrix}\begin{bmatrix} r \\ d \end{bmatrix}.$

Using technology to carry out the multiplication yields:

$$\begin{bmatrix} 0.5 \\ 0.5 \end{bmatrix} = \begin{bmatrix} 1 & 0 \\ 0 & 1 \end{bmatrix}\begin{bmatrix} r \\ d \end{bmatrix}$$

$$\begin{bmatrix} r \\ d \end{bmatrix} = \begin{bmatrix} 0.5 \\ 0.5 \end{bmatrix}$$

In last year's election 50% of voters were Republicans and 50% were Democrats.

29. a. Let x represent the largest loan, y represent the medium size loan, and z represent the smallest loan.

$$\begin{cases} x + y + z = 400{,}000 \\ x = (y + z) + 100{,}000 \\ z = \dfrac{1}{2}y \end{cases}$$

or

$$\begin{cases} x + y + z = 400{,}000 \\ x - y - z = 100{,}000 \\ -\dfrac{1}{2}y + z = 0 \end{cases}$$

$$\begin{bmatrix} 1 & 1 & 1 \\ 1 & -1 & -1 \\ 0 & -\dfrac{1}{2} & 1 \end{bmatrix}\begin{bmatrix} x \\ y \\ z \end{bmatrix} = \begin{bmatrix} 400{,}000 \\ 100{,}000 \\ 0 \end{bmatrix}$$

b.
$$\begin{bmatrix} 1 & 1 & 1 \\ 1 & -1 & -1 \\ 0 & -\frac{1}{2} & 1 \end{bmatrix} \begin{bmatrix} x \\ y \\ z \end{bmatrix} = \begin{bmatrix} 400,000 \\ 100,000 \\ 0 \end{bmatrix}$$

Let $A = \begin{bmatrix} 1 & 1 & 1 \\ 1 & -1 & -1 \\ 0 & -\frac{1}{2} & 1 \end{bmatrix}$. Using technology to calculate A^{-1} yields $A^{-1} = \begin{bmatrix} \frac{1}{2} & \frac{1}{2} & 0 \\ \frac{1}{3} & -\frac{1}{3} & -\frac{2}{3} \\ \frac{1}{6} & -\frac{1}{6} & \frac{2}{3} \end{bmatrix}$.

$$\begin{bmatrix} \frac{1}{2} & \frac{1}{2} & 0 \\ \frac{1}{3} & -\frac{1}{3} & -\frac{2}{3} \\ \frac{1}{6} & -\frac{1}{6} & \frac{2}{3} \end{bmatrix} \begin{bmatrix} 1 & 1 & 1 \\ 1 & -1 & -1 \\ 0 & -\frac{1}{2} & 1 \end{bmatrix} \begin{bmatrix} x \\ y \\ z \end{bmatrix} = \begin{bmatrix} \frac{1}{2} & \frac{1}{2} & 0 \\ \frac{1}{3} & -\frac{1}{3} & -\frac{2}{3} \\ \frac{1}{6} & -\frac{1}{6} & \frac{2}{3} \end{bmatrix} \begin{bmatrix} 400,000 \\ 100,000 \\ 0 \end{bmatrix}$$

Using technology to carry out the multiplication:
$$\begin{bmatrix} 1 & 0 & 0 \\ 0 & 1 & 0 \\ 0 & 0 & 1 \end{bmatrix} \begin{bmatrix} x \\ y \\ z \end{bmatrix} = \begin{bmatrix} 250,000 \\ 100,000 \\ 50,000 \end{bmatrix}$$
$$\begin{bmatrix} x \\ y \\ z \end{bmatrix} = \begin{bmatrix} 250,000 \\ 100,000 \\ 50,000 \end{bmatrix}$$
$x = 250,000, \ y = 100,000, \ z = 50,000$

The largest loan is $250,000. The next medium size is $100,000. The smallest loan is $50,000.

31. Let x represent the amount invested at 6%, y represent the amount invested at 8%, and z represent the amount invested at 10%.

$$\begin{cases} x + y + z = 400{,}000 \\ 2x = y \\ 0.06x + 0.08y + 0.10z = 36{,}000 \end{cases} \quad \text{or} \quad \begin{cases} x + y + z = 400{,}000 \\ 2x - y = 0 \\ 0.06x + 0.08y + 0.10z = 36{,}000 \end{cases}$$

$$\begin{bmatrix} 1 & 1 & 1 \\ 2 & -1 & 0 \\ 0.06 & 0.08 & 0.10 \end{bmatrix} \begin{bmatrix} x \\ y \\ z \end{bmatrix} = \begin{bmatrix} 400{,}000 \\ 0 \\ 36{,}000 \end{bmatrix}$$

Let $A = \begin{bmatrix} 1 & 1 & 1 \\ 2 & -1 & 0 \\ 0.06 & 0.08 & 0.10 \end{bmatrix}$. Using technology to calculate A^{-1} yields $A^{-1} = \begin{bmatrix} \dfrac{5}{4} & \dfrac{1}{4} & -\dfrac{25}{2} \\[2mm] \dfrac{5}{2} & -\dfrac{1}{2} & -25 \\[2mm] -\dfrac{11}{4} & \dfrac{1}{4} & \dfrac{75}{2} \end{bmatrix}$.

$$\begin{bmatrix} \dfrac{5}{4} & \dfrac{1}{4} & -\dfrac{25}{2} \\[2mm] \dfrac{5}{2} & -\dfrac{1}{2} & -25 \\[2mm] -\dfrac{11}{4} & \dfrac{1}{4} & \dfrac{75}{2} \end{bmatrix} \begin{bmatrix} 1 & 1 & 1 \\ 2 & -1 & 0 \\ 0.06 & 0.08 & 0.10 \end{bmatrix} \begin{bmatrix} x \\ y \\ z \end{bmatrix} = \begin{bmatrix} \dfrac{5}{4} & \dfrac{1}{4} & -\dfrac{25}{2} \\[2mm] \dfrac{5}{2} & -\dfrac{1}{2} & -25 \\[2mm] -\dfrac{11}{4} & \dfrac{1}{4} & \dfrac{75}{2} \end{bmatrix} \begin{bmatrix} 400{,}000 \\ 0 \\ 36{,}000 \end{bmatrix}$$

Using technology to carry out the multiplication yields

$$\begin{bmatrix} 1 & 0 & 0 \\ 0 & 1 & 0 \\ 0 & 0 & 1 \end{bmatrix} \begin{bmatrix} x \\ y \\ z \end{bmatrix} = \begin{bmatrix} 50{,}000 \\ 100{,}000 \\ 250{,}000 \end{bmatrix}$$

$$\begin{bmatrix} x \\ y \\ z \end{bmatrix} = \begin{bmatrix} 50{,}000 \\ 100{,}000 \\ 250{,}000 \end{bmatrix}$$

$x = 50{,}000$, $y = 100{,}000$, $z = 250{,}000$

$50,000 is invested in the 6% account, $100,000 is invested in the 8% account, and $250,000 is invested in the 10% account.

33. Let x represent the percentage of venture capital from business loans, y represent the percentage of venture capital from auto loans, and z represent the percentage of venture capital from home loans.

$$\begin{cases} 532x + 58y + 682z = 483.94 \\ 562x + 62y + 695z = 503.28 \\ 578x + 69y + 722z = 521.33 \end{cases}$$

$$\begin{bmatrix} 532 & 58 & 682 \\ 562 & 62 & 695 \\ 578 & 69 & 722 \end{bmatrix} \begin{bmatrix} x \\ y \\ z \end{bmatrix} = \begin{bmatrix} 483.94 \\ 503.28 \\ 521.33 \end{bmatrix}$$

Let $A = \begin{bmatrix} 532 & 58 & 682 \\ 562 & 62 & 695 \\ 578 & 69 & 722 \end{bmatrix}$. Using technology to calculate A^{-1} and applying A^{-1} to both sides

of the equation yields $A^{-1} \begin{bmatrix} 532 & 58 & 682 \\ 562 & 62 & 695 \\ 578 & 69 & 722 \end{bmatrix} \begin{bmatrix} x \\ y \\ z \end{bmatrix} = A^{-1} \begin{bmatrix} 483.94 \\ 503.28 \\ 521.33 \end{bmatrix}$.

Using technology to carry out the multiplication yields

```
[B]

 [[483.94]
  [503.28]
  [521.33]]
```

```
[A]⁻¹[B]

      [[.47]
       [.27]
       [.32]]
```

$$\begin{bmatrix} 1 & 0 & 0 \\ 0 & 1 & 0 \\ 0 & 0 & 1 \end{bmatrix} \begin{bmatrix} x \\ y \\ z \end{bmatrix} = \begin{bmatrix} 0.47 \\ 0.27 \\ 0.32 \end{bmatrix}$$

$$\begin{bmatrix} x \\ y \\ z \end{bmatrix} = \begin{bmatrix} 0.47 \\ 0.27 \\ 0.32 \end{bmatrix}$$

$x = 0.47, \ y = 0.27, \ z = 0.32$

The percentage of venture capital from business loans is 47%, the percentage of venture capital from auto loans is 27%, and the percentage of venture capital from home loans is 32%.

35. a. "Just do it" is represented by the 10, 21, 19, 20, 27, 4, 15, 27, 9, and 20. Recall that spaces are coded as 27.

b. Multiplying by the encoding matrix and applying technology to generate the solutions yields:

$$\begin{bmatrix} 4 & 4 \\ 1 & 2 \end{bmatrix}\begin{bmatrix} 10 \\ 21 \end{bmatrix} = \begin{bmatrix} 124 \\ 52 \end{bmatrix}$$

$$\begin{bmatrix} 4 & 4 \\ 1 & 2 \end{bmatrix}\begin{bmatrix} 19 \\ 20 \end{bmatrix} = \begin{bmatrix} 156 \\ 59 \end{bmatrix}$$

$$\begin{bmatrix} 4 & 4 \\ 1 & 2 \end{bmatrix}\begin{bmatrix} 27 \\ 4 \end{bmatrix} = \begin{bmatrix} 124 \\ 35 \end{bmatrix}$$

$$\begin{bmatrix} 4 & 4 \\ 1 & 2 \end{bmatrix}\begin{bmatrix} 15 \\ 27 \end{bmatrix} = \begin{bmatrix} 168 \\ 69 \end{bmatrix}$$

$$\begin{bmatrix} 4 & 4 \\ 1 & 2 \end{bmatrix}\begin{bmatrix} 9 \\ 20 \end{bmatrix} = \begin{bmatrix} 116 \\ 49 \end{bmatrix}$$

The pairs of coded numbers are 124, 52, 156, 59, 124, 35, 168, 69, 116, and 49.

37 "Neatness counts" is represented by the 14, 5, 1, 20, 14, 5, 19, 19, 27, 3, 15, 21, 14, 20, and 19. Recall that spaces are coded as 27.

Multiplying by the encoding matrix and applying technology to generate the solutions yields:

$$\begin{bmatrix} 4 & 4 & 4 \\ 1 & 2 & 3 \\ 2 & 4 & 2 \end{bmatrix}\begin{bmatrix} 14 \\ 5 \\ 1 \end{bmatrix} = \begin{bmatrix} 80 \\ 27 \\ 50 \end{bmatrix}$$

$$\begin{bmatrix} 4 & 4 & 4 \\ 1 & 2 & 3 \\ 2 & 4 & 2 \end{bmatrix}\begin{bmatrix} 20 \\ 14 \\ 5 \end{bmatrix} = \begin{bmatrix} 156 \\ 63 \\ 106 \end{bmatrix}$$

$$\begin{bmatrix} 4 & 4 & 4 \\ 1 & 2 & 3 \\ 2 & 4 & 2 \end{bmatrix}\begin{bmatrix} 19 \\ 19 \\ 27 \end{bmatrix} = \begin{bmatrix} 260 \\ 138 \\ 168 \end{bmatrix}$$

$$\begin{bmatrix} 4 & 4 & 4 \\ 1 & 2 & 3 \\ 2 & 4 & 2 \end{bmatrix}\begin{bmatrix} 3 \\ 15 \\ 21 \end{bmatrix} = \begin{bmatrix} 156 \\ 96 \\ 108 \end{bmatrix}$$

$$\begin{bmatrix} 4 & 4 & 4 \\ 1 & 2 & 3 \\ 2 & 4 & 2 \end{bmatrix}\begin{bmatrix} 14 \\ 20 \\ 19 \end{bmatrix} = \begin{bmatrix} 212 \\ 111 \\ 146 \end{bmatrix}$$

The triples of coded numbers are 80, 27, 50, 156, 63, 106, 260, 138, 168, 156, 96, 108, 212, 111, and 146.

39. [A]⁻¹▸Frac
$$\begin{bmatrix} 1 & 1 \\ 2 & 3 \end{bmatrix}$$

Multiplying pairs of codes by the inverse of the encoding matrix decodes the message. Multiplying by A^{-1} and using technology to simplify yields:

$$\begin{bmatrix} 1 & 1 \\ 2 & 3 \end{bmatrix}\begin{bmatrix} 51 \\ -29 \end{bmatrix} = \begin{bmatrix} 22 \\ 15 \end{bmatrix}$$

$$\begin{bmatrix} 1 & 1 \\ 2 & 3 \end{bmatrix}\begin{bmatrix} 55 \\ -35 \end{bmatrix} = \begin{bmatrix} 20 \\ 5 \end{bmatrix}$$

$$\begin{bmatrix} 1 & 1 \\ 2 & 3 \end{bmatrix}\begin{bmatrix} 76 \\ -49 \end{bmatrix} = \begin{bmatrix} 27 \\ 5 \end{bmatrix}$$

$$\begin{bmatrix} 1 & 1 \\ 2 & 3 \end{bmatrix}\begin{bmatrix} -15 \\ 16 \end{bmatrix} = \begin{bmatrix} 1 \\ 18 \end{bmatrix}$$

$$\begin{bmatrix} 1 & 1 \\ 2 & 3 \end{bmatrix}\begin{bmatrix} 11 \\ 1 \end{bmatrix} = \begin{bmatrix} 12 \\ 25 \end{bmatrix}$$

The decoded message is "Vote early."

41. [A]⁻¹▸Frac
$$\begin{bmatrix} 1 & 1 & 1 \\ 1 & 2 & 1 \\ 2 & 1 & 1 \end{bmatrix}$$

Multiplying triples of codes by the inverse of the encoding matrix decodes the message. Multiplying by A^{-1} and using technology to simplify yields:

$$\begin{bmatrix}1&1&1\\1&2&1\\2&1&1\end{bmatrix}\begin{bmatrix}1\\-4\\16\end{bmatrix}=\begin{bmatrix}13\\9\\14\end{bmatrix}$$

$$\begin{bmatrix}1&1&1\\1&2&1\\2&1&1\end{bmatrix}\begin{bmatrix}21\\23\\-40\end{bmatrix}=\begin{bmatrix}4\\27\\25\end{bmatrix}$$

$$\begin{bmatrix}1&1&1\\1&2&1\\2&1&1\end{bmatrix}\begin{bmatrix}3\\6\\6\end{bmatrix}=\begin{bmatrix}15\\21\\18\end{bmatrix}$$

$$\begin{bmatrix}1&1&1\\1&2&1\\2&1&1\end{bmatrix}\begin{bmatrix}-26\\-14\\67\end{bmatrix}=\begin{bmatrix}27\\13\\1\end{bmatrix}$$

$$\begin{bmatrix}1&1&1\\1&2&1\\2&1&1\end{bmatrix}\begin{bmatrix}-9\\0\\23\end{bmatrix}=\begin{bmatrix}14\\14\\5\end{bmatrix}$$

$$\begin{bmatrix}1&1&1\\1&2&1\\2&1&1\end{bmatrix}\begin{bmatrix}9\\1\\8\end{bmatrix}=\begin{bmatrix}18\\19\\27\end{bmatrix}$$

The decoded message is "Mind your manners."

43.
```
[A]⁻¹▶Frac
...14/3 -8/3 -5/3...
...-1/3 1/3  1/3 ...
...3    -2   -1  ...
```

Multiplying triples of codes by the inverse of the encoding matrix decodes the message. Multiplying by A^{-1} and using technology to simplify yields:

$$\begin{bmatrix}-\frac{14}{3}&-\frac{8}{3}&\frac{5}{3}\\-\frac{1}{3}&\frac{1}{3}&\frac{1}{3}\\3&-2&-1\end{bmatrix}\begin{bmatrix}29\\-1\\75\end{bmatrix}=\begin{bmatrix}13\\15\\14\end{bmatrix}$$

$$\begin{bmatrix}-\frac{14}{3}&-\frac{8}{3}&-\frac{5}{3}\\-\frac{1}{3}&\frac{1}{3}&\frac{1}{3}\\3&-2&-1\end{bmatrix}\begin{bmatrix}-19\\-66\\50\end{bmatrix}=\begin{bmatrix}4\\1\\25\end{bmatrix}$$

$$\begin{bmatrix}-\frac{14}{3}&-\frac{8}{3}&-\frac{5}{3}\\-\frac{1}{3}&\frac{1}{3}&\frac{1}{3}\\3&-2&-1\end{bmatrix}\begin{bmatrix}46\\41\\47\end{bmatrix}=\begin{bmatrix}27\\14\\9\end{bmatrix}$$

$$\begin{bmatrix}-\frac{14}{3}&-\frac{8}{3}&\frac{5}{3}\\-\frac{1}{3}&\frac{1}{3}&\frac{1}{3}\\3&-2&-1\end{bmatrix}\begin{bmatrix}3\\-38\\65\end{bmatrix}=\begin{bmatrix}7\\8\\20\end{bmatrix}$$

The decoded message is "Monday night."

45. Answers will vary.

Section 7.5 Skills Check

1. Given: $\begin{vmatrix} 3 & 2 \\ 4 & 1 \end{vmatrix}$

$D = (1)(3) - (4)(2)$

$D = 3 - 8$

$D = -5$

3. Given: $\begin{vmatrix} 5 & 1 \\ 5 & 1 \end{vmatrix}$

$D = (5)(1) - (5)(1)$

$D = 5 - 5$

$D = 0$

5. Given: $\begin{vmatrix} 3 & -2 & 1 \\ -1 & 0 & 2 \\ 0 & 1 & 1 \end{vmatrix}$

$D = 3\begin{bmatrix} 0 & 2 \\ 1 & 1 \end{bmatrix} - (-1)\begin{bmatrix} -2 & 1 \\ 1 & 1 \end{bmatrix} + 0\begin{bmatrix} -2 & 1 \\ 0 & 2 \end{bmatrix}$

$D = 3(0 - 2) + 1(-2 - 1) + 0(-4 - 0)$

$D = -6 - 3 + 0$

$D = -9$

7. Given: $\begin{vmatrix} 0 & -1 & 2 \\ 3 & 1 & -1 \\ 4 & -1 & 3 \end{vmatrix}$

$D = 0\begin{bmatrix} 1 & -1 \\ -1 & 3 \end{bmatrix} - (3)\begin{bmatrix} -1 & 2 \\ -1 & 3 \end{bmatrix} + 4\begin{bmatrix} -1 & 2 \\ 1 & -1 \end{bmatrix}$

$D = 0(3 - 1) - 3(-3 + 2) + 4(1 - 2)$

$D = 0 + 3 - 4$

$D = -1$

9. Given: $\begin{bmatrix} 2 & 3 \\ -1 & 4 \end{bmatrix}$

$D = (2)(4) - (3)(-1)$

$D = 8 + 3$

$D = 11$

Since the determinant is not 0, the matrix has an inverse.

11. Given: $\begin{bmatrix} 1 & 3 & -2 \\ 2 & -1 & 5 \\ 3 & 2 & 3 \end{bmatrix}$

$D = 1\begin{bmatrix} -1 & 5 \\ 2 & 3 \end{bmatrix} - 2\begin{bmatrix} 3 & -2 \\ 2 & 3 \end{bmatrix} + 3\begin{bmatrix} 3 & -2 \\ -1 & 5 \end{bmatrix}$

$D = 1(-3 - 10) - 2(9 + 4) + 3(15 - 2)$

$D = -13 - 26 + 39$

$D = 0$

Since the determinant is 0, the matrix does not have an inverse.

13. The coefficient matrix is: $\begin{bmatrix} 1 & 2 \\ 3 & 4 \end{bmatrix}$

$D = (4)(1) - (2)(3)$

$D = 4 - 6$

$D = -2$

The x matrix is: $\begin{bmatrix} 4 & 2 \\ 10 & 4 \end{bmatrix}$

$D_x = (4)(4) - (2)(10)$

$D_x = 16 - 20$

$D_x = -4$

The y matrix is: $\begin{bmatrix} 1 & 4 \\ 3 & 10 \end{bmatrix}$

$D_y = (1)(10) - (4)(3)$

$D_y = 10 - 12$

$D_y = -2$

Therefore,

$x = \dfrac{D_x}{D} = \dfrac{-4}{-2} = 2$

$y = \dfrac{D_y}{D} = \dfrac{-2}{-2} = 1$

The solution is $(2,1)$.

15. The coefficient matrix is: $\begin{bmatrix} 2 & -1 \\ 3 & 1 \end{bmatrix}$

$D = (2)(1) - (-1)(3)$

$D = 2 + 3$

$D = 5$

The x matrix is: $\begin{bmatrix} 4 & -1 \\ 5 & 1 \end{bmatrix}$

$D_x = (4)(1) - (-1)(5)$

$D_x = 4 + 5$

$D_x = 9$

The y matrix is: $\begin{bmatrix} 2 & 4 \\ 3 & 5 \end{bmatrix}$

$D_y = (5)(2) - (3)(4)$

$D_y = 10 - 12$

$D_y = -2$

Therefore,

$x = \dfrac{D_x}{D} = \dfrac{9}{5} = 1.8$

$y = \dfrac{D_y}{D} = \dfrac{-2}{5} = -0.4$

The solution is $(1.8, -0.4)$.

17. The coefficient matrix is: $\begin{bmatrix} 2 & -1 \\ -4 & 2 \end{bmatrix}$

$D = (2)(2) - (-1)(-4)$

$D = 4 - 4$

$D = 0$

The x matrix is: $\begin{bmatrix} 2 & -1 \\ 1 & 2 \end{bmatrix}$

$D_x = (2)(2) - (-1)(1)$

$D_x = 4 + 1$

$D_x = 5$

The y matrix is: $\begin{bmatrix} 2 & 2 \\ -4 & 1 \end{bmatrix}$

$D_y = (2)(1) - (2)(-4)$

$D_y = 2 + 8$

$D_y = 10$

Therefore, the equations are inconsistent because the coefficient matrix determinant is 0 and the x (and y) determinant is not.

19. The coefficient matrix is: $\begin{bmatrix} 2 & -4 \\ 3 & -6 \end{bmatrix}$

$D = (2)(-6) - (3)(-4)$

$D = -12 + 12$

$D = 0$

The x matrix is: $\begin{bmatrix} 5 & -4 \\ 7.5 & -6 \end{bmatrix}$

$D_x = (5)(-6) - (-4)(7.5)$

$D_x = -30 + 30$

$D_x = 0$

The y matrix is: $\begin{bmatrix} 2 & 5 \\ 3 & 7.5 \end{bmatrix}$

$D_y = (2)(7.5) - (3)(5)$

$D_y = 15 - 15$

$D_y = 0$

Therefore, the equations have infinitely many solutions because all of the determinants are 0.

21. The coefficient matrix is: $\begin{bmatrix} 1 & 1 & 1 \\ 2 & 1 & 1 \\ 2 & 2 & 1 \end{bmatrix}$

$D = 1\begin{bmatrix} 1 & 1 \\ 2 & 1 \end{bmatrix} - 2\begin{bmatrix} 1 & 1 \\ 2 & 1 \end{bmatrix} + 2\begin{bmatrix} 1 & 1 \\ 1 & 1 \end{bmatrix}$

$D = 1(1-2) - 2(1-2) + 2(1-1)$

$D = -1 + 2 + 0$

$D = 1$

The x matrix is: $\begin{bmatrix} 3 & 1 & 1 \\ 4 & 1 & 1 \\ 5 & 2 & 1 \end{bmatrix}$

$D_x = 3\begin{bmatrix} 1 & 1 \\ 2 & 1 \end{bmatrix} - 4\begin{bmatrix} 1 & 1 \\ 2 & 1 \end{bmatrix} + 5\begin{bmatrix} 1 & 1 \\ 1 & 1 \end{bmatrix}$

$D_x = 3(1-2) - 4(1-2) + 5(1-1)$

$D_x = -3 + 4 + 0$

$D_x = 1$

The y matrix is: $\begin{bmatrix} 1 & 3 & 1 \\ 2 & 4 & 1 \\ 2 & 5 & 1 \end{bmatrix}$

$D_y = 1\begin{bmatrix} 4 & 1 \\ 5 & 1 \end{bmatrix} - 2\begin{bmatrix} 3 & 1 \\ 5 & 1 \end{bmatrix} + 2\begin{bmatrix} 3 & 1 \\ 4 & 1 \end{bmatrix}$

$D_y = 1(4-5) - 2(3-5) + 2(3-4)$

$D_y = -1 + 4 - 2$

$D_y = 1$

The z matrix is: $\begin{bmatrix} 1 & 1 & 3 \\ 2 & 1 & 4 \\ 2 & 2 & 5 \end{bmatrix}$

$D_z = 1\begin{bmatrix} 1 & 4 \\ 2 & 5 \end{bmatrix} - 2\begin{bmatrix} 1 & 3 \\ 2 & 5 \end{bmatrix} + 2\begin{bmatrix} 1 & 3 \\ 1 & 4 \end{bmatrix}$

$D_z = 1(5-8) - 2(5-6) + 2(4-3)$

$D_z = -3 + 2 + 2$

$D_z = 1$

Therefore,

$x = \dfrac{D_x}{D} = \dfrac{1}{1} = 1$

$y = \dfrac{D_y}{D} = \dfrac{1}{1} = 1$

$z = \dfrac{D_z}{D} = \dfrac{1}{1} = 1$

The solution is $(1,1,1)$.

23. The coefficient matrix is: $\begin{bmatrix} 1 & 1 & 2 \\ 2 & 1 & 1 \\ 2 & 2 & 1 \end{bmatrix}$

$D = 1\begin{bmatrix} 1 & 1 \\ 2 & 1 \end{bmatrix} - 2\begin{bmatrix} 1 & 2 \\ 2 & 1 \end{bmatrix} + 2\begin{bmatrix} 1 & 2 \\ 1 & 1 \end{bmatrix}$

$D = 1(1-2) - 2(1-4) + 2(1-2)$

$D = -1 + 6 - 2$

$D = 3$

The x matrix is: $\begin{bmatrix} 8 & 1 & 2 \\ 7 & 1 & 1 \\ 10 & 2 & 1 \end{bmatrix}$

$D_x = 8\begin{bmatrix} 1 & 1 \\ 2 & 1 \end{bmatrix} - 7\begin{bmatrix} 1 & 2 \\ 2 & 1 \end{bmatrix} + 10\begin{bmatrix} 1 & 2 \\ 1 & 1 \end{bmatrix}$

$D_x = 8(1-2) - 7(1-4) + 10(1-2)$

$D_x = -8 + 21 - 10$

$D_x = 3$

The y matrix is: $\begin{bmatrix} 1 & 8 & 2 \\ 2 & 7 & 1 \\ 2 & 10 & 1 \end{bmatrix}$

$D_y = 1\begin{bmatrix} 7 & 1 \\ 10 & 1 \end{bmatrix} - 2\begin{bmatrix} 8 & 2 \\ 10 & 1 \end{bmatrix} + 2\begin{bmatrix} 8 & 2 \\ 7 & 1 \end{bmatrix}$

$D_y = 1(7-10) - 2(8-20) + 2(8-14)$

$D_y = -3 + 24 - 12$

$D_y = 9$

The z matrix is: $\begin{bmatrix} 1 & 1 & 8 \\ 2 & 1 & 7 \\ 2 & 2 & 10 \end{bmatrix}$

$D_z = 1\begin{bmatrix} 1 & 7 \\ 2 & 10 \end{bmatrix} - 2\begin{bmatrix} 1 & 8 \\ 2 & 10 \end{bmatrix} + 2\begin{bmatrix} 1 & 8 \\ 1 & 7 \end{bmatrix}$

$D_z = 1(10-14) - 2(10-16) + 2(7-8)$

$D_z = -4 + 12 - 2$

$D_z = 6$

Therefore,

$x = \dfrac{D_x}{D} = \dfrac{3}{3} = 1$

$y = \dfrac{D_y}{D} = \dfrac{9}{3} = 3$

$z = \dfrac{D_z}{D} = \dfrac{6}{3} = 2$

The solution is $(1,3,2)$.

25. Given: $\begin{vmatrix} -2 & -11 \\ x & 6 \end{vmatrix} = 32$

$32 = (-2)(6) - (-11)(x)$

$32 = -12 + 11x$

$44 = 11x$

$4 = x$

27. Given: $\begin{vmatrix} x+3 & -6 \\ x-2 & -4 \end{vmatrix} = 18$

$18 = (x+3)(-4) - (x-2)(-6)$

$18 = -4x - 12 + 6x - 12$

$18 = 2x - 24$

$42 = 2x$

$21 = x$

Section 7.5 Exercises

29. a. Let $x =$ amount of medication A and $y =$ the amount of medication B. Based on the information given, solve the following set of equations:

$6x + 2y = 50.6$

$8x - 5y = 0$

The coefficient matrix is: $\begin{bmatrix} 6 & 2 \\ 8 & -5 \end{bmatrix}$

$D = (6)(-5) - (8)(2)$

$D = -30 - 16$

$D = -46$

The x matrix is: $\begin{bmatrix} 50.6 & 2 \\ 0 & -5 \end{bmatrix}$

$D_x = (50.6)(-5) - (0)(2)$

$D_x = -253 - 0$

$D_x = -253$

The y matrix is: $\begin{bmatrix} 6 & 50.6 \\ 8 & 0 \end{bmatrix}$

$D_y = (6)(0) - (8)(50.6)$

$D_y = 0 - 404.8$

$D_y = -404.8$

Therefore,

$x = \dfrac{D_x}{D} = \dfrac{-253}{-46} = 5.5$

$y = \dfrac{D_y}{D} = \dfrac{-404.8}{-46} = 8.8$

Patient I should get 5.5 mg of medication A and 8.8 mg of medication B.

b. Let $x =$ amount of medication A and $y =$ the amount of medication B. Based on the information given, solve the following set of equations:

$6x + 2y = 92$

$8x - 5y = 0$

The coefficient matrix is: $\begin{bmatrix} 6 & 2 \\ 8 & -5 \end{bmatrix}$

$D = (6)(-5) - (8)(2)$

$D = -30 - 16$

$D = -46$

The x matrix is: $\begin{bmatrix} 92 & 2 \\ 0 & -5 \end{bmatrix}$

$D_x = (92)(-5) - (0)(2)$

$D_x = -460 - 0$

$D_x = -460$

The y matrix is: $\begin{bmatrix} 6 & 92 \\ 8 & 0 \end{bmatrix}$

$D_y = (6)(0) - (8)(92)$

$D_y = 0 - 736$

$D_y = -736$

Therefore,

$x = \dfrac{D_x}{D} = \dfrac{-460}{-46} = 10$

$y = \dfrac{D_y}{D} = \dfrac{-736}{-46} = 16$

Patient II should get 10 mg of medication A and 16 mg of medication B.

31. Let x = the number of deluxe table saws, y = the number of premium table saws, and z = the number of ultimate table saws. Based on the information given, solve the following set of equations:

$1.6x + 2y + 2.4z = 96$

$2x + 3y + 4z = 156$

$0.5x + 0.5y + z = 37$

The coefficient matrix is: $\begin{bmatrix} 1.6 & 2 & 2.4 \\ 2 & 3 & 4 \\ 0.5 & 0.5 & 1 \end{bmatrix}$

$D = 1.6\begin{bmatrix} 3 & 4 \\ .5 & 1 \end{bmatrix} - 2\begin{bmatrix} 2 & 2.4 \\ .5 & 1 \end{bmatrix} + .5\begin{bmatrix} 2 & 2.4 \\ 3 & 4 \end{bmatrix}$

$D = 1.6(3 - 2) - 2(2 - 1.2) + 0.5(8 - 7.2)$

$D = 1.6 - 1.6 + 0.4$

$D = 0.4$

The x matrix is: $\begin{bmatrix} 96 & 2 & 2.4 \\ 156 & 3 & 4 \\ 37 & 0.5 & 1 \end{bmatrix}$

$D_x = 96\begin{bmatrix} 3 & 4 \\ .5 & 1 \end{bmatrix} - 156\begin{bmatrix} 2 & 2.4 \\ .5 & 1 \end{bmatrix} + 37\begin{bmatrix} 2 & 2.4 \\ 3 & 4 \end{bmatrix}$

$D_x = 96(3 - 2) - 156(2 - 1.2) + 37(8 - 7.2)$

$D_x = 96 - 124.8 + 29.6$

$D_x = 0.8$

The y matrix is: $\begin{bmatrix} 1.6 & 96 & 2.4 \\ 2 & 156 & 4 \\ 0.5 & 37 & 1 \end{bmatrix}$

$D_y = 1.6\begin{bmatrix} 156 & 4 \\ 37 & 1 \end{bmatrix} - 2\begin{bmatrix} 96 & 2.4 \\ 37 & 1 \end{bmatrix} + .5\begin{bmatrix} 96 & 2.4 \\ 156 & 4 \end{bmatrix}$

$D_y = 1.6(156 - 148) - 2(96 - 88.8)$
$ + 0.5(384 - 374.4)$

$D_y = 12.8 - 14.4 + 4.8$

$D_y = 3.2$

The z matrix is: $\begin{bmatrix} 1.6 & 2 & 96 \\ 2 & 3 & 156 \\ 0.5 & 0.5 & 37 \end{bmatrix}$

$D_z = 1.6\begin{bmatrix} 3 & 156 \\ .5 & 37 \end{bmatrix} - 2\begin{bmatrix} 2 & 96 \\ .5 & 37 \end{bmatrix} + .5\begin{bmatrix} 2 & 96 \\ 3 & 156 \end{bmatrix}$

$D_z = 1.6(111 - 78) - 2(74 - 48) + 0.5(312 - 288)$

$D_z = 52.8 - 52 + 12$

$D_z = 12.8$

Therefore,

$x = \dfrac{D_x}{D} = \dfrac{0.8}{0.4} = 2$

$y = \dfrac{D_y}{D} = \dfrac{3.2}{0.4} = 8$

$z = \dfrac{D_z}{D} = \dfrac{12.8}{0.4} = 32$

Therefore, the manufacturer should produce 2 deluxe table saws, 8 premium, and 32 ultimate.

33. Let x = the number of large cars, y = the number of medium cars, and z = the number of compact cars. Based on the information given, solve the following set of equations:

$x + y + z = 120$

$48x + 28y + 18z = 3160$

$90x + 70y + 50z = 7600$

(Note: equation 2 has been divided by 1000 to make the calculations easier)

The coefficient matrix is: $\begin{bmatrix} 1 & 1 & 1 \\ 48 & 28 & 18 \\ 90 & 70 & 50 \end{bmatrix}$

$D = 1\begin{bmatrix} 28 & 18 \\ 70 & 50 \end{bmatrix} - 48\begin{bmatrix} 1 & 1 \\ 70 & 50 \end{bmatrix} + 90\begin{bmatrix} 1 & 1 \\ 28 & 18 \end{bmatrix}$

$D = 1(1400 - 1260) - 48(50 - 70) + 90(18 - 28)$

$D = 140 + 960 - 900$

$D = 200$

The *x* matrix is: $\begin{bmatrix} 120 & 1 & 1 \\ 3160 & 28 & 18 \\ 7600 & 70 & 50 \end{bmatrix}$

$D_x = 120\begin{bmatrix} 28 & 18 \\ 70 & 50 \end{bmatrix} - 3160\begin{bmatrix} 1 & 1 \\ 70 & 50 \end{bmatrix} + 7600\begin{bmatrix} 1 & 1 \\ 28 & 18 \end{bmatrix}$

$D_x = 120(1400 - 1260) - 3160(50 - 70) + 7600(18 - 28)$

$D_x = 16,800 + 63,200 - 76,000$

$D_x = 4000$

The *y* matrix is: $\begin{bmatrix} 1 & 120 & 1 \\ 48 & 3160 & 18 \\ 90 & 7600 & 50 \end{bmatrix}$

$D_y = 1\begin{bmatrix} 3160 & 18 \\ 7600 & 50 \end{bmatrix} - 48\begin{bmatrix} 120 & 1 \\ 7600 & 50 \end{bmatrix} + 90\begin{bmatrix} 120 & 1 \\ 3160 & 18 \end{bmatrix}$

$D_y = 1(158,000 - 136,800) - 48(6000 - 7600)$
$\qquad + 90(2160 - 3160)$

$D_y = 21,200 + 76,800 - 90,000$

$D_y = 8000$

The *z* matrix is: $\begin{bmatrix} 1 & 1 & 120 \\ 48 & 28 & 3160 \\ 90 & 70 & 7600 \end{bmatrix}$

$D_z = 1\begin{bmatrix} 28 & 3160 \\ 70 & 7600 \end{bmatrix} - 48\begin{bmatrix} 1 & 120 \\ 70 & 7600 \end{bmatrix} + 90\begin{bmatrix} 1 & 120 \\ 28 & 3160 \end{bmatrix}$

$D_z = 1(212,800 - 221,200) - 48(7600 - 8400)$
$\qquad + 90(3160 - 3360)$

$D_z = -8400 + 38,400 - 18,000$

$D_z = 12,000$

Therefore,

$x = \dfrac{D_x}{D} = \dfrac{4000}{200} = 20$

$y = \dfrac{D_y}{D} = \dfrac{8000}{200} = 40$

$z = \dfrac{D_z}{D} = \dfrac{12,000}{200} = 60$

Therefore, they should buy 20 large cars, 40 midsize cars, and 60 compact cars.

35. Let x = species I, y = species II, and z = species III. Based on the information given, solve the following set of equations:

$x + 2y + 2z = 5100$
$x \qquad + 3z = 6900$
$2x + 2y + 5z = 12,000$

The coefficient matrix is: $\begin{bmatrix} 1 & 2 & 2 \\ 1 & 0 & 3 \\ 2 & 2 & 5 \end{bmatrix}$

$D = 1\begin{bmatrix} 0 & 3 \\ 2 & 5 \end{bmatrix} - 1\begin{bmatrix} 2 & 2 \\ 2 & 5 \end{bmatrix} + 2\begin{bmatrix} 2 & 2 \\ 0 & 3 \end{bmatrix}$

$D = 1(0 - 6) - 1(10 - 4) + 2(6 - 0)$

$D = -6 - 6 + 12$

$D = 0$

The *x* matrix is: $\begin{bmatrix} 5100 & 2 & 2 \\ 6900 & 0 & 3 \\ 12,000 & 2 & 5 \end{bmatrix}$

$D_x = 5100\begin{bmatrix} 0 & 3 \\ 2 & 5 \end{bmatrix} - 6900\begin{bmatrix} 2 & 2 \\ 2 & 5 \end{bmatrix} + 12,000\begin{bmatrix} 2 & 2 \\ 0 & 3 \end{bmatrix}$

$D_x = 5100(0 - 6) - 6900(10 - 4) + 12,000(6 - 0)$

$D_x = -30,600 - 41,400 + 72,000$

$D_x = 0$

The *y* matrix is: $\begin{bmatrix} 1 & 5100 & 2 \\ 1 & 6900 & 3 \\ 2 & 12,000 & 5 \end{bmatrix}$

$D_y = 1\begin{bmatrix} 6900 & 3 \\ 12,000 & 5 \end{bmatrix} - 1\begin{bmatrix} 5100 & 2 \\ 12,000 & 5 \end{bmatrix}$
$\qquad + 2\begin{bmatrix} 5100 & 2 \\ 6900 & 3 \end{bmatrix}$

$D_y = 1(34,500 - 36,000) - 1(25,500 - 24,000)$
$\qquad + 2(15,300 - 13,800)$

$D_y = -1500 - 1500 + 3000$

$D_y = 0$

The z matrix is: $\begin{bmatrix} 1 & 2 & 5100 \\ 1 & 0 & 6900 \\ 2 & 2 & 12{,}000 \end{bmatrix}$

$D_z = 1\begin{bmatrix} 0 & 6900 \\ 2 & 12{,}000 \end{bmatrix} - 1\begin{bmatrix} 2 & 5100 \\ 2 & 12{,}000 \end{bmatrix} + 2\begin{bmatrix} 2 & 5100 \\ 0 & 6900 \end{bmatrix}$

$D_z = 1(0 - 13{,}800) - 1(24{,}000 - 10{,}200) + 2(13{,}800 - 0)$

$D_z = -13{,}800 - 13{,}800 + 27{,}600$

$D_z = 0$

Therefore, a different way would need to be used to solve the problem because all the determinants have a value of 0. Using the rref function on a calculator, the following matrix will be shown:

$\begin{bmatrix} 1 & 0 & 3 & 6900 \\ 0 & 1 & -0.5 & -900 \\ 0 & 0 & 0 & 0 \end{bmatrix}$

Therefore, species III can be any amount between 1800 and 2300, species I will be $6900 - 3$ (species III), and species II will be 0.5(species III) $- 900$.

37. Let $x =$ the amount invested at 9%, $y =$ the amount invested at 8%, and $z =$ the amount invested at 7.5%. Based on the information given, solve the following set of equations:

$x + y + z = 800{,}000$

$0.09x + 0.08y + 0.075z = 67{,}400$

$x - y - z = 0$

The coefficient matrix is:

$\begin{bmatrix} 1 & 1 & 1 \\ 0.09 & 0.08 & 0.075 \\ 1 & -1 & -1 \end{bmatrix}$

$D = 1\begin{bmatrix} 0.08 & 0.075 \\ -1 & -1 \end{bmatrix} - 0.09\begin{bmatrix} 1 & 1 \\ -1 & -1 \end{bmatrix}$

$\qquad\qquad + 1\begin{bmatrix} 1 & 1 \\ 0.08 & 0.075 \end{bmatrix}$

$D = 1(-0.08 + 0.075) - 0.09(-1 + 1)$

$\qquad\qquad + 1(0.075 - 0.08)$

$D = -0.005 + 0 - 0.005$

$D = -0.01$

The x matrix is: $\begin{bmatrix} 800{,}000 & 1 & 1 \\ 67{,}400 & 0.08 & 0.075 \\ 0 & -1 & -1 \end{bmatrix}$

$D_x = 800{,}000\begin{bmatrix} 0.08 & 0.075 \\ -1 & -1 \end{bmatrix} - 67{,}400\begin{bmatrix} 1 & 1 \\ -1 & -1 \end{bmatrix}$

$\qquad\qquad + 0\begin{bmatrix} 1 & 1 \\ 0.08 & 0.075 \end{bmatrix}$

$D_x = 800{,}000(-0.08 + 0.075) - 67{,}400(-1 + 1)$

$\qquad\qquad + 0(0.075 - 0.08)$

$D_x = -4000 + 0 + 0$

$D_x = -4000$

The y matrix is: $\begin{bmatrix} 1 & 800{,}000 & 1 \\ 0.09 & 67{,}400 & 0.075 \\ 1 & 0 & -1 \end{bmatrix}$

$D_y = 1\begin{bmatrix} 67{,}400 & 0.075 \\ 0 & -1 \end{bmatrix} - 0.09\begin{bmatrix} 800{,}000 & 1 \\ 0 & -1 \end{bmatrix}$

$\qquad\qquad + 1\begin{bmatrix} 800{,}000 & 1 \\ 67{,}400 & 0.075 \end{bmatrix}$

$D_y = 1(-67{,}400 - 0) - 0.09(-800{,}000 - 0)$

$\qquad\qquad + 1(60{,}000 - 67{,}400)$

$D_y = -67{,}400 + 72{,}000 - 7400$

$D_y = -2800$

The z matrix is: $\begin{bmatrix} 1 & 1 & 800{,}000 \\ 0.09 & 0.08 & 67{,}400 \\ 1 & -1 & 0 \end{bmatrix}$

$D_z = 1\begin{bmatrix} 0.08 & 67{,}400 \\ -1 & 0 \end{bmatrix} - 0.09\begin{bmatrix} 1 & 800{,}000 \\ -1 & 0 \end{bmatrix}$

$\qquad\qquad + 1\begin{bmatrix} 1 & 800{,}000 \\ 0.08 & 67{,}400 \end{bmatrix}$

$D_z = 1(0 + 67{,}400) - 0.09(0 + 800{,}000)$

$\qquad\qquad + 1(67{,}400 - 64{,}000)$

$D_z = 67{,}400 - 72{,}000 + 3400$

$D_z = -1200$

Therefore,

$$x = \frac{D_x}{D} = \frac{-4000}{-0.01} = 400,000$$

$$y = \frac{D_y}{D} = \frac{-2800}{-0.01} = 280,000$$

$$z = \frac{D_z}{D} = \frac{-1200}{-0.01} = 120,000$$

Therefore, the woman should invest $400,000 at 9%, $280,000 at 8%, and $120,000 at 7.5%.

Section 7.6 Skills Check

1. Isolating y in the first equation yields
   $y = x^2$.
   Substituting into the other equation yields
   $3x + y = 0$

   $3x + x^2 = 0$

   $x(3 + x) = 0$

   $x = 0, \ x = -3$

   Back substituting to calculate y

   $x = 0 \Rightarrow y = (0)^2 = 0$

   $x = -3 \Rightarrow y = (-3)^2 = 9$

   The solutions to the system are
   $(0,0)$ and $(-3,9)$.

3. Isolating y in the second equation yields
   $2x + 3y = 4$

   $3y = 4 - 2x$

   $y = \frac{4 - 2x}{3}$

   Substituting into the other equation yields
   $x^2 - 3y = 4$

   $x^2 - 3\left(\frac{4 - 2x}{3}\right) = 4$

   $x^2 - (4 - 2x) = 4$

   $x^2 - 4 + 2x = 4$

   $x^2 + 2x - 8 = 0$

   $(x - 2)(x + 4) = 0$

   $x = 2, \ x = -4$

Back substituting to calculate y

$$x = 2 \Rightarrow y = \frac{4 - 2(2)}{3} = 0$$

$$x = -4 \Rightarrow y = \frac{4 - 2(-4)}{3} = 4$$

The solutions to the system are
$(2,0)$ and $(-4,4)$.

5. Substituting $y = 2x$ into the other equation
   yields
   $x^2 + y^2 = 80$

   $x^2 + (2x)^2 = 80$

   $x^2 + 4x^2 = 80$

   $5x^2 = 80$

   $x^2 = 16$

   $x = \pm\sqrt{16} = \pm 4$

   $x = 4, \ x = -4$

   Back substituting to calculate y

   $x = 4 \Rightarrow y = 2(4) = 8$

   $x = -4 \Rightarrow y = 2(-4) = -8$

   The solutions to the system are
   $(4,8)$ and $(-4,-8)$.

7. Isolating y in the first equation yields
$y = 8 - x.$
Substituting into the other equation yields
$xy = 12$

$x(8 - x) = 12$

$8x - x^2 = 12$

$x^2 - 8x + 12 = 0$

$(x - 6)(x - 2) = 0$

$x = 6, \; x = 2$

Back substituting to calculate y

$x = 6 \Rightarrow y = 8 - (6) = 2$

$x = 2 \Rightarrow y = 8 - (2) = 6$

The solutions to the system are
$(6, 2)$ and $(2, 6)$.

9. Isolating y in the second equation yields
$2x - y + 4 = 0$ or $y = 2x + 4$.
Substituting into the other equation yields
$x^2 + 5x - y = 6$

$x^2 + 5x - (2x + 4) = 6$

$x^2 + 5x - 2x - 4 = 6$

$x^2 + 3x - 10 = 0$

$(x - 2)(x + 5) = 0$

$x = 2, \; x = -5$

Back substituting to calculate y

$x = 2 \Rightarrow y = 2(2) + 4 = 8$

$x = -5 \Rightarrow y = 2(-5) + 4 = -6$

The solutions to the system are
$(2, 8)$ and $(-5, -6)$.

11. Isolating y in the second equation yields
$2y - x = 61$ or $y = \dfrac{61 + x}{2}$.
Substituting into the other equation yields

$2x^2 - 2y + 7x = 19$

$2x^2 - 2\left(\dfrac{61 + x}{2}\right) + 7x = 19$

$2x^2 - 61 - x + 7x = 19$

$2x^2 + 6x - 80 = 0$

$2(x^2 + 3x - 40) = 0$

$2(x - 5)(x + 8) = 0$

$x = 5, \; x = -8$

Back substituting to calculate y

$x = 5 \Rightarrow y = \dfrac{61 + (5)}{2} = 33$

$x = -8 \Rightarrow y = \dfrac{61 + (-8)}{2} = \dfrac{53}{2}$

The solutions to the system are
$(5, 33)$ and $\left(-8, \dfrac{53}{2}\right)$.

13. Isolating y in both equations:
$4y = 28 - x^2$

$y = \dfrac{28 - x^2}{4}$

and

$y = 1 + \sqrt{x}$

Solving graphically by applying the intersection of graphs method:

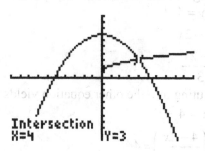

Intersection
X=4   Y=3

$[-10, 10]$ by $[-10, 10]$

The solution to the system is $(4, 3)$.

15. Isolating y in the second equation:

$$4y = x + 8$$

$$y = \frac{x+8}{4}$$

Substituting into the other equation:

$$4xy + x = 10$$

$$4x\left(\frac{x+8}{4}\right) + x = 10$$

$$x^2 + 8x + x = 10$$

$$x^2 + 9x - 10 = 0$$

$$(x+10)(x-1) = 0$$

$$x = -10, \ x = 1$$

Back substituting to calculate y

$$x = -10 \Rightarrow y = \frac{(-10)+8}{4} = -\frac{1}{2}$$

$$x = 1 \Rightarrow y = \frac{(1)+8}{4} = \frac{9}{4}$$

The solutions to the system are

$$\left(-10, -\frac{1}{2}\right) \text{ and } \left(1, \frac{9}{4}\right).$$

17. Solving graphically by applying the intersection of graphs method yields:

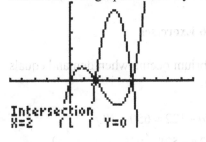

Intersection
X=2   Y=0

$[-5, 10]$ by $[-15, 25]$

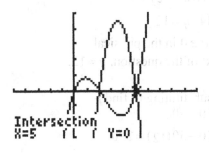

Intersection
X=5   Y=0

$[-5, 10]$ by $[-15, 25]$

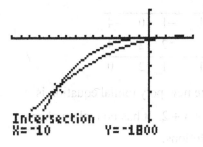

Intersection
X=-10   Y=-1800

$[-15, 5]$ by $[-3500, 1000]$

The solutions to the system are $(-10, -1800), (2, 0),$ and $(5, 0)$.

19. Isolating y in both equations:

$$y = 15 - x^2 - x^3$$

and

$$y = 11 - 2x^2$$

Solving graphically by applying the intersection of graphs method:

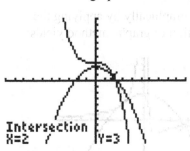

Intersection
X=2   Y=3

$[-10, 10]$ by $[-50, 50]$

The solution to the system is $(2, 3)$.

21. a. $y = 11 - 2x^2$

Substituting into the other equation

$$x^3 + x^2 + (11 - 2x^2) = 15$$

$$x^3 - x^2 - 4 = 0$$

b.

$$\begin{array}{r} 2)\overline{\begin{array}{rrrr} 1 & -1 & 0 & -4 \\ & 2 & 2 & 4 \end{array}} \\ \hline \begin{array}{rrrr} 1 & 1 & 2 & 0 \end{array} \end{array}$$

The new polynomial equation is $x^2 + x + 2$. It has no real number solutions.

Therefore, the only solution to the system is $(2,3)$.

23. Isolating y in both equations:

$$4y = 17 - x^2$$

$$y = \frac{17 - x^2}{4}$$

and

$$x^2 y = 3 + 5x$$

$$y = \frac{3 + 5x}{x^2}$$

Solving graphically by applying the intersection of graphs method yields:

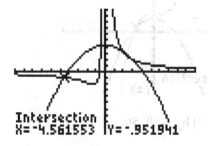

Intersection
X=-4.561553 Y=-.951941

[−10, 10] by [−10, 10]

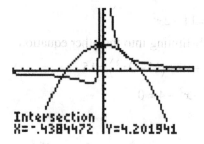

Intersection
X=-.4384472 Y=4.201941

[−10, 10] by [−10, 10]

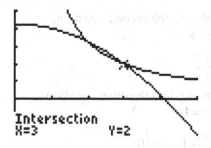

Intersection
X=3   Y=2

[0, 5] by [−2, 5]

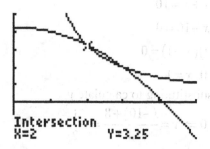

Intersection
X=2   Y=3.25

[0, 5] by [−2, 5]

Pick any two of the following four solutions to the system:
Exact solutions

$$(2, 3.25), (3, 2)$$

Approximate solutions

$$(-0.438, 4.20), (-4.562, -0.952)$$

Section 7.6 Exercises

25. Equilibrium occurs when demand equals supply.

$$q^2 + 2q + 122 = 650 - 30q$$

$$q^2 + 32q - 528 = 0$$

$$(q + 44)(q - 12) = 0$$

$$q = -44, \quad q = 12$$

Since $q \geq 0$ in the physical context of the question, $q = 12$.

Back substituting to find p

$$p = 650 - 30q$$

$$p = 650 - 30(12) = 290$$

Equilibrium occurs when the price is $290 and the demand is 1200 units.

27. Equilibrium occurs when demand equals supply.

$$0.1q^2 + 50q + 1027.50 = 6000 - 20q$$

$$0.1q^2 + 70q - 4972.50 = 0$$

$$q^2 + 700q - 49,725 = 0$$

$$q = \frac{-b \pm \sqrt{b^2 - 4ac}}{2a}$$

$$q = \frac{-700 \pm \sqrt{(700)^2 - 4(1)(-49,725)}}{2(1)}$$

$$q = \frac{-700 \pm \sqrt{490,000 + 198,900}}{2}$$

$$q = \frac{-700 \pm 830}{2}$$

$$q = \frac{-700 - 830}{2} = -765,$$

$$q = \frac{-700 + 830}{2} = 65$$

Since $q \geq 0$ in the physical context of the question, $q = 65$.

Back substituting to find p
$$p = 6000 - 20q$$
$$p = 6000 - 20(65) = 4700$$

Equilibrium occurs when the price is $4700 and the demand is 6500 units.

29. Break-even occurs when cost equals revenue.

$$C(x) = R(x)$$

$$2000x + 18,000 + 60x^2 = 4620x - 12x^2 - x^3$$

$$x^3 + 72x^2 - 2620x + 18,000 = 0$$

Solving graphically by applying the x-intercept method:

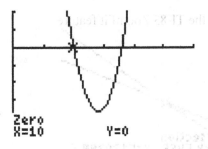

Zero
X=10   Y=0

[0, 30] by [−2500, 1000]

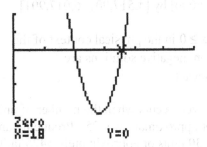

Zero
X=18   Y=0

[0, 30] by [−2500, 1000]

Since $x \geq 0$ in the physical context of the question, negative solutions are not relevant.

Break-even occurs when the number of thousands of units is 10 or 18. Producing and selling 10,000 units or 18,000 units results in revenue equaling cost.

31. Break-even occurs when cost equals revenue. $C(x) = R(x)$

Solving graphically by applying the intersection of graphs method yields:

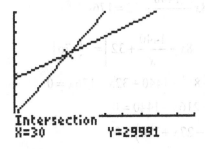

Intersection
X=30   Y=29991

[0, 100] by [−10,000, 50,000]

Using the TI-83 ZoomFit feature:

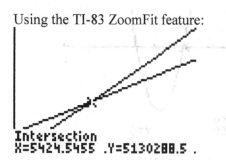

Intersection
X=5424.5455 .Y=5130288.5 .

[5000, 6000] by [4,517,901, 6,017,901]

Since $x \geq 0$ in the physical context of the question, negative solutions are not relevant.

Break-even occurs when the number of units is 30 or approximately 5425. Producing and selling 30 units or approximately 5425 units results in revenue equaling cost.

33. Let x = length and y = width.
$$\begin{cases} xy = 180 \\ 2(x-4)(y-4) = 176 \end{cases}$$

Solving the first equation for y
$$y = \frac{180}{x}$$
Substituting
$$2(x-4)(y-4) = 176$$
$$2(x-4)\left(\left[\frac{180}{x}\right]-4\right) = 176$$
$$(2x-8)\left(\frac{180}{x}-4\right) = 176$$
$$360 - 8x - \frac{1440}{x} + 32 = 176$$
$$x\left[360 - 8x - \frac{1440}{x} + 32\right] = x[176]$$
$$360x - 8x^2 - 1440 + 32x - 176x = 0$$
$$-8x^2 + 216x - 1440 = 0$$
$$-8(x^2 - 27x + 180) = 0$$
$$-8(x-15)(x-12) = 0$$
$$x = 15, \ x = 12$$

The dimensions are 15 inches by 12 inches.

35. Let x = length of the shorter side and y = length of the longer side. Assuming that the box is open,
$$V = (x)(x)(y) = x^2 y.$$
$$\text{Surface Area} = xy + 2xy + 2x^2$$
$$= 3xy + 2x^2$$
$$\begin{cases} x^2 y = 2000 \\ 3xy + 2x^2 = 800 \end{cases}$$

Solving the first equation for y
$$y = \frac{2000}{x^2}$$
Substituting
$$3xy + 2x^2 = 800$$
$$3x\left(\frac{2000}{x^2}\right) + 2x^2 = 800$$
$$\frac{6000}{x} + 2x^2 = 800$$
$$x\left[\frac{6000}{x} + 2x^2\right] = x[800]$$
$$6000 + 2x^3 = 800x$$
$$2x^3 - 800x + 6000 = 0$$

Solving graphically by applying the x-intercept method with $x \geq 0$ yields:

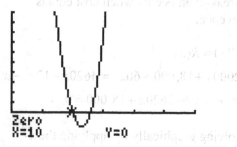

Zero
X=10   Y=0

[0, 30] by [−250, 1000]

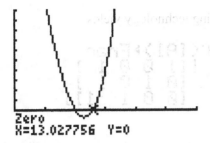

Zero
X=13.027756 Y=0

[0, 30] by [–250, 1000]

Substituting to find y

$$x = 10 \Rightarrow y = \frac{2000}{(10)^2} = 20$$

$$x \approx 13.03 \Rightarrow y \approx \frac{2000}{(13.03)^2} = 11.78$$

In the second case, x is not the smaller side. Therefore, the solution is a box with dimension 10 cm by 10 cm by 20 cm.

37. Let t = time in years and y = amount in dollars.

$$\begin{cases} y = 50,000(1.10)^t \\ y = 12,968.72t \end{cases}$$

$$50,000(1.10)^t = 12,968.72t$$

Solving graphically by applying the intersections of graphs method

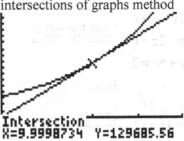

Intersection
X=9.9998734 Y=129685.56

[0, 20] by [–50,000, 250,000]

In approximately ten years the trust fund will equal the amount of money received from the second account.

Chapter 7 Skills Check

1. $\begin{cases} 2x - 3y + z = 2 \\ 3x + 2y - z = 6 \\ x - 4y + 2z = 2 \end{cases} \xrightarrow[\substack{-3Eq3+Eq2 \to Eq2 \\ -2Eq3+Eq1 \to Eq3}]{}$

$\begin{cases} 2x - 3y + z = 2 \\ 14y - 7z = 0 \\ 5y - 3z = -2 \end{cases} \xrightarrow{-5Eq2+14Eq3 \to Eq3}$

$\begin{cases} 2x - 3y + z = 2 \\ 14y - 7z = 0 \\ -7z = -28 \end{cases} \xrightarrow{\frac{1}{7}Eq3 \to Eq3}$

$\begin{cases} 2x - 3y + z = 2 \\ 14y - 7z = 0 \\ z = 4 \end{cases}$

Since z is isolated, back substitution yields

$$14y - 7(4) = 0$$
$$14y - 28 = 0$$
$$14y = 28$$
$$y = 2$$

and

$$2x - 3(2) + (4) = 2$$
$$2x - 6 + 4 = 2$$
$$2x = 4$$
$$x = 2$$

The solutions are $x = 2$, $y = 2$, $z = 4$.

Applying technology yields

```
rref([A])▶Frac
        [[1 0 0 2]
         [0 1 0 2]
         [0 0 1 4]]
```

3. $\begin{cases} 3x+2y-z=6 \\ 2x-4y-2z=0 \\ 5x+3y+6z=2 \end{cases} \xrightarrow[\;-5Eq2+2Eq3\rightarrow Eq3\;]{-3Eq2+2Eq1\rightarrow Eq2}$

$\begin{cases} 3x+2y-z=6 \\ 16y+4z=12 \\ 26y+22z=4 \end{cases} \xrightarrow[\;\frac{1}{2}Eq3\rightarrow Eq3\;]{\frac{1}{4}Eq2\rightarrow Eq2}$

$\begin{cases} 3x+2y-z=6 \\ 4y+z=3 \\ 13y+11z=2 \end{cases} \xrightarrow{\;-13Eq2+4Eq3\rightarrow Eq3\;}$

$\begin{cases} 3x+2y-z=6 \\ 4y+z=3 \\ 31z=-31 \end{cases} \xrightarrow{\;\frac{1}{31}Eq3\rightarrow Eq3\;}$

$\begin{cases} 3x+2y-z=6 \\ 4y+z=3 \\ z=-1 \end{cases}$

Since z is isolated, back substitution yields

$4y+(-1)=3$

$4y=4$

$y=1$

and

$3x+2y-z=6$

$3x+2(1)-(-1)=6$

$3x+3=6$

$3x=3$

$x=1$

The solutions are $x=1$, $y=1$, $z=-1$.

5. Writing the system as an augmented matrix and reducing yields

$$\begin{bmatrix} 1 & 2 & -2 & | & 1 \\ 2 & -1 & 5 & | & 15 \\ 3 & -4 & 1 & | & 7 \end{bmatrix} \xrightarrow[\;-3R_1+R_3\rightarrow R_3\;]{-2R_1+R_2\rightarrow R_2}$$

$$\begin{bmatrix} 1 & 2 & -2 & | & 1 \\ 0 & -5 & 9 & | & 13 \\ 0 & -10 & 7 & | & 4 \end{bmatrix} \xrightarrow{\;-2R_2+R_3\rightarrow R_3\;}$$

$$\begin{bmatrix} 1 & 2 & -2 & | & 1 \\ 0 & -5 & 9 & | & 13 \\ 0 & 0 & -11 & | & -22 \end{bmatrix} \xrightarrow[\;\left(-\frac{1}{11}\right)R_3\rightarrow R_3\;]{\left(-\frac{1}{5}\right)R_2\rightarrow R_2}$$

$$\begin{bmatrix} 1 & 2 & -2 & | & 1 \\ 0 & 1 & -\dfrac{9}{5} & | & -\dfrac{13}{5} \\ 0 & 0 & 1 & | & 2 \end{bmatrix}$$

$z=2$

and

$y-\dfrac{9}{5}(2)=-\dfrac{13}{5}$

$y-\dfrac{18}{5}=-\dfrac{13}{5}$

$y=\dfrac{5}{5}=1$

and

$x+2(1)-2(2)=1$

$x+2-4=1$

$x-2=1$

$x=3$

The solutions are $x=3$, $y=1$, $z=2$.

Applying technology yields

```
rref([A])▶Frac
        [[1 0 0 1 ]
         [0 1 0 1 ]
         [0 0 1 -1]]
```

Applying technology yields

```
rref([A])▶Frac
      [[1 0 0 3]
       [0 1 0 1],
       [0 0 1 2]]
```

7. Writing the system as an augmented matrix and reducing yields

$$\begin{bmatrix} 2 & 5 & 8 & | & 30 \\ 18 & 42 & 18 & | & 60 \end{bmatrix} \xrightarrow{-9R_1+R_2\to R_2}$$

$$\begin{bmatrix} 2 & 5 & 8 & | & 30 \\ 0 & -3 & -54 & | & -210 \end{bmatrix} \xrightarrow{\left(-\frac{1}{3}\right)R_2\to R_2}$$

$$\begin{bmatrix} 2 & 5 & 8 & | & 30 \\ 0 & 1 & 18 & | & 70 \end{bmatrix}$$

Since the system has more variables than equations, the system is dependent and has infinitely many solutions.

Let $z = z =$ any real number

$y + 18z = 70$

$y = 70 - 18z$

and

$2x + 5y + 8z = 30$

$2x + 5(70 - 18z) + 8z = 30$

$2x + 350 - 90z + 8z = 30$

$2x + 350 - 82z = 30$

$2x = 82z - 320$

$x = 41z - 160$

The solution is

$x = 41z - 160, \ y = 70 - 18z, \ z = z.$

Applying technology yields

```
rref([B])▶Frac
[[1 0 -41 -160]
 [0 1 18   70 ]]
```

See the back substitution process above.

9. Writing the system as an augmented matrix and using technology to produce the reduced row-echelon form yields

$$\begin{bmatrix} 1 & 3 & 2 & | & 5 \\ 9 & 12 & 15 & | & 6 \\ 2 & 1 & 3 & | & -10 \end{bmatrix}$$

```
rref([A])
    [[1 0 1.4 0]
     [0 1 .2  0]
     [0 0 0   1]]
```

Since the system has a last row of all zeros with a 1 as the augment, the system is inconsistent and has no solution.

11. Writing the system as an augmented matrix and using technology to produce the reduced row-echelon form yields

$$\begin{bmatrix} 3 & 2 & -1 & 1 & | & 12 \\ 1 & -4 & 3 & -1 & | & -18 \\ 1 & 1 & 3 & 2 & | & 0 \\ 2 & -1 & 3 & -3 & | & -10 \end{bmatrix}$$

```
rref([A])▶Frac
    [[1 0 0 0 1 ]
     [0 1 0 0 3 ]
     [0 0 1 0 -2]
     [0 0 0 1 1 ]]
```

The solutions are
$x = 1, \ y = 3, \ z = -2,$ and $w = 1..$

13. $B + D$

$$= \begin{bmatrix} 1 & 2 & 1 \\ 2 & -1 & 3 \end{bmatrix} + \begin{bmatrix} -2 & 3 & 1 \\ -3 & 2 & 2 \end{bmatrix}$$

$$= \begin{bmatrix} 1-2 & 2+3 & 1+1 \\ 2-3 & -1+2 & 3+2 \end{bmatrix}$$

$$= \begin{bmatrix} -1 & 5 & 2 \\ -1 & 1 & 5 \end{bmatrix}$$

15. $5C$

$$= 5\begin{bmatrix} 2 & 3 \\ -1 & 2 \\ 3 & -2 \end{bmatrix}$$

$$= \begin{bmatrix} 5(2) & 5(3) \\ 5(-1) & 5(2) \\ 5(3) & 5(-2) \end{bmatrix}$$

$$= \begin{bmatrix} 10 & 15 \\ -5 & 10 \\ 15 & -10 \end{bmatrix}$$

17. BA

$$= \begin{bmatrix} 1 & 2 & 1 \\ 2 & -1 & 3 \end{bmatrix}\begin{bmatrix} 1 & 3 & -3 \\ 2 & 4 & 1 \\ -1 & 3 & 2 \end{bmatrix}$$

$$= \begin{bmatrix} 1+4-1 & 3+8+3 & -3+2+2 \\ 2-2-3 & 6-4+9 & -6-1+6 \end{bmatrix},$$

$$= \begin{bmatrix} 4 & 14 & 1 \\ -3 & 11 & -1 \end{bmatrix}$$

21. A^{-1}

$$\begin{bmatrix} 1 & 3 & -3 & | & 1 & 0 & 0 \\ 2 & 4 & 1 & | & 0 & 1 & 0 \\ -1 & 3 & 2 & | & 0 & 0 & 1 \end{bmatrix} \xrightarrow[R_1+R_3 \to R_3]{-2R_1+R_2 \to R_2}$$

$$= \begin{bmatrix} 1 & 3 & -3 & | & 1 & 0 & 0 \\ 0 & -2 & 7 & | & -2 & 1 & 0 \\ 0 & 6 & -1 & | & 1 & 0 & 1 \end{bmatrix} \xrightarrow{3R_2+R_3 \to R_3}$$

$$= \begin{bmatrix} 1 & 3 & -3 & | & 1 & 0 & 0 \\ 0 & -2 & 7 & | & -2 & 1 & 0 \\ 0 & 0 & 20 & | & -5 & 3 & 1 \end{bmatrix} \xrightarrow{\left(\frac{1}{20}\right)R_3 \to R_3}$$

$$= \begin{bmatrix} 1 & 3 & -3 & | & 1 & 0 & 0 \\ 0 & -2 & 7 & | & -2 & 1 & 0 \\ 0 & 0 & 1 & | & -\dfrac{1}{4} & \dfrac{3}{20} & \dfrac{1}{20} \end{bmatrix} \xrightarrow[3R_3+R_1 \to R_1]{-7R_3+R_2 \to R_2}$$

19. DC

$$= \begin{bmatrix} -2 & 3 & 1 \\ -3 & 2 & 2 \end{bmatrix}\begin{bmatrix} 2 & 3 \\ -1 & 2 \\ 3 & -2 \end{bmatrix}$$

$$= \begin{bmatrix} -4-3+3 & -6+6-2 \\ -6-2+6 & -9+4-4 \end{bmatrix}$$

$$= \begin{bmatrix} -4 & -2 \\ -2 & -9 \end{bmatrix}$$

$$= \begin{bmatrix} 1 & 3 & 0 & \dfrac{1}{4} & \dfrac{9}{20} & \dfrac{3}{20} \\ 0 & -2 & 0 & -\dfrac{1}{4} & -\dfrac{1}{20} & -\dfrac{7}{20} \\ 0 & 0 & 1 & -\dfrac{1}{4} & \dfrac{3}{20} & \dfrac{1}{20} \end{bmatrix} \xrightarrow{\left(-\frac{1}{2}\right)R_2 \to R_2}$$

$$= \begin{bmatrix} 1 & 3 & 0 & \dfrac{1}{4} & \dfrac{9}{20} & \dfrac{3}{20} \\ 0 & 1 & 0 & \dfrac{1}{8} & \dfrac{1}{40} & \dfrac{7}{40} \\ 0 & 0 & 1 & -\dfrac{1}{4} & \dfrac{3}{20} & \dfrac{1}{20} \end{bmatrix} \xrightarrow{(-3)R_2 + R_1 \to R_1}$$

$$= \begin{bmatrix} 1 & 0 & 0 & -\dfrac{1}{8} & \dfrac{3}{8} & -\dfrac{3}{8} \\ 0 & 1 & 0 & \dfrac{1}{8} & \dfrac{1}{40} & \dfrac{7}{40} \\ 0 & 0 & 1 & -\dfrac{1}{4} & \dfrac{3}{20} & \dfrac{1}{20} \end{bmatrix}$$

$$A^{-1} = \begin{bmatrix} -\dfrac{1}{8} & \dfrac{3}{8} & -\dfrac{3}{8} \\ \dfrac{1}{8} & \dfrac{1}{40} & \dfrac{7}{40} \\ -\dfrac{1}{4} & \dfrac{3}{20} & \dfrac{1}{20} \end{bmatrix}$$

23. Applying technology to calculate A^{-1} yields

```
[A]-1▶Frac
     [[1   0   -1]
      [-1  1   1 ]
      [2   -2  -1]]
```

25. $\begin{bmatrix} 1 & 1 & -3 \\ 2 & 4 & 1 \\ -1 & 3 & 2 \end{bmatrix} \begin{bmatrix} x \\ y \\ z \end{bmatrix} = \begin{bmatrix} 8 \\ 15 \\ 5 \end{bmatrix}$

Let $A = \begin{bmatrix} 1 & 1 & -3 \\ 2 & 4 & 1 \\ -1 & 3 & 2 \end{bmatrix}$. Applying technology to calculate A^{-1} yields $A^{-1} = \begin{bmatrix} -\dfrac{1}{6} & \dfrac{11}{30} & -\dfrac{13}{30} \\ \dfrac{1}{6} & \dfrac{1}{30} & \dfrac{7}{30} \\ -\dfrac{1}{3} & \dfrac{2}{15} & -\dfrac{1}{15} \end{bmatrix}$.

Multiplying both sides of the matrix equation by A^{-1} yields

$$\begin{bmatrix} -\dfrac{1}{6} & \dfrac{11}{30} & -\dfrac{13}{30} \\[6pt] \dfrac{1}{6} & \dfrac{1}{30} & \dfrac{7}{30} \\[6pt] -\dfrac{1}{3} & \dfrac{2}{15} & -\dfrac{1}{15} \end{bmatrix} \begin{bmatrix} 1 & 1 & -3 \\ 2 & 4 & 1 \\ -1 & 3 & 2 \end{bmatrix} \begin{bmatrix} x \\ y \\ z \end{bmatrix} = \begin{bmatrix} -\dfrac{1}{6} & \dfrac{11}{30} & -\dfrac{13}{30} \\[6pt] \dfrac{1}{6} & \dfrac{1}{30} & \dfrac{7}{30} \\[6pt] -\dfrac{1}{3} & \dfrac{2}{15} & -\dfrac{1}{15} \end{bmatrix} \begin{bmatrix} 8 \\ 15 \\ 5 \end{bmatrix}.$$

Using technology to carry out the multiplication yields:

$$\begin{bmatrix} 1 & 0 & 0 \\ 0 & 1 & 0 \\ 0 & 0 & 1 \end{bmatrix} \begin{bmatrix} x \\ y \\ z \end{bmatrix} = \begin{bmatrix} 2 \\ 3 \\ -1 \end{bmatrix}$$

$$\begin{bmatrix} x \\ y \\ z \end{bmatrix} = \begin{bmatrix} 2 \\ 3 \\ -1 \end{bmatrix}$$

$x = 2, \; y = 3, \; z = -1$

27. $\begin{bmatrix} -1 & 0 & 1 & 1 \\ 0 & 1 & 1 & 1 \\ 2 & 5 & 1 & 0 \\ 0 & 3 & 0 & 1 \end{bmatrix} \begin{bmatrix} x_1 \\ x_2 \\ x_3 \\ x_4 \end{bmatrix} = \begin{bmatrix} 6 \\ 12 \\ 20 \\ 24 \end{bmatrix}$

Let $A = \begin{bmatrix} -1 & 0 & 1 & 1 \\ 0 & 1 & 1 & 1 \\ 2 & 5 & 1 & 0 \\ 0 & 3 & 0 & 1 \end{bmatrix}$. Applying technology to calculate A^{-1} yields

$A^{-1} = \begin{bmatrix} -1.4 & 1.6 & -0.2 & -0.2 \\ 0.4 & -0.6 & 0.2 & 0.2 \\ 0.8 & -0.2 & 0.4 & -0.6 \\ -1.2 & 1.8 & -0.6 & 0.4 \end{bmatrix}.$

Multiplying both sides of the matrix equation by A^{-1} yields

$$\begin{bmatrix} -1.4 & 1.6 & -0.2 & -0.2 \\ 0.4 & -0.6 & 0.2 & 0.2 \\ 0.8 & -0.2 & 0.4 & -0.6 \\ -1.2 & 1.8 & -0.6 & 0.4 \end{bmatrix} \begin{bmatrix} -1 & 0 & 1 & 1 \\ 0 & 1 & 1 & 1 \\ 2 & 5 & 1 & 0 \\ 0 & 3 & 0 & 1 \end{bmatrix} \begin{bmatrix} x_1 \\ x_2 \\ x_3 \\ x_4 \end{bmatrix} = \begin{bmatrix} -1.4 & 1.6 & -0.2 & -0.2 \\ 0.4 & -0.6 & 0.2 & 0.2 \\ 0.8 & -0.2 & 0.4 & -0.6 \\ -1.2 & 1.8 & -0.6 & 0.4 \end{bmatrix} \begin{bmatrix} 6 \\ 12 \\ 20 \\ 24 \end{bmatrix}.$$

Using technology to carry out the multiplication yields

$$\begin{bmatrix} 1 & 0 & 0 & 0 \\ 0 & 1 & 0 & 0 \\ 0 & 0 & 1 & 0 \\ 0 & 0 & 0 & 1 \end{bmatrix}\begin{bmatrix} x_1 \\ x_2 \\ x_3 \\ x_4 \end{bmatrix} = \begin{bmatrix} 2 \\ 4 \\ -4 \\ 12 \end{bmatrix}$$

$$\begin{bmatrix} x_1 \\ x_2 \\ x_3 \\ x_4 \end{bmatrix} = \begin{bmatrix} 2 \\ 4 \\ -4 \\ 12 \end{bmatrix}$$

$$x_1 = 2, \ x_2 = 4, \ x_3 = -4, \ x_4 = 12$$

29. Given: $\begin{vmatrix} -1 & 4 \\ 2 & -3 \end{vmatrix}$

$$D = (-1)(-3) - (4)(2)$$
$$D = 3 - 8$$
$$D = -5$$

31. The coefficient matrix is: $\begin{bmatrix} 2 & -3 \\ -1 & 1 \end{bmatrix}$

$$D = (2)(1) - (-1)(-3)$$
$$D = 2 - 3$$
$$D = -1$$

The x matrix is: $\begin{bmatrix} 7 & -3 \\ 4 & 1 \end{bmatrix}$

$$D_x = (7)(1) - (4)(-3)$$
$$D_x = 7 + 12$$
$$D_x = 19$$

The y matrix is: $\begin{bmatrix} 2 & 7 \\ -1 & 4 \end{bmatrix}$

$$D_y = (2)(4) - (7)(-1)$$
$$D_y = 8 + 7$$
$$D_y = 15$$

Therefore,

$$x = \frac{D_x}{D} = \frac{19}{-1} = -19$$

$$y = \frac{D_y}{D} = \frac{15}{-1} = -15$$

The solution is $(-19, -15)$.

33. Isolating y in the second equation
$$4x - y = 4$$
$$y = 4x - 4$$

Substituting into the other equation
$$x^2 - y = x$$
$$x^2 - (4x - 4) = x$$
$$x^2 - 4x + 4 = x$$
$$x^2 - 5x + 4 = 0$$
$$(x - 4)(x - 1) = 0$$
$$x = 4, \ x = 1$$

Back substituting to calculate y
$$x = 4 \Rightarrow y = 4(4) - 4 = 12$$
$$x = 1 \Rightarrow y = 4(1) - 4 = 0$$

The solutions to the system are
$(4, 12)$ and $(1, 0)$.

Chapter 7 Review Exercises

35. Let x represent the number of $40 tickets, y represent the number of $60 tickets, and z represent the number of $100 tickets.

$$\begin{cases} x+y+z=4000 \\ 40x+60y+100z=200{,}000 \\ y=\frac{1}{4}(x+z) \end{cases} \xrightarrow{4Eq3\to Eq3}$$

$$\begin{cases} x+y+z=4000 \\ 40x+60y+100z=200{,}000 \\ x-4y+z=0 \end{cases} \xrightarrow{\substack{-40Eq1+Eq2\to Eq2 \\ -1Eq1+Eq3\to Eq3}}$$

$$\begin{cases} x+y+z=4000 \\ 20y+60z=40{,}000 \\ -5y=-4000 \end{cases}$$

$-5y=-4000$
$y=800$

Substituting to find z Substituting to find x
$20(800)+60z=40{,}000$ $x+(800)+(400)=4000$
$16{,}000+60z=40{,}000$ $x=2800$
$60z=24{,}000$
$z=400$

To generate $200,000, the concert promoter needs to sell 2800 $40 tickets, 800 $60 tickets, and 400 $100 tickets.

36. Let x represent the daily dosage of medication A, y represent the daily dosage of medication B, and z represent the daily dosage of medication C.

$$\begin{cases} 6x+2y+z=28.7 \\ z=\frac{1}{2}(x+y) \\ \frac{x}{y}=\frac{2}{3} \end{cases}$$

or

$$\begin{cases} 6x+2y+z=28.7 \\ x+y-2z=0 \\ 3x-2y=0 \end{cases}$$

$$\begin{cases} 6x+2y+z=28.7 \\ x+y-2z=0 \\ 3x-2y=0 \end{cases} \xrightarrow[\substack{Eq1-6Eq2\to Eq2 \\ -3Eq2+Eq3\to Eq3}]{}$$

$$\begin{cases} 6x+2y+z=28.7 \\ -4y+13z=28.7 \\ -5y+6z=0 \end{cases} \xrightarrow{-5Eq2+4Eq3\to Eq3}$$

$$\begin{cases} 6x+2y+z=28.7 \\ -4y+13z=28.7 \\ -41z=-143.5 \end{cases} \xrightarrow{-\frac{1}{41}Eq3\to Eq3}$$

$$\begin{cases} 6x+2y+z=28.7 \\ -4y+13z=28.7 \\ z=3.5 \end{cases}$$

Substituting to find y Substituting to find x

$-4y+13(3.5)=28.7$ $6x+2(4.2)+(3.5)=28.7$

$-4y+45.5=28.7$ $6x+8.4+3.5=28.7$

$-4y=-16.8$ $6x=16.8$

$y=4.2$ $x=2.8$

Each dosage of medication A contains 2.8 mg, each dosage of medication B contains 4.2 mg, and each dosage of medication C contains 3.5 mg.

37. Let x represent the amount invested in property I (12%), y represent the amount invested in property II (15%), and z represent the amount invested in property III (10%).

$$\begin{cases} x+y+z=750,000 \\ 0.12x+0.15y+0.10z=89,500 \\ z=\frac{1}{2}(x+y) \end{cases} \text{ or } \begin{cases} x+y+z=750,000 \\ 0.12x+0.15y+0.10z=89,500 \\ x+y-2z=0 \end{cases}$$

or

$$\begin{bmatrix} 1 & 1 & 1 & 750,000 \\ 0.12 & 0.15 & 0.10 & 89,500 \\ 1 & 1 & -2 & 0 \end{bmatrix}$$

Using technology to solve the augmented matrix yields

```
rref([A])▶Frac
[[1 0 0 350000]
 [0 1 0 150000]
 [0 0 1 250000]]
```

To generate an annual return of $89,500, $350,000 needs to be invested in property I, $150,000 needs to be invested in property II, and $250,000 needs to be invested property III.

38. Let $x =$ the amount in 12% fund, $y =$ the amount in the 16% fund, and $z =$ the amount in the 8% fund.

$$\begin{cases} x + y + z = 360,000 \\ 0.12x + 0.16y + 0.08z = 35,200 \\ z = 2(x + y) \end{cases} \quad \text{or} \quad \begin{cases} x + y + z = 360,000 \\ 0.12x + 0.16y + 0.08z = 35,200 \\ 2x + 2y - z = 0 \end{cases}$$

or

$$\begin{bmatrix} 1 & 1 & 1 & | & 360,000 \\ 0.12 & 0.16 & 0.08 & | & 35,200 \\ 2 & 2 & -1 & | & 0 \end{bmatrix}$$

Using technology to solve the augmented matrix yields

```
rref([A])▶Frac
[[1 0 0 80000 ]
 [0 1 0 40000 ]
 [0 0 1 240000]]
```

$80,000 is invested in the 12% fund, $40,000 is invested in the 16% fund, and $240,000 is invested in the 8% fund.

39. Let x represent the cost of property I, y represent the cost of property II, and z represent the cost of property III.

$$\begin{cases} x + y + z = 1,180,000 \\ x = y + 75,000 \\ z = 3(x + y) \end{cases} \quad \text{or} \quad \begin{cases} x + y + z = 1,180,000 \\ x - y = 75,000 \\ 3x + 3y - z = 0 \end{cases}$$

or

$$\begin{bmatrix} 1 & 1 & 1 & | & 1,180,000 \\ 1 & -1 & 0 & | & 75,000 \\ 3 & 3 & -1 & | & 0 \end{bmatrix}$$

Using technology to solve the augmented matrix yields

```
rref([A])▶Frac
[[1 0 0 185000]
 [0 1 0 110000]
 [0 0 1 885000]]
```

Property I costs $185,000, property II costs $110,000, and property III costs $885,000.

40. Let x = the number of Portfolio I units, y = the number of Portfolio II units, and z = the number of Portfolio III units.

$$\begin{cases}10x+12y+10z=290 \\ 2x+8y+4z=138 \\ 3x+5y+8z=161\end{cases} \quad \text{or} \quad \begin{bmatrix}10 & 12 & 10 & | & 290 \\ 2 & 8 & 4 & | & 138 \\ 3 & 5 & 8 & | & 161\end{bmatrix}$$

Using the calculator to generate the reduced row-echelon form of the matrix yields

```
rref([A])
[[1 0 0 5 ]
 [0 1 0 10]
 [0 0 1 12]]
```

The solution to the system is $x=5$, $y=10$, $z=12$. The client needs to purchase 5 units of Portfolio I, 10 units of Portfolio II, and 12 units of Portfolio III to achieve the investment objectives.

41. $\begin{cases}x+y+z=375,000 \\ x=y+50,000 \\ z=\frac{1}{2}(x+y)\end{cases}$

Note that $x=y+50,000$ implies $x-y=50,000$. Likewise $z=\frac{1}{2}(x+y)$ implies that

$$2z=2\left(\frac{1}{2}(x+y)\right)$$
$$2z=x+y$$
$$x+y-2z=0.$$

Therefore, the system can be written as follows:

$$\begin{cases}x+y+z=375,000 \\ x-y=50,000 \\ x+y-2z=0\end{cases}$$

Writing the system as an augmented matrix yields

$$\begin{bmatrix} 1 & 1 & 1 & | & 375{,}000 \\ 1 & -1 & 0 & | & 50{,}000 \\ 1 & 1 & -2 & | & 0 \end{bmatrix}$$

Applying technology to produce the reduced row-echelon form of the matrix yields

```
rref([A])
[[1 0 0 150000]
 [0 1 0 100000]
 [0 0 1 125000]]
```

Therefore, $x = 150{,}000$, $y = 100{,}000$, and $z = 125{,}000$. The first property costs $150,000, the second property costs $100,000, and the third property costs $125,000.

42. Let x = grams of Food I, y = grams of Food II, and z = grams of Food III.

$$\begin{cases} 10\%x + 11\%y + 18\%z = 12.5 \\ 12\%x + 9\%y + 10\%z = 9.1 \\ 14\%x + 12\%y + 8\%z = 9.6 \end{cases} \quad \text{or} \quad \begin{cases} 0.10x + 0.11y + 0.18z = 12.5 \\ 0.12x + 0.09y + 0.10z = 9.1 \\ 0.14x + 0.12y + 0.08z = 9.6 \end{cases}$$

Converting the system to an augmented matrix yields

$$\begin{bmatrix} 0.10 & 0.11 & 0.18 & | & 12.5 \\ 0.12 & 0.09 & 0.10 & | & 9.1 \\ 0.14 & 0.12 & 0.08 & | & 9.6 \end{bmatrix}$$

Using the calculator to generate the reduced row-echelon form of the matrix yields

```
rref([A])
[[1 0 0 20]
 [0 1 0 30]
 [0 0 1 40]]
```

The solution to the system is $x = 20$, $y = 30$, $z = 40$. The nutritionist recommends 20 grams of Food I, 30 grams of Food II, and 40 grams of Food III.

43. Let x = the number of passenger aircraft, y = the number of transport aircraft, and z = the number of jumbo aircraft.

$$\begin{cases} 200x + 200y + 200z = 2200 \\ 300x + 40y + 700z = 3860 \\ 40x + 130y + 70z = 920 \end{cases}$$

Converting the system to an augmented matrix yields

$$\begin{bmatrix} 200 & 200 & 200 & | & 2200 \\ 300 & 40 & 700 & | & 3860 \\ 40 & 130 & 70 & | & 920 \end{bmatrix}$$

Using the calculator to generate the reduced row-echelon form of the matrix yields

```
rref([A])
      [[1 0 0 3]
       [0 1 0 4]
       [0 0 1 4]]
```

The solution to the system is $x = 3$, $y = 4$, $z = 4$. The delivery service needs to schedule 3 passenger planes, 4 transport planes, and 4 jumbo planes.

44. Let $x =$ amount invested in the tech fund, $y =$ amount invested in the balanced fund, and $z =$ amount invested in the utility fund.

$$\begin{cases} 180x + 210y + 120z = 210,000 \\ 18x + 42y + 18z = 12\%(210,000) \end{cases} \quad \text{or} \quad \begin{cases} 180x + 210y + 120z = 210,000 \\ 18x + 42y + 18z = 0.12(210,000) \end{cases}$$

Converting the system to an augmented matrix yields

$$\begin{bmatrix} 180 & 210 & 120 & | & 210,000 \\ 18 & 42 & 18 & | & 25,200 \end{bmatrix}$$

Note that the system has more variables than equations. The system is dependent. Using the calculator to generate the reduced row-echelon form of the matrix yields

```
rref([A])▶Frac
...1 0 1/3 2800/3...
...0 1 2/7 200    ...
```

Let $z = z$.

$$y + \frac{2}{7}z = 200$$

$$y = 200 - \frac{2}{7}z$$

and

$$x + \frac{1}{3}z = \frac{2800}{3}$$

$$x = \frac{2800}{3} - \frac{1}{3}z$$

$$x = \frac{2800 - z}{3}$$

The solution is $x = \dfrac{2800 - z}{3}$, $y = 200 - \dfrac{2}{7}z$, $z = z$.

Since y must be greater than or equal to zero, z must not exceed 700. Therefore, $0 \le z \le 700$.

45. Let x = the number of units of Product A, y = the number of units of Product B, and
z = the number of units of Product C.

$$\begin{cases} 25x + 30y + 40z = 9260 \\ 30x + 36y + 60z = 12{,}000 \\ 150x + 180y + 200z = 52{,}600 \end{cases}$$

Converting the system to an augmented matrix yields

$$\begin{bmatrix} 25 & 30 & 40 & | & 9260 \\ 30 & 36 & 60 & | & 12{,}000 \\ 150 & 180 & 200 & | & 52{,}600 \end{bmatrix}$$

Using the calculator to generate the reduced row-echelon form of the matrix yields

```
rref([A])
  [[1 1.2 0 252]
   [0 0   1 74 ]
   [0 0   0 0  ]]
```

Since the last row consists entirely of zeros with zero as the augment, the system is dependent and has infinitely many solutions.

$z = 74$

Let $y = y$.

$x + 1.2y = 252$

$x = 252 - 1.2y$

The solution is $x = 252 - 1.2y$, $y = y$, $z = 74$.

Note that $x \geq 0$. Furthermore, $252 - 1.2y \geq 0$.

$-1.2y \geq -252$

$y \leq \dfrac{-252}{-1.2}$

$y \leq 210$

Therefore, $0 \leq y \leq 210$.

46. Let $x =$ the number of type A slugs, $y =$ the number of type B slugs, and $z =$ the number of type C slugs.

$$\begin{cases} 2x + 2y + 4z = 4000 & (\text{nutrient I}) \\ 6x + 8y + 20z = 16{,}000 & (\text{nutrient II}) \\ 2x + 4y + 12z = 8000 & (\text{nutrient III}) \end{cases}$$

Converting the system to an augmented matrix yields

$$\begin{bmatrix} 2 & 2 & 4 & 4000 \\ 6 & 8 & 20 & 16{,}000 \\ 2 & 4 & 12 & 8000 \end{bmatrix}$$

Using the calculator to generate the reduced row-echelon form of the matrix yields

```
rref([A])
  [[1  0  -2  0    ]
   [0  1  4   2000]
   [0  0  0   0    ]]
```

Since the last row consists entirely of zeros with zero as the augment, the system is dependent and has infinitely many solutions.

Let $z = z$

$y + 4z = 2000$

$y = 2000 - 4z$

and

$x - 2z = 0$

$x = 2z$

The solution is $x = 2z$, $y = 2000 - 4z$, $z = z$.

Note that $y \geq 0$. Furthermore, $2000 - 4z \geq 0$.

$-4z \geq -2000$

$z \leq \dfrac{-2000}{-4}$

$z \leq 500$

Therefore, $0 \leq z \leq 500$.

47. a.

$$A = \begin{bmatrix} 240,326 & 312,032 \\ 226,079 & 301,610 \\ 215,907 & 292,651 \\ 198,289 & 281,292 \\ 163,665 & 249,257 \\ 128,892 & 204,658 \end{bmatrix}$$

b.

$$B = \begin{bmatrix} 294,158 & 346,063 \\ 280,529 & 332,553 \\ 277,594 & 324,264 \\ 262,874 & 315,325 \\ 229,986 & 277,637 \\ 176,654 & 226,248 \end{bmatrix}$$

c.

Trade balance = Exports − Imports

$$A - B = \begin{bmatrix} 240,326 & 312,032 \\ 226,079 & 301,610 \\ 215,907 & 292,651 \\ 198,289 & 281,292 \\ 163,665 & 249,257 \\ 128,892 & 204,658 \end{bmatrix} - \begin{bmatrix} 294,158 & 346,063 \\ 280,529 & 332,553 \\ 277,594 & 324,264 \\ 262,874 & 315,325 \\ 229,986 & 277,637 \\ 176,654 & 226,248 \end{bmatrix}$$

$$= \begin{bmatrix} 240,326 - 294,158 & 312,032 - 346,063 \\ 226,079 - 280,529 & 301,610 - 332,553 \\ 215,907 - 277,594 & 292,651 - 324,264 \\ 198,289 - 262,874 & 281,292 - 315,325 \\ 163,665 - 229,986 & 249,257 - 277,637 \\ 128,892 - 176,654 & 204,658 - 226,248 \end{bmatrix}$$

$$= \begin{bmatrix} -53,832 & -34,031 \\ -54,450 & -30,943 \\ -61,687 & -31,613 \\ -64,585 & -34,033 \\ -66,321 & -28,380 \\ -47,762 & -21,590 \end{bmatrix}$$

d.   No. The trend of the trade balances is getting worse for the U.S. over time.

e.   In 2014, the trade balance is worse with Mexico.

48. a. Let x = the amount invested at 6%,
y = the amount invested at 7%, and
z = the amount invested at 10%.

$$\begin{cases} x + y + z = 800,000 \\ 6\%x + 7\%y + 10\%z = 71,600 \end{cases}$$

or

$$\begin{cases} x + y + z = 800,000 \\ 0.06x + 0.07y + 0.10z = 71,600 \end{cases}$$

b.
$$\begin{bmatrix} 1 & 1 & 1 & | & 800,000 \\ 0.06 & 0.07 & 0.10 & | & 71,600 \end{bmatrix}$$

Using the calculator to generate the reduced row-echelon form of the matrix yields:

$$\begin{bmatrix} 1 & 0 & -3 & | & -1,560,000 \\ 0 & 1 & 4 & | & 2,360,000 \end{bmatrix}$$

Let $z = z$.

$y + 4z = 2,360,000$

$y = 2,360,000 - 4z$

and

$x - 3z = -1,560,000$

$x = 3z - 1,560,000$

The solution is

$x = 3z - 1,560,000,$

$y = 2,360,000 - 4z,$

$z = z$

with all nonnegative values
and $520,000 \le z \le 590,000$.

c. Let $y = 0$ and solve for z.

$y = 2,360,000 - 4z$

$0 = 2,360,000 - 4z$

$-2,360,000 = -4z$

$590,000 = z$

Let $z = 590,000$ and solve for x.

$x = 3z - 1,560,000$

$x = 3(590,000) - 1,560,000$

$x = 1,770,000 - 1,560,000$

$x = 210,000$

Therefore, the investor should invest $210,000 at 6%, $0 at 7%, and $590,000 at 10%.

d. Let $x = 0$ and solve for z.

$x = 3z - 1,560,000$

$0 = 3z - 1,560,000$

$1,560,000 = 3z$

$520,000 = z$

Let $z = 520,000$ and solve for y.

$y = 2,360,000 - 4z$

$y = 2,360,000 - 4(520,000)$

$y = 2,360,000 - 2,080,000$

$y = 280,000$

Therefore, the investor should invest $0 at 6%, $280,000 at 7%, and $520,000 at 10%.

49. a. $y = 0.586x + 450.086$

[0, 220] by [300, 600]

b. $y = 0.578x + 464.216$

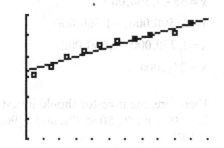

[0, 220] by [300, 600]

c. The rate of change of critical reading SAT scores as a function of family income is an increase of 0.586 points per $1000 increase in family income, while the rate of change for math SAT scores is 0.578 points per $1000 increase in family income.

50. a. $x + y = 10$

b. $20\%x + 5\%y = 15.5\%(10)$

or

$0.20x + 0.05y = 1.55$

c. Applying the substitution method

$x + y = 10$

$x = 10 - y$

Substituting into the other equation

$0.20(10 - y) + 0.05y = 1.55$

$2 - 0.20y + 0.05y = 1.55$

$-0.15y + 2 = 1.55$

$-0.15y = -0.45$

$y = \dfrac{-0.45}{-0.15} = 3$

Substituting to find x

$x + y = 10$

$x + 3 = 10$

$x = 7$

To administer the desired concentration requires 7 cc of the 20% medication and 3 cc of the 5% medication.

51. $\begin{cases} p = q + 578 \\ p = 396 + q^2 \end{cases}$

$q + 578 = 396 + q^2$

$q^2 - q - 182 = 0$

$(q - 14)(q + 13) = 0$

$q = 14, q = -13$

Since $q \geq 0$ in the context of the question, then $q = 14$.

Substituting to calculate p

$p = 14 + 578$

$p = 592$

Equilibrium occurs when 14 units are produced and sold at a price of $592 per unit.

52. $C(x) = R(x)$

$2500x + x^2 + 27,540 = 3899x - 0.1x^2$

$1.1x^2 - 1399x + 27,540 = 0$

Solving graphically by applying the x-intercept method

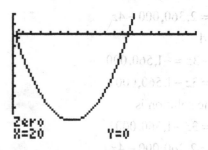

[0, 2000] by [−500,000, 100,000]

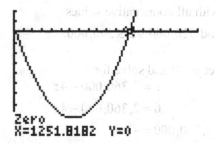

[0, 2000] by [−500,000, 100,000]

Break-even occurs when 20 units are produced and sold or when approximately 1252 units are produced and sold.

Group Activity/Extended Application I

1. $A = \begin{bmatrix} 108,481 & 138,921 & 121,312 \\ 82,834 & 93,020 & 84,493 \\ 78,390 & 91,563 & 69,969 \end{bmatrix}$

$B = \begin{bmatrix} 98,552 & 127,542 & 109,720 \\ 79,493 & 86,476 & 78,355 \\ 73,372 & 87,570 & 66,427 \end{bmatrix}$

2. $B - A$

$= \begin{bmatrix} -9,929 & -11,379 & -11,592 \\ -3,341 & -6,544 & -6,138 \\ -5,018 & -3,993 & -3,542 \end{bmatrix}$

3. Based on the matrix in part 2, at all institutions female professors are paid less than male professors. Gender bias seems to exist at the institutions.

4.

[20,000, 150,000] by [20,000, 130,000]

5. $y = 0.8637x + 6325.335$

6.

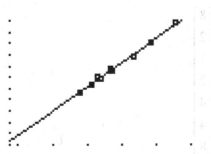

[20,000, 150,000] by [20,000, 130,000]

7.

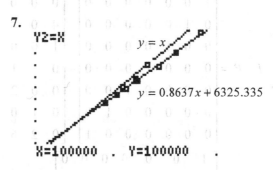

[20,000, 150,000] by [20,000, 130,000]

8. Based on the graph in part 7, as salaries increase the gap between male and female salaries also increases.

9. Using the intersection of graphs method, it appears that the salary level at which male salaries become greater than female salaries is at $46,422.49.

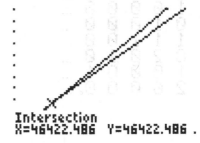

[20,000, 150,000] by [20,000, 130,000]

Extended Application II

1. $D = \begin{bmatrix} 8 \\ 2 \\ 1 \\ 2 \\ 4 \\ 4 \\ 8 \end{bmatrix}$

2. $I - P = \begin{bmatrix} 1 & 0 & 0 & 0 & 0 & 0 & 0 \\ 0 & 1 & 0 & 0 & 0 & 0 & 0 \\ 0 & 0 & 1 & 0 & 0 & 0 & 0 \\ 0 & 0 & 0 & 1 & 0 & 0 & 0 \\ 0 & 0 & 0 & 0 & 1 & 0 & 0 \\ 0 & 0 & 0 & 0 & 0 & 1 & 0 \\ 0 & 0 & 0 & 0 & 0 & 0 & 1 \end{bmatrix} - \begin{bmatrix} 0 & 0 & 0 & 0 & 0 & 0 & 0 \\ 4 & 0 & 0 & 0 & 0 & 0 & 0 \\ 1 & 0 & 0 & 0 & 0 & 0 & 0 \\ 0 & 1 & 1 & 0 & 0 & 0 & 0 \\ 0 & 0 & 2 & 0 & 0 & 0 & 0 \\ 0 & 1 & 0 & 0 & 0 & 0 & 0 \\ 0 & 2 & 6 & 0 & 0 & 0 & 0 \end{bmatrix}$

$= \begin{bmatrix} 1 & 0 & 0 & 0 & 0 & 0 & 0 \\ -4 & 1 & 0 & 0 & 0 & 0 & 0 \\ -1 & 0 & 1 & 0 & 0 & 0 & 0 \\ 0 & -1 & -1 & 1 & 0 & 0 & 0 \\ 0 & 0 & -2 & 0 & 1 & 0 & 0 \\ 0 & -1 & 0 & 0 & 0 & 1 & 0 \\ 0 & -2 & -6 & 0 & 0 & 0 & 1 \end{bmatrix}$

3. Applying technology to find the inverse of $I - P$:

```
[ [1    0  0  0  0  0  ...
  [4    1  0  0  0  0  ...
  [1    0  1  0  0  0  ...
  [5    1  1  1  0  0  ...
  [2    0  2  0  1  0  ...
  [4    1  0  0  0  1  ...
  [14   2  6  0  0  0  ...
```

```
... 0 0 0 0 0 0]
... 1 0 0 0 0 0]
... 0 1 0 0 0 0]
... 1 1 1 0 0 0]
... 0 2 0 1 0 0]
... 1 0 0 0 1 0]
...4 2 6 0 0 0 1]]
```

Note that the inverse matrix is shown in two graphics with several repeated columns. The inverse matrix is

$$(I-P)^{-1} = \begin{bmatrix} 1 & 0 & 0 & 0 & 0 & 0 & 0 \\ 4 & 1 & 0 & 0 & 0 & 0 & 0 \\ 1 & 0 & 1 & 0 & 0 & 0 & 0 \\ 5 & 1 & 1 & 1 & 0 & 0 & 0 \\ 2 & 0 & 2 & 0 & 1 & 0 & 0 \\ 4 & 1 & 0 & 0 & 0 & 1 & 0 \\ 14 & 2 & 6 & 0 & 0 & 0 & 1 \end{bmatrix}$$

4. $(I-P)X = D$

$(I-P)^{-1}(I-P)X = (I-P)^{-1}D$

$$X = \begin{bmatrix} 1 & 0 & 0 & 0 & 0 & 0 & 0 \\ 4 & 1 & 0 & 0 & 0 & 0 & 0 \\ 1 & 0 & 1 & 0 & 0 & 0 & 0 \\ 5 & 1 & 1 & 1 & 0 & 0 & 0 \\ 2 & 0 & 2 & 0 & 1 & 0 & 0 \\ 4 & 1 & 0 & 0 & 0 & 1 & 0 \\ 14 & 2 & 6 & 0 & 0 & 0 & 1 \end{bmatrix} \begin{bmatrix} 8 \\ 2 \\ 1 \\ 2 \\ 4 \\ 4 \\ 8 \end{bmatrix}$$

Applying technology to carry out the multiplication yields

$$X = \begin{bmatrix} 8 \\ 34 \\ 9 \\ 45 \\ 22 \\ 38 \\ 130 \end{bmatrix}$$

5. Based on the calculations in part 4, the required number of pipes is 45, the required number of clamps is 22, the required number of braces is 38, and the required number of bolts is 130.

Chapter 8 - Special Topics in Algebra

Section 8.1 Skills Check

1. $y \le 5x - 4$

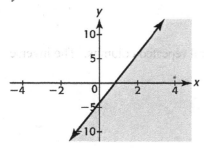

Test (0, 0).

$0 \le 5(0) - 4$

$0 \le -4$

Since the statement is false, the region not containing (0, 0) is the solution to the inequality. Note that the line is solid because of the "equal to" in the given inequality.

3. $6x - 3y \ge 12$

$-3y \ge -6x + 12$

$\dfrac{-3y}{-3} \le \dfrac{-6x + 12}{-3}$

$y \le 2x - 4$

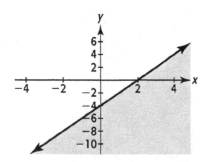

Test (0, 0).

$0 \le 2(0) - 4$

$0 \le -4$

Since the statement is false, the region not containing (0, 0) is the solution to the inequality. Note that the line is solid because of the "equal to" in the given inequality.

5. $4x + 5y \le 20$

$5y \le 20 - 4x$

$y \le \dfrac{20 - 4x}{5}$

$y \le 4 - \dfrac{4}{5}x$

Test (0, 0).

$y \le 4 - \dfrac{4}{5}x$

$0 \le 4 - \dfrac{4}{5}(0)$

$0 \le 4$

Since the statement is true, the region containing (0, 0) is the solution to the inequality. Note that the line is solid because of the "equal to" in the given inequality.

7. The answer is C since the slopes of both lines are positive and the first line has a y-intercept which is negative. Also there is no restriction to Quadrant I.

9. The answer is D since the slopes of both lines will be negative, with y-intercepts of +5 and +2. Quadrant I restriction, and test point (0, 0) works in neither equation.

11. To determine the solution region and the corners of the solution region, pick a point in a potential solution region and test it. For example, pick $(2,2)$.

$$x + y \leq 5$$
$$2 + 2 \leq 5$$
$$4 \leq 5$$

  True statement

$$2x + y \leq 8$$
$$2(2) + 2 \leq 8$$
$$4 + 2 \leq 8$$
$$6 \leq 8$$

  True statement

$$x \geq 0$$
$$2 \geq 0$$

  True statement

$$y \geq 0$$
$$2 \geq 0$$

  True statement

Since all the inequalities are true at the point $(2,2)$, the region that contains $(2,2)$ is the solution region. The corners of the region are $(0,0),(4,0),(3,2),(0,5)$.

13. To determine the solution region and the corners of the solution region, pick a point in a potential solution region and test it. Pick $(2,4)$.

$$4x + 2y > 8$$
$$4(2) + 2(4) > 8$$
$$16 > 8$$

True statement

$$3x + y > 5$$
$$3(2) + 4 > 5$$
$$10 > 5$$

True statement

$$x \geq 0$$
$$2 \geq 0$$

True statement

$$y \geq 0$$
$$4 \geq 0$$

True statement

Since all the inequalities are true at the point $(2,4)$, the region that contains $(2,4)$ is the solution region. The corners of the region are $(2,0),(0,5),(1,2)$.

15. To determine the solution region and the corners of the solution region, pick a point in a potential solution region and test it. Pick $(3,3)$.

$$2x + 6y \geq 12$$
$$2(3) + 6(3) \geq 12$$
$$6 + 18 \geq 12$$
$$24 \geq 12$$

True statement

$$3x + y \geq 5$$
$$3(3) + 3 \geq 5$$
$$12 \geq 5$$

True statement

$$x + 2y \geq 5$$
$$3 + 2(3) \geq 5$$
$$9 \geq 5$$

True statement

$$x \geq 0$$
$$3 \geq 0$$

True statement

$$y \geq 0$$
$$3 \geq 0$$

True statement

Since all the inequalities are true at the point $(3,3)$, the region that contains $(3,3)$ is the solution region. The corners of the region are $(0,5),(1,2),(3,1),(6,0)$.

17. The graph of the system is

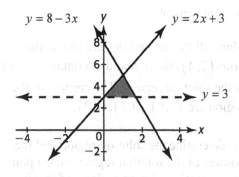

Note that $y=3$ is a dashed line. The other two lines are solid. The intersection point between $y=8-3x$ and $y=2x+3$ is $(1,5)$. The intersection point between $y=3$ and $y=2x+3$ is $(0,3)$.

The intersection point between $y=3$ and $y=8-3x$ is $\left(\frac{5}{3},3\right)$.

To determine the solution region, pick a point to test. Pick $(1,4)$.

$y \le 8-3x$

$4 \le 8-3(1)$

$4 \le 5$

True statement

$y \le 2x+3$

$4 \le 2(1)+3$

$4 \le 5$

True statement

$y > 3$

$4 > 3$

True statement

Since all the inequalities that form the system are true at the point $(1,4)$, the region that contains $(1,4)$ is the solution region. The corners of the region are

$(0,3),(1,5),\left(\frac{5}{3},3\right)$.

19. Rewriting the inequalities:
$2x+y<5$

$y<-2x+5$

and

$2x-y>-1$

$-y>-1-2x$

$\dfrac{-y}{-1}<\dfrac{-1-2x}{-1}$

$y<2x+1$

The new system is
$$\begin{cases} y<-2x+5 \\ y<2x+1 \\ x \ge 0, y \ge 0 \end{cases}$$

The graph of the system is

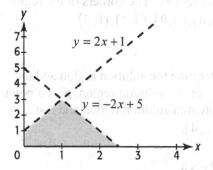

Note that both lines are dashed because there is no "equal to" in either first two inequality statements. The solution is also restricted to Quadrant I. The intersection point between the two lines is $(1,3)$.

To determine the solution region, pick a point to test. Pick $(1, 2)$.

$y < -2x + 5$

$2 < -2(1) + 5$

$2 < 3$

True statement

$y < 2x + 1$

$2 < 2(1) + 1$

$2 < 3$

True statement

$x \geq 0$

$1 \geq 0$

True statement

$y \geq 0$

$2 \geq 0$

True statement

Since all the inequalities that form the system are true at the point $(1, 2)$, the region that contains $(1, 2)$ is the solution region. Considering the graph, the corners of the region are $(0, 0), (0, 1), (2.5, 0), (1, 3)$.

21. Rewriting the system:

$x + 2y \geq 4$

$2y \geq 4 - x$

$y \geq \dfrac{4 - x}{2}$

and

$x + y \leq 5$

$y \leq 5 - x$

and

$2x + y \leq 8$

$y \leq 8 - 2x$

The new system is

$$\begin{cases} y \geq \dfrac{4 - x}{2} \\ y \leq 5 - x \\ y \leq 8 - 2x \\ x \geq 0, y \geq 0 \end{cases}$$

The graph of the system is

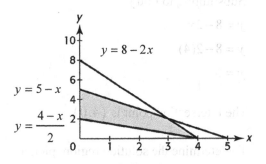

Note that all the lines are solid, and Quadrant I restriction.

Determining the intersection point between $y = 8 - 2x$ and $y = 5 - x$:

$8 - 2x = 5 - x$

$-x = -3$

$x = 3$

Substituting to find y

$y = 5 - x$

$y = 5 - 3$

$y = 2$

The intersection point is $(3, 2)$.

Determining the intersection point between $y = 8 - 2x$ and $y = \dfrac{4 - x}{2}$:

$$8 - 2x = \frac{4 - x}{2}$$

$$2(8 - 2x) = 2\left(\frac{4 - x}{2}\right)$$

$$16 - 4x = 4 - x$$

$$-3x = -12$$

$$x = 4$$

Substituting to find y

$$y = 8 - 2x$$

$$y = 8 - 2(4)$$

$$y = 0$$

The intersection point is $(4, 0)$.

To determine the solution region, pick a point to test. Pick $(1, 3)$.

$$y \geq \frac{4 - x}{2}$$

$$3 \geq \frac{4 - 1}{2}$$

$$3 \geq \frac{3}{2}$$

True statement

$$y \leq 5 - x$$

$$3 \leq 5 - 1$$

$$3 \leq 4$$

True statement

$$y \leq 8 - 2x$$

$$3 \leq 8 - 2(1)$$

$$3 \leq 6$$

True statement

$$x \geq 0$$

$$1 \geq 0$$

True statement

$$y \geq 0$$

$$3 \geq 0$$

True statement

Since all the inequalities that form the system are true at the point $(1, 3)$, the region that contains $(1, 3)$ is the solution region. Considering the graph of the system, the corners of the region are $(0, 2), (0, 5), (3, 2), (4, 0)$.

23. Rewriting the system:

$$x^2 - 3y < 4$$

$$-3y < 4 - x^2$$

$$y > \frac{4 - x^2}{-3}$$

$$y > \frac{x^2 - 4}{3}$$

and

$$2x + 3y < 4$$

$$3y < 4 - 2x$$

$$y < \frac{4 - 2x}{3}$$

The new system is

$$\begin{cases} y > \dfrac{x^2 - 4}{3} \\ y < \dfrac{4 - 2x}{3} \end{cases}$$

The graph of the system is

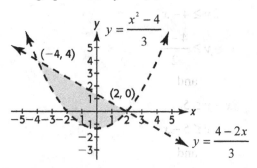

Note that all the lines are dashed.

Determining the intersection points between the two functions graphically results in intersection points of $(-4, 4)$ and $(2, 0)$.

To determine the solution region, pick a point to test. Pick $(0,0)$.

$$y > \frac{x^2 - 4}{3}$$

$$0 > \frac{(0)^2 - 4}{3}$$

$$0 > -\frac{4}{3}$$

True statement

$$y < \frac{4 - 2x}{3}$$

$$0 < \frac{4 - 2(0)}{3}$$

$$0 < \frac{4}{3}$$

True statement

Since all the inequalities that form the system are true at the point $(0,0)$, the region that contains $(0,0)$ is the solution region. Considering the graph of the system, the corners of the region are $(-4,4),(2,0)$.

25. Rewriting the system:

$$x^2 - y - 8x \le -6$$
$$-y \le -x^2 + 8x - 6$$
$$y \ge x^2 - 8x + 6$$

and

$$y + 9x \le 18$$
$$y \le 18 - 9x$$

The new system is
$$\begin{cases} y \ge x^2 - 8x + 6 \\ y \le 18 - 9x \end{cases}$$

The graph of the system is

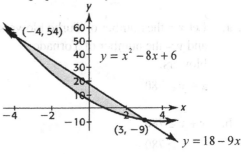

Note that all the lines are solid.

Determining the intersection points between the two functions graphically results in intersection points of $(3,-9)$ and $(-4,54)$.

To determine the solution region, pick a point to test. Pick $(0,10)$.

$$y \ge x^2 - 8x + 6$$
$$10 \ge (0)^2 - 8(0) + 6$$
$$10 \ge 6$$

True statement

$$y \le 18 - 9x$$
$$10 \le 18 - 9(0)$$
$$10 \le 18$$

True statement

Since all the inequalities that form the system are true at the point $(0,10)$, the region that contains $(0,10)$ is the solution region. Considering the graph of the system, the corners of the region are $(3,-9),(-4,54)$.

Section 8.1 Exercises

27. a. Let x = the number of Turbo blowers, and y = the number of Tornado blowers.

$x + y \geq 780$

b. $x + y \geq 780$

$y \geq 780 - x$

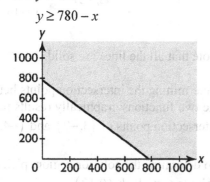

Note that the line is solid since an "equal to" is part of the inequality.

29. a. Let x = minutes of cable television time, and y = minutes of radio time.

$240x + 150y \leq 36,000$

b. $240x + 150y \leq 36,000$

$150y \leq 36,000 - 240x$

$y \leq \dfrac{36,000 - 240x}{150}$

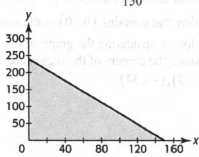

Note that the line is solid since an "equal to" is part of the inequality.

31. a. Let x = minutes of television time, and y = minutes of radio time. Then, $0.12x$ represents the number, in millions, of eligible voters reached by television advertising, and $0.009y$ represents the number, in millions, of eligible voters reached by radio advertising. The corresponding inequalities are

$$\begin{cases} x + y \geq 100 \\ 0.12x + 0.009y \geq 7.56 \\ x \geq 0, y \geq 0 \end{cases}$$

b. Rewriting the system:

$x + y \geq 100$

$y \geq 100 - x$

and

$0.12x + 0.009y \geq 7.56$

$0.009y \geq 7.56 - 0.12x$

$y \geq \dfrac{7.56 - 0.12x}{0.009}$

The new system is

$$\begin{cases} y \geq 100 - x \\ y \geq \dfrac{7.56 - 0.12x}{0.009} \\ x \geq 0, y \geq 0 \end{cases}$$

The graph of the system is

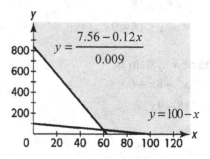

The intersection point between the two lines is $(60,40)$. Both lines are solid since there is an "equal to" in both inequalities.

To determine the solution region, pick a point to test. Pick $(60,100)$.

$x + y \geq 100$

$60 + 100 \geq 100$

$160 \geq 100$

True statement

$0.12x + 0.009y \geq 7.56$

$0.12(60) + 0.009(100) \geq 7.56$

$7.2 + 0.9 \geq 7.56$

$8.1 \geq 7.56$

True statement

$x \geq 0$

$60 \geq 0$

True statement

$y \geq 0$

$100 \geq 0$

True statement

Since all the inequalities that form the system are true at the point $(60,100)$, the region that contains $(60,100)$ is the solution region.

One of the corner points is the intersection point of the two lines, $(60,40)$. Another corner point occurs where $y = \dfrac{7.56 - 0.12x}{0.009}$ crosses the y-axis. A third corner point occurs where $y = 100 - x$ crosses the x-axis.

To find the y-intercept, let $x = 0$.

$y = \dfrac{7.56 - 0.12(0)}{0.009}$

$y = \dfrac{7.56}{0.009}$

$y = 840$

$(0,840)$

To find the x-intercept, let $y = 0$.

$0 = 100 - x$

$x = 100$

$(100,0)$

Therefore, the corner points of the solution region are $(60,40),(100,0),(0,840)$.

33. The system of inequalities is

$$\begin{cases} x + y \geq 780 \\ 78x + 117y \leq 76{,}050 \\ x \geq 0, y \geq 0 \end{cases}$$

Rewriting the system:

$x + y \geq 780$

$y \geq 780 - x$

and

$78x + 117y \leq 76{,}050$

$117y \leq 76{,}050 - 78x$

$y \leq \dfrac{76{,}050 - 78x}{117}$

The new system is

$$\begin{cases} y \geq 780 - x \\ y \leq \dfrac{76{,}050 - 78x}{117} \\ x \geq 0, y \geq 0 \end{cases}$$

The graph of the system is

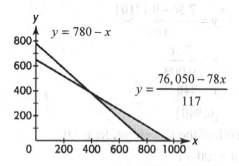

The intersection point between the two lines is $(390, 390)$. Both lines are solid since there is an "equal to" in both inequalities.

To determine the solution region, pick a point to test. Pick $(700, 100)$.

$x + y \geq 780$

$700 + 100 \geq 780$

$800 \geq 780$

True statement

$78x + 117y \leq 76,050$

$78(700) + 117(100) \leq 76,050$

$54,600 + 11,700 \leq 76,050$

$66,300 \leq 76,050$

True statement

$x \geq 0$

$700 \geq 0$

True statement

$y \geq 0$

$100 \geq 0$

True statement

Since all the inequalities that form the system are true at the point $(700, 100)$, the region that contains $(700, 100)$ is the solution region.

One of the corner points is the intersection point of the two lines, $(390, 390)$. A second corner point occurs where $y = 780 - x$ crosses the x-axis. A third corner point occurs where $y = \dfrac{76,050 - 78x}{117}$ crosses the x-axis.

To find the x-intercept, let $y = 0$.

$y = 780 - x$

$0 = 780 - x$

$x = 780$

$(780, 0)$

To find the x-intercept, let $y = 0$.

$0 = \dfrac{76,050 - 78x}{117}$

$0 = 76,050 - 78x$

$78x = 76,050$

$x = \dfrac{76,050}{78} = 975$

$(975, 0)$

Therefore, the corner points of the solution region are $(390, 390), (780, 0), (975, 0)$.

35. a. Let x = number of manufacturing days on assembly line 1, and y = number of manufacturing days on assembly line 2.

Then, the system of inequalities is
$$\begin{cases} 80x + 40y \geq 3200 \\ 20x + 20y \geq 1000 \\ 100x + 40y \geq 3400 \\ x \geq 0, y \geq 0 \end{cases}$$

Rewriting the system:

$$80x + 40y \geq 3200$$

$$40y \geq 3200 - 80x$$

$$y \geq \frac{3200 - 80x}{40}$$

$$y \geq 80 - 2x$$

and

$$20x + 20y \geq 1000$$

$$20y \geq 1000 - 20x$$

$$y \geq \frac{1000 - 20x}{20}$$

$$y \geq 50 - x$$

and

$$100x + 40y \geq 3400$$

$$40y \geq 3400 - 100x$$

$$y \geq \frac{3400 - 100x}{40}$$

$$y \geq 85 - 2.5x$$

The new system is

$$\begin{cases} y \geq 80 - 2x \\ y \geq 50 - x \\ y \geq 85 - 2.5x \\ x \geq 0, y \geq 0 \end{cases}$$

b. To solve the system, first consider the graph of the system

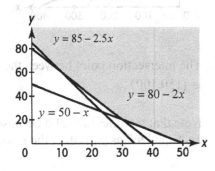

The intersection point between

$y = 80 - 2x$ and $y = 50 - x$ is $(30, 20)$.

The intersection point between

$y = 80 - 2x$ and $y = 85 - 2.5x$ is

$(10, 60)$.

The intersection point between

$y = 50 - x$ and $y = 85 - 2.5x$ is

$(23.\overline{3}, 26.\overline{6})$.

Note that all the lines are solid because there is an "equal to" as part of all the inequalities.

To determine the solution region, pick a point to test. Pick $(25, 40)$.

$$80x + 40y \geq 3200$$

$$80(25) + 40(40) \geq 3200$$

$$3600 \geq 3200$$

True statement

$$20x + 20y \geq 1000$$

$$20(25) + 20(40) \geq 1000$$

$$500 + 800 \geq 1000$$

$$1300 \geq 1000$$

True statement

$$100x + 40y \geq 3400$$

$$100(25) + 40(40) \geq 3400$$

$$4100 \geq 3400$$

True statement

$$x \geq 0$$

$$25 \geq 0$$

True statement

$$y \geq 0$$

$$40 \geq 0$$

True statement

Since all the inequalities that form the system are true at the point $(25, 40)$, the region that contains $(25, 40)$ is the solution region.

One of the corner points is the intersection point between $y = 80 - 2x$ and $y = 85 - 2.5x$, which is $(10, 60)$. A second corner point is the intersection point between $y = 80 - 2x$ and $y = 50 - x$, which is $(30, 20)$. A third corner point occurs where $y = 50 - x$ crosses the x-axis. A fourth corner point occurs where $y = 85 - 2.5x$ crosses the y-axis. Therefore, to find the x-intercept, let $y = 0$.

$0 = 50 - x$

$x = 50$

$(50, 0)$

To find the y-intercept, let $x = 0$.

$y = 85 - 2.5(0) = 85$

$(0, 85)$

Therefore, the corner points of the solution region are

$(10, 60), (30, 20), (50, 0), (0, 85)$.

37. Let x = the number of bass, and y = the number of trout.

Then, the system of inequalities is

$$\begin{cases} 4x + 10y \le 1600 \\ 6x + 7y \le 1600 \\ x \ge 0, y \ge 0 \end{cases}$$

Rewriting the system:

$4x + 10y \le 1600$

$10y \le 1600 - 4x$

$y \le \dfrac{1600 - 4x}{10}$

and

$6x + 7y \le 1600$

$7y \le 1600 - 6x$

$y \le \dfrac{1600 - 6x}{7}$

The new system is

$$\begin{cases} y \le \dfrac{1600 - 4x}{10} \\ y \le \dfrac{1600 - 6x}{7} \\ x \ge 0, y \ge 0 \end{cases}$$

To solve the system, first consider the graph of the system

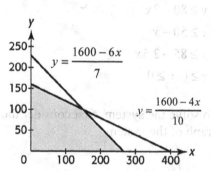

The intersection point between the two lines is $(150, 100)$.

Note that the lines are solid because there is an "equal to" as part of all the inequalities.

To determine the solution region, pick a point to test. Pick $(100, 75)$.

$$y \leq \frac{1600 - 4x}{10}$$

$$75 \leq \frac{1600 - 4(100)}{10}$$

$$75 \leq \frac{1200}{10}$$

$$75 \leq 120$$

True statement

$$y \leq \frac{1600 - 6x}{7}$$

$$75 \leq \frac{1600 - 6(100)}{7}$$

$$75 \leq \frac{1000}{7} \approx 142.857$$

$$75 \leq 142.857$$

True statement

$$x \geq 0$$

$$100 \geq 0$$

True statement

$$y \geq 0$$

$$75 \geq 0$$

True statement

Since all the inequalities that form the system are true at the point $(100, 75)$, the region that contains $(100, 75)$ is the solution region.

One of the corner points is the intersection point between the two lines, which is $(150, 100)$. A second corner point occurs where $y = \frac{1600 - 4x}{10}$ crosses the y-axis. A third corner point occurs where $y = \frac{1600 - 6x}{7}$ crosses the x-axis.

To find the y-intercept, let $x = 0$.

$$y = \frac{1600 - 4(0)}{10}$$

$$y = \frac{1600}{10}$$

$$y = 160$$

$$(0, 160)$$

To find the x-intercept, let $y = 0$.

$$0 = \frac{1600 - 6x}{7}$$

$$0 = 1600 - 6x$$

$$6x = 1600$$

$$x = \frac{1600}{6} = 266.\overline{6}$$

$$(266.\overline{6}, 0)$$

A fourth corner point occurs at the origin, $(0, 0)$. Therefore, the corner points of the solution region are

$$(0, 0), (150, 100), (0, 160), (266.\overline{6}, 0).$$

39. a. Let x = number of Standard chairs, and let y = number of Deluxe chairs.

Then, the system of inequalities is

$$\begin{cases} 4x + 6y \leq 480 \\ 2x + 6y \leq 300 \\ x \geq 0, y \geq 0 \end{cases}$$

Rewriting the system:

$$4x + 6y \leq 480$$

$$6y \leq 480 - 4x$$

$$y \leq \frac{480 - 4x}{6}$$

and

$$2x + 6y \leq 300$$

$$6y \leq 300 - 2x$$

$$y \leq \frac{300 - 2x}{6}$$

The new system is

$$\begin{cases} y \leq \dfrac{480 - 4x}{6} \\[2mm] y \leq \dfrac{300 - 2x}{6} \\[2mm] x \geq 0, y \geq 0 \end{cases}$$

b. To solve the system, first consider the graph of the system

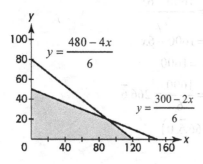

The intersection point between the two lines is $(90, 20)$.

Note that the lines are solid because there is an "equal to" as part of all the inequalities.

To determine the solution region, pick a point to test. Pick $(5,10)$.

$4x + 6y \leq 480$

$4(5) + 6(10) \leq 480$

$80 \leq 480$

True statement

$2x + 6y \leq 300$

$2(5) + 6(10) \leq 300$

$70 \leq 300$

True statement

$x \geq 0$

$5 \geq 0$

True statement

$y \geq 0$

$10 \geq 0$

True statement

Since all the inequalities that form the system are true at the point $(5,10)$, the region that contains $(5,10)$ is the solution region.

One of the corner points is the intersection point between the two lines, which is $(90, 20)$. A second corner point occurs where $y = \dfrac{480 - 4x}{6}$ crosses the x-axis. A third corner point occurs where $y = \dfrac{300 - 2x}{6}$ crosses the y-axis.

To find the x-intercept, let $y = 0$.

$0 = \dfrac{480 - 4x}{6}$

$0 = 480 - 4x$

$4x = 480$

$x = \dfrac{480}{4}$

$(120, 0)$

To find the y-intercept, let $x = 0$.

$y = \dfrac{300 - 2(0)}{6}$

$y = \dfrac{300}{6}$

$y = 50$

$(0, 50)$

A fourth corner point occurs at the origin, $(0,0)$. Therefore, the corner points of the solution region are $(0,0), (90,20), (120,0), (0,50)$.

41. a. Let x = number of commercial heating systems, and y = number of domestic heating systems.

Then, the system of inequalities is
$$\begin{cases} x + y \leq 1400 \\ x \geq 500 \\ y \geq 750 \end{cases}$$

Rewriting the system:

$x + y \leq 1400$
$\quad y \leq 1400 - x$

The new system is
$$\begin{cases} y \leq 1400 - x \\ x \geq 500 \\ y \geq 750 \end{cases}$$

b. To solve the system, first consider the graph of the system

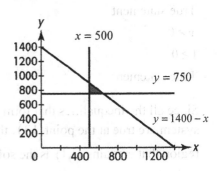

The intersection point between $x = 500$ and $y = 1400 - x$ is $(500, 900)$.

The intersection point between $y = 750$ and $y = 1400 - x$ is $(650, 750)$.

The intersection point between $x = 500$ and $y = 750$ is $(500, 750)$.

Note that the lines are solid because there is an "equal to" as part of all the inequalities.

To determine the solution region, pick a point to test. Pick $(600, 775)$.

$y \leq 1400 - x$
$775 \leq 1400 - 600$
$775 \leq 800$
True statement
$x \geq 500$
$600 \geq 500$
True statement
$y \geq 750$
$775 \geq 750$
True statement

Since all the inequalities that form the system are true at the point $(600, 775)$, the region that contains $(600, 775)$ is the solution region.

One of the corner points is the intersection point between $x = 500$ and $y = 1400 - x$, which is $(500, 900)$.

Another corner point is the intersection point between $y = 750$ and $y = 1400 - x$, which is $(650, 750)$.

A third corner point occurs at the intersection between $x = 500$ and $y = 750$, which is $(500, 750)$.

Therefore, the corner points of the solution region are
$(500, 750), (500, 900), (650, 750)$.

Section 8.2 Skills Check

1. Test the corner points of the feasible region in the objective function, $f = 4x + 9y$.

    At $(0,0)$: $f = 4(0) + 9(0) = 0$

    At $(0,40)$: $f = 4(0) + 9(40) = 360$

    At $(67,0)$: $f = 4(67) + 9(0) = 268$

    At $(10,38)$: $f = 4(10) + 9(38) = 382$

    The maximum value is 382 occurring at $(10,38)$, and the minimum value is 0 occurring at $(0,0)$.

3. Test the corner points of the feasible region in the objective function, $f = 4x + 2y$.

    At $(0,0)$: $f = 4(0) + 2(0) = 0$

    At $(0,2)$: $f = 4(0) + 2(2) = 4$

    At $(2,4)$: $f = 4(2) + 2(4) = 16$

    At $(4,3)$: $f = 4(4) + 2(3) = 22$

    At $(5,0)$: $f = 4(5) + 2(0) = 20$

    The maximum value is 22 occurring at $(4,3)$, and the minimum value is 0 occurring at $(0,0)$.

5. a. The graph of the system is

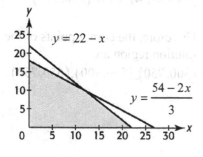

The intersection point between the two lines is $(12,10)$.

To determine the solution region, pick a point to test. Pick $(1,1)$.

$$y \le \frac{54 - 2x}{3}$$

$$1 \le \frac{54 - 2(1)}{3}$$

$$1 \le \frac{52}{3}$$

$$1 \le 17.\overline{3}$$

True statement

$$y \le 22 - x$$

$$1 \le 22 - 1$$

$$1 \le 21$$

True statement

$$x \ge 0$$

$$1 \ge 0$$

True statement

$$y \ge 0$$

$$1 \ge 0$$

True statement

Since all the inequalities that form the system are true at the point $(1,1)$, the region that contains $(1,1)$ is the solution region.

Note that since all the inequalities contain an "equal to", all the boundary lines are solid.

The graph above represents the feasible region. The corners of the region are $(0,0), (12,10), (22,0), (0,18)$.

b. Testing the corner points of the feasible region in the objective function, $f = 3x + 5y$ yields

At $(0,0)$:	$f = 3(0) + 5(0) = 0$
At $(0,18)$:	$f = 3(0) + 5(18) = 90$
At $(22,0)$:	$f = 3(22) + 5(0) = 66$
At $(12,10)$:	$f = 3(12) + 5(10) = 86$

The maximum value is 90 occurring at $(0,18)$.

7. a. To solve the system and determine the feasible region, first graph the system. To graph the system, rewrite the system by solving for y in the inequalities.

$$x + 2y \ge 15$$
$$2y \ge 15 - x$$
$$y \ge \frac{15 - x}{2}$$

and

$$x + y \ge 10$$
$$y \ge 10 - x$$

The new system is

$$\begin{cases} y \ge \dfrac{15 - x}{2} \\ y \ge 10 - x \\ x \ge 0, y \ge 0 \end{cases}$$

The graph of the system is

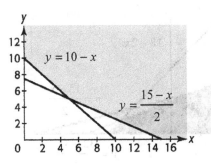

The intersection point between the two lines is $(5,5)$.

To determine the solution region, pick a point to test. Pick $(5,6)$.

$$y \ge \frac{15 - x}{2}$$
$$6 \ge \frac{15 - 5}{2}$$
$$6 \ge \frac{10}{2}$$
$$6 \ge 5$$
True statement

$$y \ge 10 - x$$
$$6 \ge 10 - 5$$
$$6 \ge 5$$
True statement

$$x \ge 0$$
$$5 \ge 0$$
True statement

$$y \ge 0$$
$$6 \ge 0$$
True statement

Since all the inequalities that form the system are true at the point $(5,6)$, the region that contains $(5,6)$ is the solution region.

Note that since all the inequalities contain an "equal to", all the boundary lines are solid.

The graph above represents the feasible region. The corners of the region are $(5,5), (0,10), (15,0)$.

b. Testing the corner points of the feasible region in the objective function, $g = 4x + 2y$ yields

At $(5,5)$: $g = 4(5) + 2(5) = 30$

At $(0,10)$: $g = 4(0) + 2(10) = 20$

At $(15,0)$: $g = 4(15) + 2(0) = 60$

The minimum value is 20 occurring at $(0,10)$.

9. Graphing the system yields

$$\begin{cases} y \le -\dfrac{1}{2}x + 4 \\ y \le -x + 6 \\ y \le -\dfrac{1}{3}x + 4 \\ x \ge 0, y \ge 0 \end{cases}$$

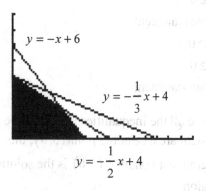

$y = -x + 6$

$y = -\dfrac{1}{3}x + 4$

$y = -\dfrac{1}{2}x + 4$

[0, 15] by [0, 8]

The intersection point between the lines $y = -x + 6$ and $y = -\dfrac{1}{2}x + 4$ is $(4,2)$.

To determine the solution region, pick a point to test. Pick $(1,1)$. When substituted into the inequalities that form the system, the point $(1,1)$ creates true statements in all cases.

Since all the inequalities that form the system are true at the point $(1,1)$, the region that contains $(1,1)$ is the solution region.

Note that since all the inequalities contain an "equal to", all the boundary lines are solid.

The corner points of the feasible region are $(0,0), (0,4), (6,0), (4,2)$.

Testing the corner points of the feasible region to maximize the objective function, $f = 20x + 30y$, yields

At $(0,0)$: $f = 20(0) + 30(0) = 0$

At $(0,4)$: $f = 20(0) + 30(4) = 120$

At $(6,0)$: $f = 20(6) + 30(0) = 120$

At $(4,2)$: $f = 20(4) + 30(2) = 140$

The maximum value is 140 occurring at $(4,2)$.

11. Rewriting the system of constraints and graphing the system yields

$$\begin{cases} y \le \dfrac{32 - 3x}{4} \\ y \le \dfrac{15 - x}{2} \\ y \le 18 - 2x \\ x \ge 0, y \ge 0 \end{cases}$$

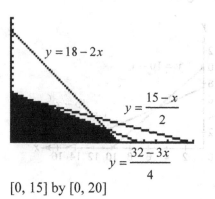

$y = 18 - 2x$

$y = \dfrac{15 - x}{2}$

$y = \dfrac{32 - 3x}{4}$

[0, 15] by [0, 20]

The intersection point between the lines

$y = \dfrac{32 - 3x}{4}$ and $y = \dfrac{15 - x}{2}$ is $(2, 6.5)$, and

between $y = 18 - 2x$ and $y = \dfrac{32 - 3x}{4}$ is

$(8, 2)$.

To determine the solution region, pick a point to test. Pick $(1, 1)$. When substituted into the inequalities that form the system, the point $(1, 1)$ creates true statements in all cases.

Since all the inequalities that form the system are true at the point $(1, 1)$, the region that contains $(1, 1)$ is the solution region.

Note that since all the inequalities contain an "equal to", all the boundary lines are solid.

The corner points of the feasible region are $(0, 0), (0, 7.5), (9, 0), (2, 6.5), (8, 2)$.

Testing the corner points of the feasible region to maximize the objective function, $f = 80x + 160y$, yields

At $(0, 0)$: $f = 80(0) + 160(0) = 0$

At $(0, 7.5)$: $f = 80(0) + 160(7.5) = 1200$

At $(9, 0)$: $f = 80(9) + 160(0) = 720$

At $(2, 6.5)$: $f = 80(2) + 160(6.5) = 1200$

At $(8, 2)$: $f = 80(8) + 160(2) = 960$

The maximum value is 1200 occurring at $(0, 7.5)$ and $(2, 6.5)$. Since the maximum occurs at two corner points, the maximum also occurs at all points along the line segment connecting $(0, 7.5)$ and $(2, 6.5)$.

13. Rewriting the system of constraints and graphing the system yields

$$\begin{cases} y \geq 6 - 2x \\ y \geq 8 - 4x \\ y \geq \dfrac{6 - x}{2} \\ x \geq 0, y \geq 0 \end{cases}$$

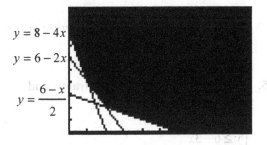

$[0, 10]$ by $[0, 10]$

The intersection point between the lines $y = 8 - 4x$ and $y = 6 - 2x$ is $(1, 4)$, and

between $y = 6 - 2x$ and $y = \dfrac{6 - x}{2}$ is $(2, 2)$.

To determine the solution region, pick a point to test. Pick $(2, 4)$. When substituted into the inequalities that form the system, the point $(2, 4)$ creates true statements in all cases.

Since all the inequalities that form the system are true at the point $(2, 4)$, the region that contains $(2, 4)$ is the solution region.

Note that since all the inequalities contain an "equal to", all the boundary lines are solid.

The corner points of the feasible region are $(0, 8), (6, 0), (1, 4), (2, 2)$.

Testing the corner points of the feasible region to minimize the objective function, $g = 40x + 30y$, yields

At $(0,8)$: $g = 40(0) + 30(8) = 240$
At $(6,0)$: $g = 40(6) + 30(0) = 240$
At $(1,4)$: $g = 40(1) + 30(4) = 160$
At $(2,2)$: $g = 40(2) + 30(2) = 140$

The minimum value is 140 occurring at $(2,2)$.

15. Rewriting the system of constraints and graphing the system yields

$$\begin{cases} y \geq 6 - 3x \\ y \geq 4 - x \\ y \leq \dfrac{8 - x}{5} \\ x \geq 0, y \geq 0 \end{cases}$$

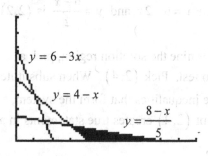

[0, 10] by [0, 8]

The intersection point between the lines $y = 4 - x$ and $y = \dfrac{8 - x}{5}$ is $(3,1)$.

To determine the solution region, pick a point to test. Pick $\left(4, \dfrac{1}{2}\right)$. When substituted into the inequalities that form the system, the point $\left(4, \dfrac{1}{2}\right)$ creates true statements in all cases.

Since all the inequalities that form the system are true at the point $\left(4, \dfrac{1}{2}\right)$, the region that contains $\left(4, \dfrac{1}{2}\right)$ is the solution region.

Note that since all the inequalities contain an "equal to", all the boundary lines are solid.

The corner points of the feasible region are $(4,0), (8,0), (3,1)$.

Testing the corner points of the feasible region to minimize the objective function, $g = 46x + 23y$, yields

At $(4,0)$: $g = 46(4) + 23(0) = 184$
At $(8,0)$: $g = 46(8) + 23(0) = 368$
At $(3,1)$: $g = 46(3) + 23(1) = 161$

The minimum value is 161 occurring at $(3,1)$.

17. Rewriting the system of constraints and graphing the system yields

$$\begin{cases} y \geq \dfrac{12 - 3x}{2} \\ y \geq 7 - 2x \\ x \geq 0, y \geq 0 \end{cases}$$

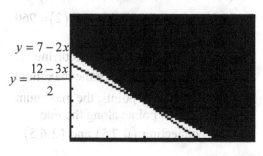

[0, 5] by [0, 10]

The intersection point between the lines
$y = 7 - 2x$ and $y = \dfrac{12 - 3x}{2}$ is $(2,3)$.

To determine the solution region, pick a point to test. Pick $(1,10)$. When substituted into the inequalities that form the system, the point $(1,10)$ creates true statements in all cases.

Since all the inequalities that form the system are true at the point $(1,10)$, the region that contains $(1,10)$ is the solution region. Note that since all the inequalities contain an "equal to", all the boundary lines are solid.

The corner points of the feasible region are $(0,7), (4,0), (2,3)$.

Testing the corner points of the feasible region to minimize the objective function, $g = 60x + 10y$, yields

At $(0,7)$: $g = 60(0) + 10(7) = 70$

At $(4,0)$: $g = 60(4) + 10(0) = 240$

At $(2,3)$: $g = 60(2) + 10(3) = 150$

The minimum value is 70 occurring at $(0,7)$.

Section 8.2 Exercises

19. See the solution to Section 8.1, exercise 33. The corner points of the feasible region are $(390,390), (780,0), (975,0)$.

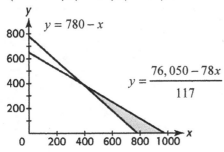

Testing the corner points of the feasible region to maximize the objective function, $f = 32x + 45y$, yields

At $(390,390)$: $f = 32(390) + 45(390)$
$= \$30,030$

At $(780,0)$: $f = 32(780) + 45(0)$
$= \$24,960$

At $(975,0)$: $f = 32(975) + 45(0)$
$= \$31,200$

The maximum is \$31,200 occurring at $(975,0)$. To maximize profit, the company should manufacture 975 Turbo models and 0 Tornado models.

21. See the solution to Section 8.1, exercise 40. The corner points of the feasible region are $(0,0), (9,6), (12,0), (0,12)$.

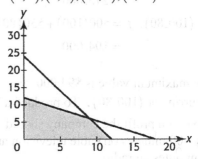

Testing the corner points of the feasible region to maximize the objective function, $f = 24x + 30y$, yields

At $(0,0)$: $f = 24(0) + 30(0) = \$0$

At $(9,6)$: $f = 24(9) + 30(6) = \$396$

At $(12,0)$: $f = 24(12) + 30(0) = \$288$

At $(0,12)$: $f = 24(0) + 30(12) = \$360$

The maximum is \$396 occurring at $(9,6)$.

To maximize profit, the company should sell 9 Safecut chainsaws and 6 Deluxe chainsaws.

23. a. See the solution to Section 8.1, exercise 34. The corner points of the feasible region are $(150,0),(132,0),(100,80)$.

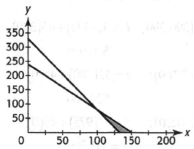

Testing the corner points of the feasible region to maximize the objective function, $f = 500x + 550y$, yields

At $(132,0)$: $f = 500(132) + 550(0)$
$= \$66,000$

At $(150,0)$: $f = 500(150) + 550(0)$
$= \$75,000$

At $(100,80)$: $f = 500(100) + 550(80)$
$= \$94,000$

The maximum value is $94,000 occurring at $(100,80)$. To produce a maximum profit the company should buy 100 minutes on cable television and 80 minutes on radio.

b. The maximum profit is $94,000.

25. See the solution to Section 8.1, exercise 35. The corner points of the feasible region are $(0,85),(50,0),(30,20),(10,60)$.

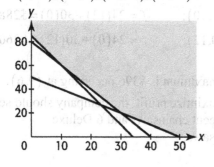

Testing the corner points of the feasible region to minimize the objective function, $g = 20,000x + 40,000y$, yields

At $(0,85)$: $g = 20,000(0) + 40,000(85)$
$= \$3,400,000$

At $(50,0)$: $g = 20,000(50) + 40,000(0)$
$= \$1,000,000$

At $(10,60)$: $g = 20,000(10) + 40,000(60)$
$= \$2,600,000$

At $(30,20)$: $g = 20,000(30) + 40,000(20)$
$= \$1,400,000$

The minimum value is $1,000,000 occurring at $(50,0)$. To minimize the cost, the company should manufacture the television sets on assembly line 1 for 50 days and on assembly line 2 for zero days.

27. See the solution to Section 8.1, exercise 32. The corner points of the feasible region are $(0,0),(15,0),(0,10),(5,8)$.

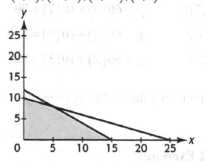

Testing the corner points of the feasible region to maximize the objective function, $f = 60,000x + 75,000y$, yields

At $(0,0)$: $f = 60,000(0) + 75,000(0)$
$$= \$0$$
At $(15,0)$: $f = 60,000(15) + 75,000(0)$
$$= \$900,000$$
At $(0,10)$: $f = 60,000(0) + 75,000(10)$
$$= \$750,000$$
At $(5,8)$: $f = 60,000(5) + 75,000(8)$
$$= \$900,000$$

The maximum profit is \$900,000 occurring at $(15,0)$ and $(5,8)$. Since there are two corner points that maximize the objective function, any point along the line segment connecting $(15,0)$ and $(5,8)$ also maximizes the objective function.

Therefore, the contractor can build 15 Van Buren models and 0 Jefferson models, or 5 Van Buren models and 8 Jefferson models, or 10 Van Buren models and 4 Jefferson models. In particular the possible solutions are $(10,4), (15,0)$ and $(5,8)$.

29. Determine the feasible region by solving and graphing the system of inequalities that represent the constraints. Let x represent the number of weeks operating facility 1, and let y represent the number of weeks operating facility 2.

$$\begin{cases} 400x + 400y \geq 4000 \\ 300x + 100y \geq 1800 \\ 200x + 400y \geq 2400 \\ x \geq 0, y \geq 0 \end{cases}$$

Rewriting yields
$$\begin{cases} y \geq 10 - x \\ y \geq 18 - 3x \\ y \geq \dfrac{12 - x}{2} \\ x \geq 0, y \geq 0 \end{cases}$$

Graphing the system yields

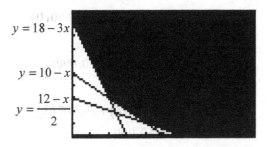

$y = 18 - 3x$

$y = 10 - x$

$y = \dfrac{12 - x}{2}$

[0, 20] by [0, 20]

The intersection point between the lines $y = 18 - 3x$ and $y = 10 - x$ is $(4,6)$, and between the lines $y = 10 - x$ and $y = \dfrac{12 - x}{2}$ is $(8,2)$.

To determine the solution region, pick a point to test. Pick $(10,5)$. When substituted into the inequalities that form the system, the point $(10,5)$ creates true statements in all cases.

Since all the inequalities that form the system are true at the point $(10,5)$, the region that contains $(10,5)$ is the solution region.

Note that since all the inequalities contain an "equal to", all the boundary lines are solid.

The corner points of the feasible region are $(0,18), (12,0), (4,6), (8,2)$.

Testing the corner points of the feasible region to minimize the objective function, $g = 15,000x + 20,000y$, yields

At $(0,18)$: $g = 15,000(0) + 20,000(18)$
$$= \$360,000$$
At $(12,0)$: $g = 15,000(12) + 20,000(0)$
$$= \$180,000$$
At $(4,6)$: $g = 15,000(4) + 20,000(6)$
$$= \$180,000$$
At $(8,2)$: $g = 15,000(8) + 20,000(2)$
$$= \$160,000$$

The minimum value is \$160,000 occurring at $(8,2)$. Operating facility 1 for 8 weeks and facility 2 for 2 weeks yields a minimum cost of \$160,000 for filling the orders.

31. Determine the feasible region by solving and graphing the system of inequalities that represent the constraints. Let x represent the number of servings of Diet A, and let y represent the number of servings of Diet B.

$$\begin{cases} 2x + 1y \geq 18 \\ 4x + 1y \geq 26 \\ x \geq 0, y \geq 0 \end{cases}$$

Rewriting yields
$$\begin{cases} y \geq 18 - 2x \\ y \geq 26 - 4x \\ x \geq 0, y \geq 0 \end{cases}$$

Graphing the system yields

$y = 26 - 4x$

$y = 18 - 2x$

[0, 10] by [0, 30]

The intersection point between the lines $y = 26 - 4x$ and $y = 18 - 2x$ is $(4,10)$.

To determine the solution region, pick a point to test. Pick $(5,10)$. When substituted into the inequalities that form the system, the point $(5,10)$ creates true statements in all cases.

Since all the inequalities that form the system are true at the point $(5,10)$, the region that contains $(5,10)$ is the solution region.

Note that since all the inequalities contain an "equal to", all the boundary lines are solid.

The corner points of the feasible region are $(0,26), (9,0), (4,10)$.

Testing the corner points of the feasible region to minimize the objective function, $g = 0.09x + 0.035y$, yields

At $(0,26)$: $g = 0.09(0) + 0.035(26) = 0.91$
At $(9,0)$: $g = 0.09(9) + 0.035(0) = 0.81$
At $(4,10)$: $g = 0.09(4) + 0.035(10) = 0.71$

The minimum value is 0.71 occurring at $(4,10)$. Four servings of Diet A and ten servings of Diet B yields the minimum amount of detrimental substance, 0.71 oz.

Section 8.3 Skills Check

1.
$$f(1) = 2(1) + 3 = 5$$
$$f(2) = 2(2) + 3 = 7$$
$$f(3) = 2(3) + 3 = 9$$
$$f(4) = 2(4) + 3 = 11$$
$$f(5) = 2(5) + 3 = 13$$
$$f(6) = 2(6) + 3 = 15$$
The first six terms are 5, 7, 9, 11, 13, and 15.

3.
$$a_1 = \frac{10}{1} = 10$$
$$a_2 = \frac{10}{2} = 5$$
$$a_3 = \frac{10}{3}$$
$$a_4 = \frac{10}{4} = \frac{5}{2}$$
$$a_5 = \frac{10}{5} = 2$$

The first five terms are $10, 5, \frac{10}{3}, \frac{5}{2},$ and 2.

5. To move from one term in the sequence to the next term, add two. Therefore the next three terms are 9, 11, and 13.

7.
$$a_n = a_1 + (n-1)d$$
$$a_n = -3 + (n-1)(4)$$
$$a_n = -3 + 4n - 4$$
$$a_n = 4n - 7$$
$$a_8 = 4(8) - 7 = 25$$

9. To move from one term in the sequence to the next term, multiply by 2. Therefore the next four terms are 24, 48, 96, and 192.

11.
$$a_n = a_1 r^{n-1}$$
$$a_n = 10(3)^{n-1}$$
$$a_6 = 10(3)^{6-1} = 10(3)^5 = 2430$$

13. $a_n = a_{n-1} - 2$ implies that the next term in the sequence is the previous term in the sequence minus 2. For example, $a_2 = a_{2-1} - 2 = a_1 - 2$. The second term is the first term minus 2. Therefore, if the first term is 5, then the first four terms will be 5, 3, 1, and -1.

15.
$$a_n = 2a_{n-1} + 3$$
$$a_1 = 2$$
$$a_2 = 2a_{2-1} + 3 = 2a_1 + 3 = 2(2) + 3 = 7$$
$$a_3 = 2a_{3-1} + 3 = 2a_2 + 3 = 2(7) + 3 = 17$$
$$a_4 = 2a_{4-1} + 3 = 2a_3 + 3 = 2(17) + 3 = 37$$
The first four terms are 2, 7, 17, and 37.

Section 8.3 Exercises

17. Let a_1 represent the starting salary
$$a_n = a_1 + (n-1)d$$
$$a_n = a_1 + (n-1)1500$$
$$a_8 = a_1 + (8-1)1500$$
$$a_8 = a_1 + (7)1500 = a_1 + 10,500$$
The salary would increase by $10,500.

19. a. $f(n) = 300 + 60n$

b.
$$f(1) = 300 + 60(1) = 360$$
$$f(2) = 300 + 60(2) = 420$$
$$f(3) = 300 + 60(3) = 480$$
$$f(4) = 300 + 60(4) = 540$$
$$f(5) = 300 + 60(5) = 600$$
$$f(6) = 300 + 60(6) = 660$$
The first six terms are 360, 420, 480, 540, 600, and 660.

21. a. Job 1

$$a_n = a_1 + (n-1)d$$

$$a_n = 40,000 + (n-1)2000$$

$$a_n = 2000n + 38,000$$

Job 2

$$a_n = a_1 + (n-1)d$$

$$a_n = 36,000 + (n-1)2400$$

$$a_n = 2400n + 33,600$$

In 6 years (after 5 years), $n = 6$.

Job 1:

$$a_6 = 2000(6) + 38,000 = 50,000$$

Job 2:

$$a_6 = 2400(6) + 33,600 = 48,000$$

In 6 years, Job 1 pays $2000 more than Job 2.

b. In 11 years (after 10 years), $n = 11$.

Job 1:

$$a_{11} = 2000(11) + 38,000 = 60,000$$

Job 2:

$$a_{11} = 2400(11) + 33,600 = 60,000$$

The salaries are the same.

c. In the 13th year (after 12 years), $n = 13$.

Job 1:

$$a_{13} = 2000(13) + 38,000 = 64,000$$

Job 2:

$$a_{13} = 2400(13) + 33,600 = 64,800$$

In the 13th year, Job 2 pays $800 more than Job 1.

23. a. Year 1: $S = P + Prt$

$$S = 1000 + 1000(5\%)(1)$$

$$S = 1000 + 1000(0.05)(1)$$

$$S = 1000 + 50$$

$$S = 1050$$

Year 2: $S = P + Prt$

$$S = 1050 + 1050(5\%)(1)$$

$$S = 1050 + 1050(0.05)(1)$$

$$S = 1050 + 52.5$$

$$S = 1102.50$$

Year 3: $S = P + Prt$

$$S = 1102.50 + 1102.50(5\%)(1)$$

$$S = 1102.50 + 1102.50(0.05)(1)$$

$$S = 1102.50 + 55.13$$

$$S = 1157.63$$

Year 4: $S = P + Prt$

$$S = 1157.63 + 1157.63(5\%)(1)$$

$$S = 1157.63 + 1157.63(0.05)(1)$$

$$S = 1157.63 + 57.88$$

$$S = 1215.51$$

The sequence is $1050, $1102.50, $1157.63, and $1215.51.

25. a.

$$20\%(50,000) = (0.20)(50,000)$$

$$= 10,000$$

$$50,000 - 3(10,000) = \$20,000$$

The value after 3 years is $20,000.

b. $a_n = 50,000 + (n)(-10,000)$

$$a_n = 50,000 - 10,000n$$

Note that n represents the number of years of ownership.

c. The value of the car depreciates as follows: $40,000, $30,000, $20,000, $10,000, and $0.

27.

$$a_n = a_1 r^{n-1}$$

$$a_n = 5000(2)^{n-1}$$

Since $a_1 = 5000$ bacteria when zero time has passed, a_7 will represent the number of bacteria after 6 hours. Thus, after six hours, $n = 7$.

$$a_7 = 5000(2)^{7-1}$$

$$a_7 = 5000(2)^6 = 320,000$$

After 6 hours, the number of bacteria in the culture is 320,000.

29. a. The number of cells in these fossils follow a geometric sequence with a common ratio of 2 (because the number doubles each time).

b. $c(n) = 2^n$, where n = the number of rounds of cell division, and c = the number of cells.

$$c(10) = 2^{10} = 1024$$

The number of cells is more than 1000 after the tenth round of division.

c. $c(n) = 2^n$, where n = the number of rounds of cell division, and c = the number of cells.

$$c(9) = 2^9 = 512$$

The number of cells is more than 500 after the ninth round of division.

31. If a company loses 2% of its profit, then 98% of its profit remains. Therefore,

$f(n) = 8,000,000(0.98)^n$, where n = the number of years and $f(n)$ = the company's profit.

$$f(5) = 8,000,000(0.98)^5$$

$$f(5) = 7,231,366.37$$

After five years the company's projected profit is $7,231,366.37.

33. $a_n = a_1 + (n-1)d$

$$a_n = 54,000 + (n-1)(3600)$$

$$a_n = 54,000 + 3600n - 3600$$

$$a_n = 50,400 + 3600n$$

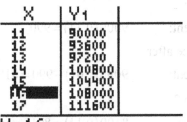

When $n = 16$ or after 15 years, the salary is twice the original amount of $54,000. Therefore, the salary doubles after 15 years.

35. $A = P\left(1 + \dfrac{r}{k}\right)^{kt}$

$$A = 10,000\left(1 + \frac{0.08}{365}\right)^{365(10)}$$

$$A = 10,000(1.000219178)^{3650}$$

$$A = 22,253.46$$

The future value of $10,000 compounded daily at 8% is $22,253.46.

37. The next number in the sequence is the sum of the previous two numbers in the sequence. Therefore, the next four numbers in the sequence are 13, 21, 34, and 55.

39. Making a payment of 1% interest plus 10% of the balance at the beginning of the month implies that she will make a payment of 11% of balance each month. However, the balance will decrease by 10%, since 1% of the payment is interest.

$$\begin{cases} \text{Month 1} \\ \text{Payment:} \qquad 10{,}000(0.11)=1100 \\ \text{Balance after} \\ \text{Payment:} \qquad 10{,}000(1.01)-1100=9000 \end{cases}$$

$$\begin{cases} \text{Month 2} \\ \text{Payment:} \qquad 9000(0.11)=990 \\ \text{Balance after} \\ \text{Payment:} \qquad 9000(1.01)-990=8100 \end{cases}$$

$$\begin{cases} \text{Month 3} \\ \text{Payment:} \qquad 8100(0.11)=891 \\ \text{Balance after} \\ \text{Payment:} \qquad 8100(1.01)-891=7290 \end{cases}$$

$$\begin{cases} \text{Month 4} \\ \text{Payment:} \qquad 7290(0.11)=801.90 \end{cases}$$

The sequence of payments is $1100, $990, $891, and $801.90.

Section 8.4 Skills Check

1. Note that $r=\dfrac{3}{9}=\dfrac{1}{3}$.

$$s_n=\frac{a_1\left(1-r^n\right)}{1-r}$$

$$s_6=\frac{9\left(1-\left(\dfrac{1}{3}\right)^6\right)}{1-\left(\dfrac{1}{3}\right)}=\frac{364}{27}$$

3. Note that $d=10-7=13-10=3$.

$$a_n=a_1+(n-1)d$$
$$a_n=7+(n-1)(3)$$
$$a_n=3n+4$$
$$a_{10}=3(10)+4=34$$

$$s_n=\frac{n(a_1+a_n)}{2}$$

$$s_{10}=\frac{10(a_1+a_{10})}{2}=\frac{10(7+34)}{2}=205$$

5. $a_n=a_1+(n-1)d$
$$a_n=-4+(n-1)(2)$$
$$a_n=2n-6$$
$$a_{15}=2(15)-6=24$$
$$s_n=\frac{n(a_1+a_n)}{2}$$
$$s_{15}=\frac{15(a_1+a_{15})}{2}=\frac{15(-4+24)}{2}=150$$

7. $s_n = \dfrac{a_1\left(1-r^n\right)}{1-r}$

$s_{15} = \dfrac{3\left(1-(2)^{15}\right)}{1-(2)} = 98{,}301$

9. Note that $r = \dfrac{10}{5} = \dfrac{20}{10} = 2$.

$s_n = \dfrac{a_1\left(1-r^n\right)}{1-r}$

$s_{10} = \dfrac{5\left(1-(2)^{10}\right)}{1-(2)} = 5115$

11. Note that $r = \dfrac{256}{1024} = \dfrac{64}{256} = \dfrac{16}{64} = \dfrac{1}{4}$.

$S = \dfrac{a_1}{1-r}$

$S = \dfrac{1024}{1-\left(\dfrac{1}{4}\right)} = \dfrac{1024}{\left(\dfrac{3}{4}\right)} = \dfrac{4096}{3}$

13. $\displaystyle\sum_{i=1}^{6} 2^i = 2^1 + 2^2 + 2^3 + 2^4 + 2^5 + 2^6$

$= 2 + 4 + 8 + 16 + 32 + 64 = 126$

15. $\displaystyle\sum_{i=1}^{4}\left(\dfrac{1+i}{i}\right) = 2 + \dfrac{3}{2} + \dfrac{4}{3} + \dfrac{5}{4} = \dfrac{73}{12}$

17. $S = \dfrac{a_1}{1-r} = \dfrac{\dfrac{3}{4}}{1-\dfrac{3}{4}} = 3$

19. Since $r = \dfrac{4}{3} \ge 1$, the infinite sum does not exist. No solution.

Section 8.4 Exercises

21. Note that $d = 400$ and $a_1 = -2000$.

$a_n = a_1 + (n-1)d$

$a_n = -2000 + (n-1)(400)$

$a_n = 400n - 2400$

$a_{12} = 400(12) - 2400 = 2400$

$s_n = \dfrac{n(a_1 + a_n)}{2}$

$s_{12} = \dfrac{12(a_1 + a_{12})}{2}$

$= \dfrac{12(-2000 + 2400)}{2} = 2400$

The profit for the year is $2400.

23. $1 + 2 + 3 + 5 = 11$. A male bee has 11 ancestors through four generations, where:
1 = the male bee's parent = 1 female
2 = the male bee's grandparents = 1 male, 1 female
3 = the male bee's great-grandparents = 1 female, and 1 male, 1 female
5 = the male bee's great-great-grandparents = 1 male, 1 female, and 1 female, and 1 male, 1 female. (whew!)

25. a. $a_n = a_1 + (n-1)d$

$a_n = 1 + (n-1)(1)$

$a_n = n$

$a_{12} = 12$

$s_n = \dfrac{n(a_1 + a_n)}{2}$

$s_{12} = \dfrac{12(1+12)}{2} = 78$

The clock chimes 78 times in a 12-hour period. (1 – 12, or 2 – 1, or 3 – 2, etc.)

b. In a 24-hour period, the clock chimes twice as many times as in a 12-hour period. Therefore the clock chimes 156 times every 24 hours.

27. Note that the sequence is geometric with a common ratio of 1.10. Therefore,

$$s_n = \frac{a_1\left(1 - r^n\right)}{1 - r}$$

$$s_{12} = \frac{2000\left(1 - \left(1.10\right)^{12}\right)}{1 - \left(1.10\right)} \approx 42,768.57$$

The total profit over the first 12 months is $42,768.57.

29. If the pump removes $\frac{1}{3}$ of the water with each stroke, then $\frac{2}{3}$ of the water remains after each stroke. Therefore, a geometric sequence can be created with a common ratio of $\frac{2}{3}$.

$$a_1 = 81\left(\frac{2}{3}\right) = 54$$

$$a_n = a_1 r^{n-1}$$

$$a_n = 54\left(\frac{2}{3}\right)^{n-1}$$

$$a_4 = 54\left(\frac{2}{3}\right)^{4-1} = 54\left(\frac{2}{3}\right)^3 = 16\,\text{cm}^3$$

The amount of water in the container after four strokes is 16 cm³.

31. a. 5

b. $5 \times 5 = 5^2 = 25$

c. $5^3 = 125$
$5^4 = 625$

d. The sequence is geometric with a common ratio of 5.

33. Since 10% of the balance is paid each month, then 90% of the balance remains. Note that the situation is modeled by a geometric series with a common ratio of 0.90.

$$a_n = 10,000\left(0.90\right)^n$$

$$a_{12} = 10,000\left(0.90\right)^{12} = 2824.30$$

35. Note that the situation is modeled by a geometric function having a common ratio of $\frac{1}{4}$.

$$a_1 = 128\left(\frac{1}{4}\right) = 32$$

$$a_n = 32\left(\frac{1}{4}\right)^{n-1}, \text{ where } n \text{ represents the}$$

number of bounces and a_n represents the height after that bounce. Note that when the ball hits the ground for the fifth time, it has bounced four times.

$$s_n = \frac{a_1\left(1 - r^n\right)}{1 - r}$$

$$s_4 = \frac{32\left(1 - \left(\frac{1}{4}\right)^4\right)}{1 - \left(\frac{1}{4}\right)}$$

$$s_4 = 42.5$$

Note that s_4 is the sum of the rebound heights. To calculate the distance the ball actually travels, double s_4 to take into consideration that the ball rises and falls with each bounce, and add in the initial distance of 128 feet before the first rebound. Therefore the total distance traveled is $128 + 2\left(42.5\right) = 213$ feet.

37. a. Note that if the car loses 16% of its value each year, then it retains 84% of its value. The value of the car is modeled by a geometric sequence with a common ratio of 0.84. The depreciated value of the car after n years is given by $a_n = 35,000\left(0.84\right)^n$. The amount of depreciation after n years is given by the original value of the car less the depreciated value. Therefore,

$$s_n = 35,000 - 35,000\left(0.84\right)^n$$

$$s_n = 35,000\left(1 - \left(0.84\right)^n\right)$$

b. $a_n = 35,000(0.84)^n$

39. $s_n = \dfrac{a_1(1 - r^n)}{1 - r}$

$$s_{96} = \dfrac{100\left[1 - \left(1 + \dfrac{0.12}{12}\right)^{96}\right]}{1 - \left(1 + \dfrac{0.12}{12}\right)}$$

$$s_{96} = \dfrac{100\left(1 - (1.01)^{96}\right)}{1 - (1.01)}$$

$s_{96} \approx 15,992.73$

The future value of the annuity is $15,992.73.

41. $s_n = \dfrac{a_1(1 - r^n)}{1 - r}$

$$s_n = \dfrac{\left(R(1 + i)^{-n}\right)\left(1 - (1 + i)^n\right)}{1 - (1 + i)}$$

$$s_n = \dfrac{R(1 + i)^{-n}\left(1 - (1 + i)^n\right)}{1 - 1 - i}$$

$$s_n = \dfrac{R\left[(1 + i)^{-n}(1) - (1 + i)^{-n}(1 + i)^n\right]}{-i}$$

$$s_n = \dfrac{R\left[(1 + i)^{-n} - (1 + i)^{-n+n}\right]}{-i}$$

$$s_n = \dfrac{R\left[(1 + i)^{-n} - (1 + i)^0\right]}{-i}$$

$$s_n = \dfrac{R\left[(1 + i)^{-n} - 1\right]}{-i}$$

$$s_n = R\left[\dfrac{1 - (1 + i)^{-n}}{i}\right]$$

Section 8.5 Skills Check

1. $f'(x) = (2x^3 + 3x + 1)(2x) +$
    $\quad (x^2 + 4)(6x^2 + 3)$
$f'(x) = (4x^4 + 6x^2 + 2x) +$
    $\quad (6x^4 + 3x^2 + 24x^2 + 12)$
$f'(x) = 10x^4 + 33x^2 + 2x + 12$

3. $f'(x) = \dfrac{x^2(3x^2) - (x^3 - 3)(2x)}{(x^2)^2}$

$f'(x) = \dfrac{3x^4 - (2x^4 - 6x)}{x^4}$

$f'(x) = \dfrac{x^4 + 6x}{x^4} = \dfrac{x(x^3 + 6)}{x^4} = \dfrac{x^3 + 6}{x^3}$

5. $f'(2) = 4(2)^3 - 3(2)^2 + 4(2) - 2 = 26$

7. $f'(3) = \dfrac{\left((3)^2 - 1\right)(2) + (2(3))(2(3))}{\left((3)^2 - 1\right)^2}$

$= \dfrac{(8)(2) + (6)(6)}{(8)^2}$

$= \dfrac{16 + 36}{64}$

$= \dfrac{52}{64} = \dfrac{13}{16}$

9. $y - y_1 = m(x - x_1)$

$y - 8 = 12(x - 2)$

$y - 8 = 12x - 24$

$y = 12x - 16$

11. a. $f(x + h) = 4(x + h) + 5$

b. $f(x + h) - f(x)$

$= \left[4(x + h) + 5\right] - \left[4x + 5\right]$

$= 4x + 4h + 5 - 4x - 5$

$= 4h$

c. $\dfrac{f(x + h) - f(x)}{h}$

$= \dfrac{4h}{h}$

$= 4$

13. $8x^2 + 4y = 12$

$4y = 12 - 8x^2$

$y = \dfrac{12 - 8x^2}{4}$

$y = \dfrac{12}{4} - \dfrac{8x^2}{4}$

$y = 3 - 2x^2$

15. $9x^3 + 5y = 18$

$5y = 18 - 9x^3$

$y = \dfrac{18 - 9x^3}{5}$

17. a. $0 = 2x - 2$

$2x = 2$

$x = 1$

b. Recall that the x-coordinate of the

vertex is given by $\dfrac{-b}{2a}$.

$\dfrac{-b}{2a} = \dfrac{-(-2)}{2(1)} = \dfrac{2}{2} = 1$

The solution to part a) and the
x-coordinate of the vertex are the same.

c. $y = (1)^2 - 2(1) + 5 = 4$

$(1, 4)$

19. a. $0 = 12x - 24$

$12x = 24$

$x = 2$

b. Recall that the x-coordinate of the

vertex is given by $\dfrac{-b}{2a}$.

$\dfrac{-b}{2a} = \dfrac{-(-24)}{2(6)} = \dfrac{24}{12} = 2$

The solution to part a) and the
x-coordinate of the vertex are the same.

c. $y = 6(2)^2 - 24(2) + 15$

$y = 24 - 48 + 15 = -9$

$(2, -9)$

21. a. $3x^2 - 18x - 48 = 0$

$3(x^2 - 6x - 16) = 0$

$3(x-8)(x+2) = 0$

$x = 8,\ x = -2$

b. $x = 8$

$y = (8)^3 - 9(8)^2 - 48(8) + 15$

$y = -433$

$(8, -433)$

$x = -2$

$y = (-2)^3 - 9(-2)^2 - 48(-2) + 15$

$y = 67$

$(-2, 67)$

c. $x = -2$ produces the maximum.

X	Y₁
-3	51
-2	67
-1	53
0	15
1	-41
2	-109
3	-183

X= -2

23. $y = 3\sqrt{x} = 3\sqrt[2]{x^1} = 3x^{\frac{1}{2}}$

25. $y = 2\sqrt[3]{x^2} = 2x^{\frac{2}{3}}$

27. $y = \sqrt[3]{x^2 + 1} = (x^2 + 1)^{\frac{1}{3}}$

29. $y = \sqrt[3]{(x^3 - 2)^2} = (x^3 - 2)^{\frac{2}{3}}$

31. $y = \sqrt{x} + \sqrt[3]{2x}$

$y = x^{\frac{1}{2}} + (2x)^{\frac{1}{3}}$

33. $y^2 + 4x - 3 = 0$

$y^2 = 3 - 4x$

$\sqrt{y^2} = \pm\sqrt{3 - 4x}$

$y = \pm\sqrt{3 - 4x}$

35. $y^2 + y - 6x = 0$

$a = 1,\ b = 1,\ c = -6x$

$y = \dfrac{-b \pm \sqrt{b^2 - 4ac}}{2a}$

$y = \dfrac{-1 \pm \sqrt{(1)^2 - 4(1)(-6x)}}{2(1)}$

$y = \dfrac{-1 \pm \sqrt{1 + 24x}}{2}$

37. $y = u^2$

$u = 4x^3 + 5$

39. $u = x^3 + x$

$y = \sqrt{x^3 + x}$

$y = \sqrt{u} = u^{\frac{1}{2}}$

41. $\dfrac{3}{2}u^{\frac{1}{2}} \cdot 2x$

$= \dfrac{3}{2}(x^2 - 1)^{\frac{1}{2}} \cdot 2x$

$= 3x\sqrt{x^2 - 1}$

43. $y = 3x^{-1} - 4x^{-2} - 6$

45. $y = (x^2 - 3)^{-3}$

47. $f'(x) = -6\left(\dfrac{1}{x^4}\right) + 4\left(\dfrac{1}{x^2}\right) + \dfrac{1}{x}$

$f'(x) = \dfrac{-6}{x^4} + \dfrac{4}{x^2} + \dfrac{1}{x}$

49. $f'(x) = \left(4x^2 - 3\right)^{-\frac{1}{2}}(8x)$

$$f'(x) = \frac{8x}{\left(4x^2 - 3\right)^{\frac{1}{2}}}$$

$$f'(x) = \frac{8x}{\sqrt{4x^2 - 3}}$$

51. $\log\left[x^3 \left(3x - 4\right)^5 \right]$

$\log\left(x^3\right) + \log\left[\left(3x - 4\right)^5\right]$

$3\log x + 5\log\left(3x - 4\right)$

53. $f'(x) = 0$

$3x^2 - 3x = 0$

$3x(x - 1) = 0$

$3x = 0,\ x - 1 = 0$

$x = 0,\ x = 1$

55. $f'(x) = 0$

$\left(x^2 - 4\right)3\left(x - 3\right)^2 + \left(x - 3\right)^3(2x) = 0$

$\left(x - 3\right)^2\left[3\left(x^2 - 4\right) + \left(x - 3\right)(2x)\right] = 0$

$\left(x - 3\right)^2\left[3x^2 - 12 + 2x^2 - 6x\right] = 0$

$\left(x - 3\right)^2\left[5x^2 - 6x - 12\right] = 0$

$\left(x - 3\right)^2 = 0$ implies $x = 3$

$5x^2 - 6x - 12 = 0$

$a = 5,\ b = -6,\ c = -12$

$$x = \frac{-b \pm \sqrt{b^2 - 4ac}}{2a}$$

$$x = \frac{-(-6) \pm \sqrt{(-6)^2 - 4(5)(-12)}}{2(5)}$$

$$x = \frac{6 \pm \sqrt{36 + 240}}{10}$$

$$x = \frac{6 \pm \sqrt{276}}{10}$$

$$x = \frac{6 \pm \sqrt{4 \cdot 69}}{10}$$

$$x = \frac{6 \pm 2\sqrt{69}}{10}$$

$$x = \frac{3 \pm \sqrt{69}}{5}$$

$$x = 3,\ x = \frac{3 + \sqrt{69}}{5},\ x = \frac{3 - \sqrt{69}}{5}$$

Section 8.6 Skills Check

1. $\dfrac{8!}{4!4!} = \dfrac{8 \cdot 7 \cdot 6 \cdot 5 \cdot 4 \cdot 3 \cdot 2 \cdot 1}{4 \cdot 3 \cdot 2 \cdot 1 \cdot 4 \cdot 3 \cdot 2 \cdot 1}$

$\quad\quad = 2 \cdot 7 \cdot 5$

$\quad\quad = 70$

3. $\dfrac{9!}{0!9!} = \dfrac{9 \cdot 8 \cdot 7 \cdot 6 \cdot 5 \cdot 4 \cdot 3 \cdot 2 \cdot 1}{1 \cdot 9 \cdot 8 \cdot 7 \cdot 6 \cdot 5 \cdot 4 \cdot 3 \cdot 2 \cdot 1}$

$\quad\quad = 1$

5. $\dfrac{9!}{3!6!} = \dfrac{9 \cdot 8 \cdot 7 \cdot 6 \cdot 5 \cdot 4 \cdot 3 \cdot 2 \cdot 1}{3 \cdot 2 \cdot 1 \cdot 6 \cdot 5 \cdot 4 \cdot 3 \cdot 2 \cdot 1}$

$\qquad = 3 \cdot 4 \cdot 7$

$\qquad = 84$

7. $\dbinom{12}{2} = \dfrac{12!}{2!10!}$

$\qquad = \dfrac{12 \cdot 11 \cdot 10 \cdot 9 \cdot 8 \cdot 7 \cdot 6 \cdot 5 \cdot 4 \cdot 3 \cdot 2 \cdot 1}{2 \cdot 1 \cdot 10 \cdot 9 \cdot 8 \cdot 7 \cdot 6 \cdot 5 \cdot 4 \cdot 3 \cdot 2 \cdot 1}$

$\qquad = 6 \cdot 11$

$\qquad = 66$

9. $\dbinom{10}{3} = \dfrac{10!}{3!7!}$

$\qquad = \dfrac{10 \cdot 9 \cdot 8 \cdot 7 \cdot 6 \cdot 5 \cdot 4 \cdot 3 \cdot 2 \cdot 1}{3 \cdot 2 \cdot 1 \cdot 7 \cdot 6 \cdot 5 \cdot 4 \cdot 3 \cdot 2 \cdot 1}$

$\qquad = 10 \cdot 3 \cdot 4$

$\qquad = 120$

11. $_{11}C_8 = \dfrac{11!}{8!3!}$

$\qquad = \dfrac{11 \cdot 10 \cdot 9 \cdot 8 \cdot 7 \cdot 6 \cdot 5 \cdot 4 \cdot 3 \cdot 2 \cdot 1}{8 \cdot 7 \cdot 6 \cdot 5 \cdot 4 \cdot 3 \cdot 2 \cdot 1 \cdot 3 \cdot 2 \cdot 1}$

$\qquad = 11 \cdot 5 \cdot 3$

$\qquad = 165$

13. A calculator is used for the combinations:

$$(a+b)^4 = a^4 + \binom{4}{1}a^3b + \binom{4}{2}a^2b^2 + \binom{4}{3}ab^3 + b^4$$

$$= a^4 + 4a^3b + 6a^2b^2 + 4ab^3 + b^4$$

15. A calculator is used for the combinations:

$$(x-y)^4 = x^4 + \binom{4}{1}x^3(-y) + \binom{4}{2}x^2(-y)^2 + \binom{4}{3}x(-y)^3 + (-y)^4$$

$$= x^4 - 4x^3y + 6x^2y^2 - 4xy^3 + y^4$$

17. A calculator is used for the combinations:

$$(x+2y)^6 = x^6 + \binom{6}{1}x^5(2y) + \binom{6}{2}x^4(2y)^2 + \binom{6}{3}x^3(2y)^3 + \binom{6}{4}x^2(2y)^4 + \binom{6}{5}x(2y)^5 + (2y)^6$$

$$= x^6 + 12x^5y + 60x^4y^2 + 160x^3y^3 + 240x^2y^4 + 192xy^5 + 64y^6$$

19. A calculator is used for the combinations:

$$(3x+4y)^5 = (3x)^5 + \binom{5}{1}(3x)^4(4y) + \binom{5}{2}(3x)^3(4y)^2 + \binom{5}{3}(3x)^2(4y)^3 + \binom{5}{4}3x(4y)^4 + (4y)^5$$

$$= 243x^5 + 1620x^4y + 4320x^3y^2 + 5760x^2y^3 + 3840xy^4 + 1024y^5$$

21. A calculator is used for the combinations:

$$(3a+b)^4 = (3a)^4 + \binom{4}{1}(3a)^3b + \binom{4}{2}(3a)^2b^2 + \binom{4}{3}(3a)b^3 + b^4$$

$$= 81a^4 + 108a^3b + 54a^2b^2 + 12ab^3 + b^4$$

23. A calculator is used for the combinations:

$$(x+y)^7 = x^7 + \binom{7}{1}x^6 y + \binom{7}{2}x^5 y^2 + \binom{7}{3}x^4 y^3 + \binom{7}{4}x^3 y^4 + \binom{7}{5}x^2 y^5 + \binom{7}{6}xy^6 + y^7$$

$$= x^7 + 7x^6 y + 21x^5 y^2 + 35x^4 y^3 + 35x^3 y^4 + 21x^2 y^5 + 7xy^6 + y^7$$

25. The fifth term of $(m+n)^6$ will be:

$$\binom{6}{4}m^2 n^4 = 15m^2 n^4$$

27. The eighth term of $(x+y)^{10}$ will be:

$$\binom{10}{7}x^3 y^7 = 120x^3 y^7$$

29. The third term of $(3m-n)^5$ will be:

$$\binom{5}{2}(3m)^3 (-n)^2 = 10 \cdot 27m^3 n^2$$

$$= 270m^3 n^2$$

31. The fourteenth term of $(m-n)^{16}$ will be:

$$\binom{16}{13}m^3 (-n)^{13} = 560m^3 (-n^{13})$$

$$= -560m^3 n^{13}$$

33. The middle (or the sixth) term of $(x+y)^{10}$

will be: $\binom{10}{5}x^5 y^5 = 252x^5 y^5$

Section 8.7 Skills Check

1. a. The equation of the circle is:

$$(x-h)^2 + (y-k)^2 = r^2$$

$$(x-(-1))^2 + (y-2)^2 = 3^2$$

$$(x+1)^2 + (y-2)^2 = 9$$

b. The graph is:

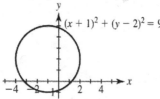

3. The equation in standard form is:

$$(x-0)^2 + (y-0)^2 = 6^2$$

Therefore, the center of the circle is $(0,0)$ and the radius is 6.

5. The equation in standard form is:

$$(x-2)^2 + (y+1)^2 = 14$$

$$(x-2)^2 + (y-(-1))^2 = (\sqrt{14})^2$$

Therefore, the center of the circle is $(2,-1)$ and the radius is $\sqrt{14}$.

7. The equation in standard form is:

$$(x-0.5)^2 + (y+3.2)^2 = 7$$

$$(x-0.5)^2 + (y-(-3.2))^2 = (\sqrt{7})^2$$

Therefore, the center of the circle is $(0.5,-3.2)$ and the radius is $\sqrt{7}$.

9. The equation in standard form is:

$x^2 + y^2 - 49 = 0$

$(x - 0)^2 + (y - 0)^2 = 7^2$

Therefore, the center of the circle is $(0,0)$ and the radius is 7 and the graph is:

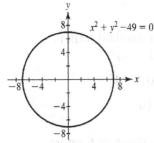

11. The equation in standard form is:

$x^2 + y^2 - 6 = 8$

$x^2 + y^2 = 14$

$(x - 0)^2 + (y - 0)^2 = \left(\sqrt{14}\right)^2$

Therefore, the center of the circle is $(0,0)$ and the radius is $\sqrt{14}$ and the graph is:

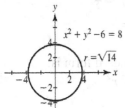

13. The equation in standard form is:

$(x + 1)^2 + (y + 3)^2 = 1$

$(x - (-1))^2 + (y - (-3))^2 = 1^2$

Therefore, the center of the circle is $(-1,-3)$ and the radius is 1 and the graph is:

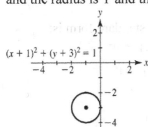

15. The equation in standard form is:

$\left(x + \tfrac{1}{2}\right)^2 + (y - 2)^2 = 6$

$\left(x - \left(-\tfrac{1}{2}\right)\right)^2 + (y - 2)^2 = \left(\sqrt{6}\right)^2$

Therefore, the center of the circle is $\left(-\tfrac{1}{2}, 2\right)$ and the radius is $\sqrt{6}$ and the graph is:

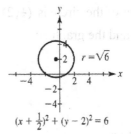

17. The equation in standard form is:

$x^2 + (y + 4)^2 = 16$

$(x - 0)^2 + (y - (-4))^2 = 4^2$

Therefore, the center of the circle is $(0,-4)$ and the radius is 4 and the graph is:

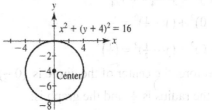

19. The equation in standard form is:

$x^2 + y^2 + 4x - 2y - 4 = 0$

$x^2 + 4x + 4 + y^2 - 2y + 1 = 4 + 4 + 1$

$(x + 2)^2 + (y - 1)^2 = 9$

$(x - (-2))^2 + (y - 1)^2 = 3^2$

Therefore, the center of the circle is $(-2,1)$ and the radius is 3 and the graph is:

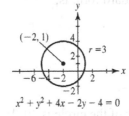

21. The equation in standard form is:

$$x^2 + y^2 - x - 4y + 4 = 0$$

$$x^2 - x + \tfrac{1}{4} + y^2 - 4y + 4 = -4 + \tfrac{1}{4} + 4$$

$$\left(x - \tfrac{1}{2}\right)^2 + (y - 2)^2 = \tfrac{1}{4}$$

$$\left(x - \tfrac{1}{2}\right)^2 + (y - 2)^2 = \left(\tfrac{1}{2}\right)^2$$

Therefore, the center of the circle is $\left(\tfrac{1}{2}, 2\right)$ and the radius is $\tfrac{1}{2}$ and the graph is:

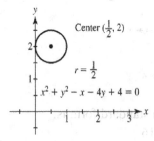

23. The equation in standard form is:

$$x^2 + y^2 - y = 0$$

$$x^2 + y^2 - y + \tfrac{1}{4} = 0 + \tfrac{1}{4}$$

$$(x - 0)^2 + \left(y - \tfrac{1}{2}\right)^2 = \tfrac{1}{4}$$

$$(x - 0)^2 + \left(y - \tfrac{1}{2}\right)^2 = \left(\tfrac{1}{2}\right)^2$$

Therefore, the center of the circle is $\left(0, \tfrac{1}{2}\right)$ and the radius is $\tfrac{1}{2}$ and the graph is:

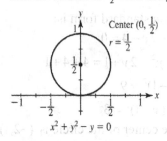

25. The equation in standard form is:

$$y = \sqrt{9 - x^2}$$

$$y^2 = 9 - x^2$$

$$x^2 + y^2 = 9$$

$$(x - 0)^2 + (y - 0)^2 = 3^2$$

Therefore, the center of the circle is $(0, 0)$ and the radius is 3 and the graph is:

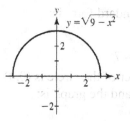

27. The equation in standard form is:

$$(x - h)^2 + (y - k)^2 = r^2$$

$$(x - 4)^2 + (y - 0)^2 = 4^2$$

$$(x - 4)^2 + y^2 = 16$$

29. The equation in standard form is:

$$(x - h)^2 + (y - k)^2 = r^2$$

$$(x - (-3))^2 + (y - (-1))^2 = 2^2$$

$$(x + 3)^2 + (y + 1)^2 = 4$$

31. The equation in standard form is:

$$(x - h)^2 + (y - k)^2 = r^2$$

$$(x - 5)^2 + (y - (-2))^2 = \tfrac{1}{2}^2$$

$$(x - 5)^2 + (y + 2)^2 = \tfrac{1}{4}$$

33. The equation in standard form is:

$$(x - h)^2 + (y - k)^2 = r^2$$

$$x^2 + y^2 = 9$$

Now solve for the positive y (because you are looking for the upper semicircle).

$$x^2 + y^2 = 9$$

$$y^2 = 9 - x^2$$

$$y = \sqrt{9 - x^2}$$

35. The equation in standard form is:

$$(x - h)^2 + (y - k)^2 = r^2$$

$$(x - 0)^2 + (y + 1)^2 = 1^2$$

$$x^2 + y^2 + 2y + 1 = 1$$

$$x^2 + y^2 + 2y = 0$$

Now solve for the negative x (because you are looking for the left semicircle).

$$x^2 + y^2 + 2y = 0$$

$$x^2 = -y^2 - 2y$$

$$x = -\sqrt{-y^2 - 2y}$$

37. The center is the midpoint of the given point:

$$\left(\frac{x_1+x_2}{2},\frac{y_1+y_2}{2}\right)=\left(\frac{1+4}{2},\frac{6+(-1)}{2}\right)$$
$$=\left(\frac{5}{2},\frac{5}{2}\right)$$

The radius is the distance from the center to either given point:

$$r=\sqrt{(x_1-x_2)^2+(y_1-y_2)^2}$$
$$r=\sqrt{(1-\tfrac{5}{2})^2+(6-\tfrac{5}{2})^2}$$
$$r=\sqrt{(-\tfrac{3}{2})^2+(\tfrac{7}{2})^2}$$
$$r=\sqrt{\tfrac{9}{4}+\tfrac{49}{4}}$$
$$r=\sqrt{\tfrac{58}{4}}$$
$$r=\frac{\sqrt{58}}{2}$$

Using the above information, the equation in standard form is:

$$(x-h)^2+(y-k)^2=r^2$$
$$(x-\tfrac{5}{2})^2+(y-\tfrac{5}{2})^2=\left(\tfrac{\sqrt{58}}{2}\right)^2$$
$$(x-\tfrac{5}{2})^2+(y-\tfrac{5}{2})^2=\tfrac{58}{4}$$

39. Since the circle is tangent to the x axis, the center of the circle will be $(-4,k)$. Using this center, find the radius from both given points:

$$r=\sqrt{(x_1-x_2)^2+(y_1-y_2)^2}$$
$$r=\sqrt{(-4+4)^2+(k-0)^2}$$
$$r=\sqrt{(0)^2+(k)^2}$$
$$r=\sqrt{k^2}$$
$$r=k$$

$$r=\sqrt{(x_1-x_2)^2+(y_1-y_2)^2}$$
$$r=\sqrt{(-4-0)^2+(k-2)^2}$$
$$r=\sqrt{16+k^2-4k+4}$$
$$r=\sqrt{k^2-4k+20}$$

Set these two equations equal to each other because they are both the radius and solve for k.

$$k=\sqrt{k^2-4k+20}$$
$$k^2=k^2-4k+20$$
$$4k=20$$
$$k=5$$

Therefore the center of the circle is $(-4,5)$ and the radius is 5 (because $r=k$) and the equation in standard form is:

$$(x-h)^2+(y-k)^2=r^2$$
$$(x-(-4))^2+(y-5)^2=5^2$$
$$(x+4)^2+(y-5)^2=25$$

41. The equation in standard form is:

$$y^2-12x=0$$
$$y^2=12x$$
$$y^2=4(3x)$$

Therefore, the vertex of the parabola is $(0,0)$, the focus is at the point $(3,0)$, the directix is $x=-3$, and the graph opens to the positive x axis and the graph is:

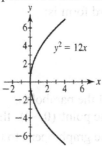

43. The equation in standard form is:

$$y^2+4x=0$$
$$y^2=-4x$$
$$y^2=4(-x)$$

Therefore, the vertex of the parabola is $(0,0)$, the focus is at the point $(-1,0)$, the directix is $x=1$, and the graph opens to the negative x axis and the graph is:

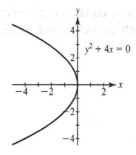

45. The equation in standard form is:
$$x^2 - 4y = 0$$
$$x^2 = 4y$$
$$x^2 = 4(y)$$
Therefore, the vertex of the parabola is $(0,0)$, the focus is at the point $(0,1)$, the directix is $y = -1$, and the graph opens to the positive y axis and the graph is:

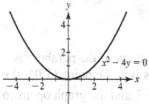

47. The equation in standard form is:
$$x^2 + 12y = 0$$
$$x^2 = -12y$$
$$x^2 = 4(-3y)$$
Therefore, the vertex of the parabola is $(0,0)$, the focus is at the point $(0,-3)$, the directix is $y = 3$, and the graph opens to the negative y axis and the graph is:

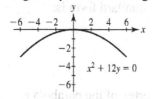

49. The equation in standard form is:
$$y = \sqrt{x-3}$$
$$y^2 = x - 3$$
$$y^2 = 4(\tfrac{1}{4}(x-3))$$

Therefore, the vertex of the parabola is $(3,0)$, the focus is at the point $(\frac{13}{4},0)$, the directix is $x = \frac{11}{4}$, the graph opens to the positive x axis, and the graph is only the top portion and the graph is:

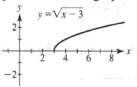

51. The equation in standard form is:
$$y = \sqrt{3-x}$$
$$y^2 = 3 - x$$
$$y^2 = 4(-\tfrac{1}{4}(x-3))$$
Therefore, the vertex of the parabola is $(3,0)$, the focus is at the point $(\frac{11}{4},0)$, the directix is $x = \frac{13}{4}$, the graph opens to the negative x axis, and the graph is only the top portion and the graph is:

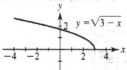

53. The equation in standard form is:
$$y^2 - 6y + 4x + 5 = 0$$
$$y^2 - 6y + 9 = -4x - 5 + 9$$
$$(y-3)^2 = -4x + 4$$
$$(y-3)^2 = 4(-(x-1))$$
Therefore, the vertex of the parabola is $(1,3)$, the focus is at the point $(0,3)$, the directix is $x = 2$, and the graph opens to the negative x axis and the graph is:

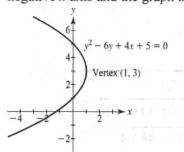

55. The equation in standard form is:

$$4x^2 + 12x + 32y + 25 = 0$$

$$x^2 + 3x + 8y + \tfrac{25}{4} = 0$$

$$x^2 + 3x + \tfrac{9}{4} = -8y - \tfrac{25}{4} + \tfrac{9}{4}$$

$$(x + \tfrac{3}{2})^2 = -8y - 4$$

$$(x + \tfrac{3}{2})^2 = 4(-2(y + \tfrac{1}{2}))$$

Therefore, the vertex of the parabola is $(-\tfrac{3}{2}, -\tfrac{1}{2})$, the focus is at the point $(-\tfrac{3}{2}, -\tfrac{5}{2})$, the directix is $y = \tfrac{3}{2}$, and the graph opens to the negative y axis and the graph is:

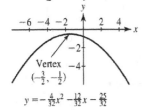

57. The equation in standard form is:

$$y = x^2 + 10x + 25$$

$$y = (x + 5)^2$$

$$(x + 5)^2 = 4(\tfrac{1}{4}y)$$

Therefore, the vertex of the parabola is $(-5, 0)$, the focus is at the point $(-5, \tfrac{1}{4})$, the directix is $y = -\tfrac{1}{4}$, and the graph opens to the positive y axis and the graph is:

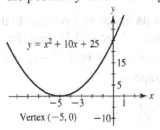

59. The equation in standard form is:

$$y = 3 + x - 2x^2$$

$$y - 3 = x - 2x^2$$

$$-\tfrac{1}{2}y + \tfrac{3}{2} = x^2 - \tfrac{1}{2}x$$

$$-\tfrac{1}{2}y + \tfrac{3}{2} + \tfrac{1}{16} = x^2 - \tfrac{1}{2}x + \tfrac{1}{16}$$

$$-\tfrac{1}{2}y + \tfrac{25}{16} = (x - \tfrac{1}{4})^2$$

$$(x - \tfrac{1}{4})^2 = 4(-\tfrac{1}{8}(y - \tfrac{25}{8}))$$

Therefore, the vertex of the parabola is $(\tfrac{1}{4}, \tfrac{25}{8})$, the focus is at the point $(\tfrac{1}{4}, 3)$, the directix is $y = \tfrac{13}{4}$, and the graph opens to the negative y axis and the graph is:

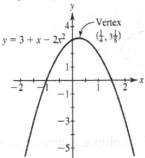

Section 8.7 Exercises

61. If the cables low point is above the origin, the right side of the cable that is connected to the tower would be considered the point $(150, 80)$. Use this point and the low point of the cable at $(0, 20)$ to find the equation:

$$x^2 = a(y - 20)$$

$$(150)^2 = a(80 - 20)$$

$$22{,}500 = 60a$$

$$a = 375$$

Therefore, the equation of this cable will be $x^2 = 375(y - 20)$.

63. Answers will vary.

Section 8.8 Skills Check

1. The equation of the ellipse in general form will be:
    $$\frac{x^2}{9}+\frac{y^2}{4}=1$$
    $$\frac{x^2}{3^2}+\frac{y^2}{2^2}=1$$

    The ellipse has its major axis on the x axis, the center will be $(0,0)$, and the major axis endpoints will be $(3,0)$ and $(-3,0)$. The graph is:

    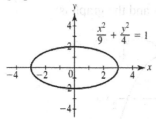

3. The equation of the ellipse in general form will be:
    $$4x^2+y^2=16$$
    $$\frac{x^2}{4}+\frac{y^2}{16}=1$$
    $$\frac{x^2}{2^2}+\frac{y^2}{4^2}=1$$

    The ellipse has its major axis on the y axis, the center will be $(0,0)$, and the major axis endpoints will be $(0,4)$ and $(0,-4)$. The graph is:

    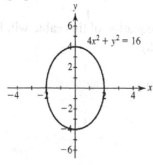

5. The equation of the ellipse in general form will be:
    $$x^2+4y^2=1$$
    $$\frac{x^2}{1}+\frac{y^2}{0.25}=1$$
    $$\frac{x^2}{1^2}+\frac{y^2}{(0.5)^2}=1$$

    The ellipse has its major axis on the x axis, the center will be $(0,0)$, and the major axis endpoints will be $(1,0)$ and $(-1,0)$. The graph is:

    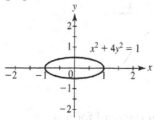

7. The equation of the ellipse in general form will be:
    $$\frac{(x+2)^2}{4}+\frac{(y-1)^2}{16}=1$$
    $$\frac{(x+2)^2}{2^2}+\frac{(y-1)^2}{4^2}=1$$

    The ellipse has its major axis parallel to the y axis, the center will be $(-2,1)$, and the major axis endpoints will be $(-2,-3)$ and $(-2,5)$. The graph is:

    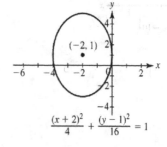

9. The equation of the ellipse in general form will be:

$$x^2 + 4y^2 + 4x - 8y + 4 = 0$$

$$x^2 + 4x + 4 + 4(y^2 - 2y + 1) = -4 + 4 + 4$$

$$(x+2)^2 + 4(y-1)^2 = 4$$

$$\frac{(x+2)^2}{4} + \frac{(y-1)^2}{1} = 1$$

$$\frac{(x+2)^2}{2^2} + \frac{(y-1)^2}{1^2} = 1$$

The ellipse has its major axis parallel to the x axis, the center will be $(-2,1)$, and the major axis endpoints will be $(0,1)$ and $(-4,1)$. The graph is:

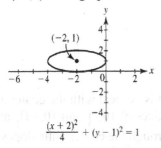

$$\frac{(x+2)^2}{4} + (y-1)^2 = 1$$

11. The equation of the ellipse in general form will be:

$$9x^2 + y^2 + 6y = 0$$

$$9x^2 + y^2 + 6y + 9 = 0 + 9$$

$$9x^2 + (y+3)^2 = 9$$

$$\frac{(x-0)^2}{1} + \frac{(y+3)^2}{9} = 1$$

$$\frac{(x-0)^2}{1^2} + \frac{(y+3)^2}{3^2} = 1$$

The ellipse has its major axis on the y axis, the center will be $(0,-3)$, and the major axis endpoints will be $(0,0)$ and $(0,-6)$. The graph is:

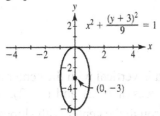

13. The equation of the ellipse in general form will be:

$$y = \frac{\sqrt{1-4x^2}}{3}$$

$$y^2 = \frac{1-4x^2}{9}$$

$$9y^2 = 1 - 4x^2$$

$$4x^2 + 9y^2 = 1$$

$$\frac{x^2}{\frac{1}{4}} + \frac{y^2}{\frac{1}{9}} = 1$$

$$\frac{x^2}{\left(\frac{1}{2}\right)^2} + \frac{y^2}{\left(\frac{1}{3}\right)^2} = 1$$

The graph will only be the top half because the original equation only has y equal to a positive value:

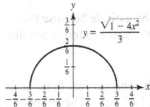

15. The equation of the ellipse in general form will be:

$$x = \frac{-2\sqrt{225-y^2}}{3}$$

$$x^2 = \frac{4(225-y^2)}{9}$$

$$9x^2 = 900 - 4y^2$$

$$9x^2 + 4y^2 = 900$$

$$\frac{x^2}{100} + \frac{y^2}{225} = 1$$

$$\frac{x^2}{10^2} + \frac{y^2}{15^2} = 1$$

The graph will only be the left side because the original equation only has x equal to a negative value:

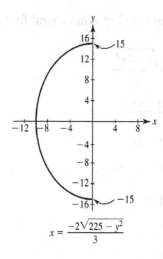

$$x = \frac{-2\sqrt{225 - y^2}}{3}$$

17. The graph will be an ellipse because the coefficients of the squared terms are different.

19. The graph will be a circle because the coefficients of the squared terms are the same.

21. The endpoints for the vertical major axis will be $(0,5)$ and $(0,-5)$. The endpoints for the horizontal minor axis will be $(1,0)$ and $(-1,0)$. The graph of the ellipse will be:

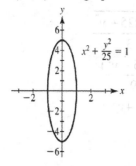

23. The equation of the hyperbola in general form will be:

$$\frac{x^2}{16} - \frac{y^2}{25} = 1$$

$$\frac{x^2}{4^2} - \frac{y^2}{5^2} = 1$$

The hyperbola is horizontal with the center at $(0,0)$, the vertices of $(4,0)$ and $(-4,0)$, and asymptotes through the center with slopes of $\frac{5}{4}$ and $-\frac{5}{4}$. The graph is:

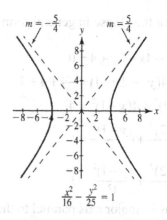

$$\frac{x^2}{16} - \frac{y^2}{25} = 1$$

25. The equation of the hyperbola in general form will be:

$$y^2 - x^2 = 1$$

$$\frac{y^2}{1} - \frac{x^2}{1} = 1$$

$$\frac{y^2}{1^2} - \frac{x^2}{1^2} = 1$$

The hyperbola is vertical with the center at $(0,0)$, the vertices of $(0,1)$ and $(0,-1)$, and asymptotes through the center with slopes of 1 and -1. The graph is:

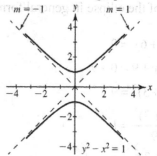

27. The equation of the hyperbola in general form will be:

$$y^2 - 4x^2 = 4$$

$$\frac{y^2}{4} - \frac{x^2}{1} = 1$$

$$\frac{y^2}{2^2} - \frac{x^2}{1^2} = 1$$

The hyperbola is vertical with the center at $(0,0)$, the vertices of $(0,2)$ and $(0,-2)$, and asymptotes through the center with slopes of 2 and -2. The graph is:

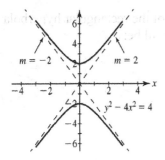

29. The equation of the hyperbola in general form will be:
$$\frac{(x+1)^2}{36} - \frac{(y+4)^2}{36} = 1$$
$$\frac{(x+1)^2}{6^2} - \frac{(y+4)^2}{6^2} = 1$$

The hyperbola is horizontal with the center at $(-1,-4)$, the vertices of $(-7,-4)$ and $(5,-4)$, and asymptotes through the center with slopes of 1 and -1. The graph is:

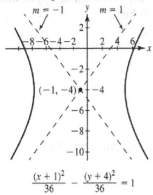

$$\frac{(x+1)^2}{36} - \frac{(y+4)^2}{36} = 1$$

31. The equation of the hyperbola in general form will be:
$$x^2 - 4y^2 - 24y - 40 = 0$$
$$x^2 - 4(y^2 + 6y + 9) = 40 - 36$$
$$x^2 - 4(y+3)^2 = 4$$
$$\frac{(x-0)^2}{4} - \frac{(y+3)^2}{1} = 1$$
$$\frac{(x-0)^2}{2^2} - \frac{(y+3)^2}{1^2} = 1$$

The hyperbola is horizontal with the center at $(0,-3)$, the vertices of $(-2,-3)$ and $(2,-3)$, and asymptotes through the center with slopes of $\frac{1}{2}$ and $-\frac{1}{2}$. The graph is:

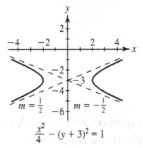

$$\frac{x^2}{4} - (y+3)^2 = 1$$

33. The equation of the hyperbola in general form will be:
$$5x^2 - 4y^2 + 20x - 16y + 24 = 0$$
$$5(x^2 + 4x + 4) - 4(y^2 + 4y + 4) = -24 + 20 - 16$$
$$5(x+2)^2 - 4(y+2)^2 = -20$$
$$-\frac{(x+2)^2}{4} + \frac{(y+2)^2}{5} = 1$$
$$\frac{(y+2)^2}{\sqrt{5}^2} - \frac{(x+2)^2}{2^2} = 1$$

The hyperbola is vertical with the center at $(-2,-2)$, the vertices of $(-2,-2+\sqrt{5})$ and $(-2,-2-\sqrt{5})$, and asymptotes through the center with slopes of $\frac{\sqrt{5}}{2}$ and $-\frac{\sqrt{5}}{2}$. The graph is:

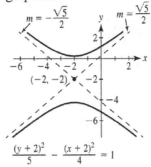

$$\frac{(y+2)^2}{5} - \frac{(x+2)^2}{4} = 1$$

35. The equation of the hyperbola in general form will be:

$$y = -\sqrt{1 + x^2}$$
$$y^2 = 1 + x^2$$
$$y^2 - x^2 = 1$$
$$\frac{y^2}{1} - \frac{x^2}{1} = 1$$
$$\frac{y^2}{1^2} - \frac{x^2}{1^2} = 1$$

The graph of the equation will be the lower branch of the graph of the equation in Exercise 25.

37. The equation of the hyperbola in general form will be:

$$x = \frac{\sqrt{9 + y^2}}{3}$$
$$x^2 = \frac{9 + y^2}{9}$$
$$9x^2 = 9 + y^2$$
$$9x^2 - y^2 = 9$$
$$\frac{x^2}{1} - \frac{y^2}{9} = 1$$
$$\frac{x^2}{1^2} - \frac{y^2}{3^2} = 1$$

The graph of the equation will be:

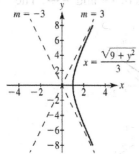

39. The equation of the rectangular hyperbola in general form will be:

$$xy = 16$$
or
$$y = \frac{16}{x}$$

The graph of the equation will be:

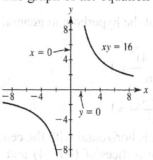

41. The equation of the rectangular hyperbola in general form will be:

$$xy = -0.5$$
or
$$y = -\frac{0.5}{x}$$

The graph of the equation will be:

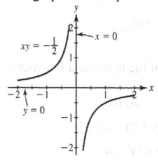

43. The equation of the rectangular hyperbola in general form will be:

$$(x-3)y = 4$$

or

$$y = \frac{4}{x-3}$$

The graph of the equation will be:

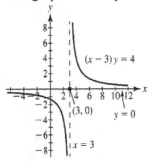

45. The equation of the hyperbola in general form will be:

$$x^2 - y^2 = 1$$

$$\frac{x^2}{1} - \frac{y^2}{1} = 1$$

$$\frac{x^2}{1^2} - \frac{y^2}{1^2} = 1$$

The graph of the left half of the equation will be:

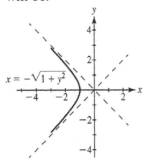

Section 8.8 Exercises

47. The arch is the upper half of an ellipse with a width of 100 feet and a height of 16 feet. The major axis is the width (or the horizontal axis) and the minor axis is the height. Therefore the equation of the ellipse is given by:

$$\frac{x^2}{a^2} + \frac{y^2}{b^2} = 1$$

$$\frac{x^2}{50^2} + \frac{y^2}{16^2} = 1$$

$$\frac{x^2}{2500} + \frac{y^2}{256} = 1$$

The distance of 19 feet from the center means that $x = 19$. Use the above equation to solve for y:

$$\frac{(19)^2}{2500} + \frac{y^2}{256} = 1$$

$$256(19)^2 + 2500y^2 = 640,000$$

$$92,416 + 2500y^2 = 640,000$$

$$2500y^2 = 547,584$$

$$y^2 = 219.0336$$

$$y \approx 14.8$$

Thus, the height of the arch is approximately 14.8 feet when you are 19 feet from the center.

49. The orbit is in the shape of an ellipse with a length of 36.18 AU and a width of 9.12 AU. The major axis is the length (or the horizontal axis) and the minor axis is the width. Therefore the equation of the ellipse is given by:

$$\frac{x^2}{a^2} + \frac{y^2}{b^2} = 1$$

$$\frac{x^2}{18.09^2} + \frac{y^2}{4.56^2} = 1$$

$$\frac{x^2}{327.2481} + \frac{y^2}{20.7936} = 1$$

51. The equation for Ohm's Law would be in the shape of a rectangular hyperbola. The equation in general form will be:

$$E = IR$$

$$120 = IR$$

or

$$I = \frac{120}{R}$$

The graph of the equation will be:

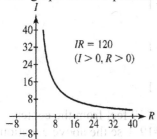

$IR = 120$
$(I > 0, R > 0)$

53.

$$\sqrt{(x+c)^2 + y^2} + \sqrt{(x-c)^2 + y^2} = 2a$$

$$\sqrt{(x+c)^2 + y^2} = 2a - \sqrt{(x-c)^2 + y^2}$$

$$\left(\sqrt{(x+c)^2 + y^2}\right)^2 = \left(2a - \sqrt{(x-c)^2 + y^2}\right)^2$$

$$(x+c)^2 + y^2 = 4a^2 - 4a\sqrt{(x-c)^2 + y^2} + \left((x-c)^2 + y^2\right)$$

$$x^2 + 2xc + c^2 + y^2 = 4a^2 - 4a\sqrt{(x-c)^2 + y^2} + x^2 - 2xc + c^2 + y^2$$

$$4xc - 4a^2 = -4a\sqrt{(x-c)^2 + y^2}$$

$$a^2 - xc = a\sqrt{(x-c)^2 + y^2}$$

$$\left(a^2 - xc\right)^2 = \left(a\sqrt{(x-c)^2 + y^2}\right)^2$$

$$a^4 - 2xca^2 + x^2c^2 = a^2\left((x-c)^2 + y^2\right)$$

$$a^4 - 2xca^2 + x^2c^2 = a^2\left(x^2 - 2xc + c^2 + y^2\right)$$

$$a^4 - 2xca^2 + x^2c^2 = a^2x^2 - 2xca^2 + a^2c^2 + a^2y^2$$

$$a^4 - a^2c^2 = a^2x^2 - c^2x^2 + a^2y^2$$

$$a^2\left(a^2 - c^2\right) = x^2\left(a^2 - c^2\right) + a^2y^2$$

$$a^2b^2 = x^2b^2 + a^2y^2$$

$$\frac{a^2b^2}{a^2b^2} = \frac{x^2b^2 + a^2y^2}{a^2b^2}$$

$$1 = \frac{x^2}{a^2} + \frac{y^2}{b^2}$$

Chapter 8 Skills Check

1. $5x + 2y \le 10$

$$2y \le 10 - 5x$$

$$y \le \frac{10 - 5x}{2}$$

Test (0, 0).

$$y \le \frac{10 - 5x}{2}$$

$$0 \le \frac{10 - 5(0)}{2}$$

$$0 \le 5$$

Since the statement is true, the region containing (0, 0) is the solution to the inequality. Note that the line is solid because of the "equal to" in the given inequality.

3. Rewriting the system yields

$$2x + y \le 3$$

$$y \le 3 - 2x$$

and

$$x + y \le 2$$

$$y \le 2 - x$$

The new system is

$$\begin{cases} y \le 3 - 2x \\ y \le 2 - x \\ x \ge 0, y \ge 0 \end{cases}$$

The graph of the system is

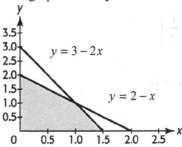

Note that all the lines are solid, and the solution is restricted to Quadrant I.

Determining the intersection point between $y = 3 - 2x$ and $y = 2 - x$:

$$3 - 2x = 2 - x$$

$$-x = -1$$

$$x = 1$$

Substituting to find y

$$y = 2 - x$$

$$y = 2 - 1$$

$$y = 1$$

The intersection point is $(1,1)$.

To determine the solution region, pick a point to test. Pick $\left(\dfrac{1}{2}, 1 \right)$.

$y \le 3 - 2x$

$1 \le 3 - 2\left(\dfrac{1}{2}\right)$

$1 \le 2$

True statement

$y \le 2 - x$

$1 \le 2 - \dfrac{1}{2}$

$1 \le \dfrac{3}{2}$

True statement

$x \ge 0$

$\dfrac{1}{2} \ge 0$

True statement

$y \ge 0$

$1 \ge 0$

True statement

Since all the inequalities that form the system are true at the point $\left(\dfrac{1}{2}, 1\right)$, the region that contains $\left(\dfrac{1}{2}, 1\right)$ is the solution region. Considering the graph of the system, the corners of the region are $(0,0), (0,2), (1,1), (1.5,0)$.

5. Rewriting the system yields
$2x + y \le 30$

$\qquad y \le 30 - 2x$

$\qquad\qquad$ and

$x + y \le 19$

$\qquad y \le 19 - x$

$\qquad\qquad$ and

$x + 2y \le 30$

$\qquad 2y \le 30 - x$

$\qquad y \le \dfrac{30 - x}{2}$

The new system is
$$\begin{cases} y \le 30 - 2x \\ y \le 19 - x \\ y \le \dfrac{30 - x}{2} \\ x \ge 0, y \ge 0 \end{cases}$$

The graph of the system is

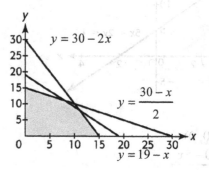

The intersection point between $y = 30 - 2x$ and $y = 19 - x$ is $(11, 8)$.

The intersection point between $y = 30 - 2x$ and $y = \dfrac{30 - x}{2}$ is $(10, 10)$.

The intersection point between $y = 19 - x$ and $y = \dfrac{30 - x}{2}$ is $(8, 11)$.

Note that all the lines are solid because there is an "equal to" as part of all the inequalities, and the solution is restricted to Quadrant I.

To determine the solution region, pick a point to test. Pick $(1, 1)$.

$2x + y \le 30$

$2(1) + (1) \le 30$

$3 \le 30$

True statement

$x+y\leq 19$

$(1)+(1)\leq 19$

$2\leq 19$

True statement

$x+2y\leq 30$

$(1)+2(1)\leq 30$

$3\leq 30$

True statement

$x\geq 0$

$1\geq 0$

True statement

$y\geq 0$

$1\geq 0$

True statement

Since all the inequalities that form the system are true at the point $(1,1)$, the region that contains $(1,1)$ is the solution region.

One of the corner points is the intersection point between $y=30-2x$ and $y=19-x$, which is $(11,8)$. A second corner point is the intersection point between $y=19-x$ and $y=\dfrac{30-x}{2}$, which is $(8,11)$. A third corner point occurs where $y=30-2x$ crosses the x-axis, which is $(15,0)$. A fourth corner point occurs where $y=\dfrac{30-x}{2}$ crosses the y-axis, which is $(0,15)$. Note that the intersection point between $y=30-2x$ and $y=\dfrac{30-x}{2}$ is not a corner point of the solution region.

Therefore, the corner points of the solution region are

$(8,11),(11,8),(15,0),(0,15),(0,0)$.

7. Rewriting the system:

$15x-x^2-y\geq 0$

$\qquad -y\geq x^2-15x$

$\qquad y\leq 15x-x^2$

and

$y-\dfrac{44x+60}{x}\geq 0$

$\qquad y\geq \dfrac{44x+60}{x}$

The new system is

$$\begin{cases} y\leq 15x-x^2 \\ y\geq \dfrac{44x+60}{x} \\ x\geq 0, y\geq 0 \end{cases}$$

The graph of the system is

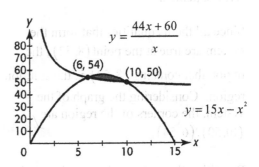

[0, 15] by [0, 80]

Note that all the lines are solid, and the solution is restricted to Quadrant I.

The intersection points between the curves are $(6,54)$ and $(10,50)$.

To determine the solution region, pick a point to test. Pick $(8,52)$.

$$y \le 15x - x^2$$

$$52 \le 15(8) - (8)^2$$

$$52 \le 56$$

True statement

$$y \ge \frac{44x + 60}{x}$$

$$52 \ge \frac{44(8) + 60}{(8)}$$

$$52 \ge 51.5$$

True statement

$$x \ge 0$$

$$8 \ge 0$$

True statement

$$y \ge 0$$

$$52 \ge 0$$

True statement

Since all the inequalities that form the system are true at the point $(8, 52)$, the region that contains $(8, 52)$ is the solution region. Considering the graph of the system, the corners of the region are $(10, 50), (6, 54)$.

9. Rewriting the system of constraints and graphing the system yields

$$\begin{cases} y \ge 8 - 3x \\ y \ge \dfrac{14 - 2x}{5} \\ x \ge 0, y \ge 0 \end{cases}$$

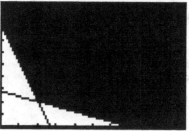

$y = \dfrac{14 - 2x}{5}$

[0, 10] by [0, 10]

The intersection point between $y = 8 - 3x$ and $y = \dfrac{14 - 2x}{5}$ is $(2, 2)$.

To determine the solution region, pick a point to test. Pick $(5, 5)$. When substituted into the inequalities that form the system, the point $(5, 5)$ creates true statements in all cases.

Since all the inequalities that form the system are true at the point $(5, 5)$, the region that contains $(5, 5)$ is the solution region.

Note that since all the inequalities contain an "equal to", all the boundary lines are solid, and the solution is restricted to Quadrant I.

The corner points of the feasible region are $(0, 8), (7, 0), (2, 2)$.

Testing the corner points of the feasible region to minimize the objective function, $g = 3x + 4y$, yields

At $(0, 8)$: $\quad g = 3(0) + 4(8) = 32$

At $(7, 0)$: $\quad g = 3(7) + 4(0) = 21$

At $(2, 2)$: $\quad g = 3(2) + 4(2) = 14$

The minimum value is 14 occurring at $(2, 2)$.

11. Rewriting the system:

$2x + y \le 30$

$\qquad y \le 30 - 2x$

$\qquad$ and

$x + y \le 19$

$\qquad y \le 19 - x$

$\qquad$ and

$x + 2y \le 30$

$\qquad 2y \le 30 - x$

$$y \le \frac{30 - x}{2}$$

The new system is

$$\begin{cases} y \le 30 - 2x \\ y \le 19 - x \\ y \le \dfrac{30 - x}{2} \\ x \ge 0, y \ge 0 \end{cases}$$

The intersection point between $y = 30 - 2x$ and $y = 19 - x$ is $(11,8)$, and between

$y = 30 - 2x$ and $y = \dfrac{30 - x}{2}$ is $(10,10)$, and

between $y = 19 - x$ and $y = \dfrac{30 - x}{2}$ is

$(8,11)$.

Note that all the lines are solid because there is an "equal to" as part of all the inequalities, and the solution is restricted to Quadrant I.

To determine the solution region, pick a point to test. Pick $(1,1)$.

$2x + y \le 30$

$2(1) + (1) \le 30$

$3 \le 30$

True statement

$x + y \le 19$

$(1) + (1) \le 19$

$2 \le 19$

True statement

$x + 2y \le 30$

$(1) + 2(1) \le 30$

$3 \le 30$

True statement

$x \ge 0$

$1 \ge 0$

True statement

$y \ge 0$

$1 \ge 0$

True statement

Since all the inequalities that form the system are true at the point $(1,1)$, the region that contains $(1,1)$ is the solution region. The graph of the solution is:

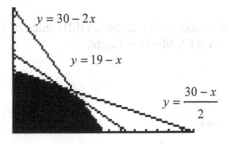

[0, 30] by [0, 30]

One of the corner points is the intersection point between $y = 30 - 2x$ and $y = 19 - x$, which is $(11,8)$. A second corner point is the intersection point between $y = 19 - x$ and $y = \dfrac{30 - x}{2}$, which is $(8,11)$. A third corner point occurs where $y = 30 - 2x$ crosses the x-axis, which is $(15,0)$. A fourth corner point occurs where $y = \dfrac{30 - x}{2}$ crosses the y-axis, which is $(0,15)$. Note that the intersection point between $y = 30 - 2x$ and $y = \dfrac{30 - x}{2}$ is not a corner point of the solution region.

The corner points of the feasible region are $(0,0),(0,15),(15,0),(8,11),(11,8)$.

Testing the corner points of the feasible region to maximize the objective function, $f = 3x + 5y$, yields

At $(0,0)$: $f = 3(0) + 5(0) = 0$
At $(0,15)$: $f = 3(0) + 5(15) = 75$
At $(15,0)$: $f = 3(15) + 5(0) = 45$
At $(8,11)$: $f = 3(8) + 5(11) = 79$
At $(11,8)$: $f = 3(11) + 5(8) = 73$

The maximum value is 79 occurring at $(8,11)$.

13. Arithmetic, with a common difference of 12.
$16 - 4 = 12, 28 - 16 = 12,$ etc.

15.

$a_n = a_1 r^{n-1}$

$a_n = 64 r^{n-1}$

$a_8 = 64 r^{8-1} = \dfrac{1}{2}$

$\dfrac{1}{64}(64 r^7) = \dfrac{1}{64}\left(\dfrac{1}{2}\right)$

$r^7 = \dfrac{1}{128}$

$r = \sqrt[7]{\dfrac{1}{128}} = \dfrac{1}{2}$

$a_n = 64\left(\dfrac{1}{2}\right)^{n-1}$

$a_5 = 64\left(\dfrac{1}{2}\right)^{5-1} = 64\left(\dfrac{1}{2}\right)^4 = 4$

17. $s_n = \dfrac{a_1(1 - r^n)}{1 - r}$

$s_{10} = \dfrac{5\left(1 - (-2)^{10}\right)}{1 - (-2)}$

$s_{10} = \dfrac{5(-1023)}{3} = -1705$

19. This is an infinite geometric series with $a_1 = \left(\dfrac{4}{5}\right)^1 = \dfrac{4}{5}$, and $r = \dfrac{4}{5}$

Therefore the sum of the series is

$S = \dfrac{a_1}{1 - r} = \dfrac{\dfrac{4}{5}}{1 - \dfrac{4}{5}} = \dfrac{\dfrac{4}{5}}{\dfrac{1}{5}} = 4$

21. $x^3 - 8x^2 - 9x = 0$

$x(x^2 - 8x - 9) = 0$

$x(x - 9)(x + 1) = 0$

$x = 0, \ x - 9 = 0, \ x + 1 = 0$

$x = 0, \ x = 9, \ x = -1$

23. $f'(x) = \dfrac{8}{x^3} + \dfrac{5}{x} + 4x$

25. $\dfrac{6!}{5!3!} = \dfrac{6 \cdot 5 \cdot 4 \cdot 3 \cdot 2 \cdot 1}{5 \cdot 4 \cdot 3 \cdot 2 \cdot 1 \cdot 3 \cdot 2 \cdot 1} = 1$

27. $_{10}C_7 = \dfrac{10!}{7!3!}$

$= \dfrac{10 \cdot 9 \cdot 8 \cdot 7 \cdot 6 \cdot 5 \cdot 4 \cdot 3 \cdot 2 \cdot 1}{7 \cdot 6 \cdot 5 \cdot 4 \cdot 3 \cdot 2 \cdot 1 \cdot 3 \cdot 2 \cdot 1}$

$= 10 \cdot 3 \cdot 4$

$= 120$

29. A calculator is used for the combinations:

$$(2a-3b)^6 = (2a)^6 + \binom{6}{1}(2a)^5(-3b) + \binom{6}{2}(2a)^4(-3b)^2$$

$$+ \binom{6}{3}(2a)^3(-3b)^3 + \binom{6}{4}(2a)^2(-3b)^4$$

$$+ \binom{6}{5}(2a)(-3b)^5 + (-3b)^6$$

$$= 64a^6 - 576a^5b + 2160a^4b^2 - 4320a^3b^3$$

$$+ 4860a^2b^4 - 2916ab^5 + 729b^6$$

31. The sixth term of $(2x+3y)^9$ will be:

$$\binom{9}{5}(2x)^4(3y)^5 = 126(16x^4)(243y^5)$$

$$= 489,888x^4y^5$$

33. The equation in standard form is:

$$x^2 - 6x + y^2 - 2y = 15$$

$$x^2 - 6x + 9 + y^2 - 2y + 1 = 15 + 9 + 1$$

$$(x-3)^2 + (y-1)^2 = 25$$

$$(x-3)^2 + (y-1)^2 = 5^2$$

Therefore, the center of the circle is $(3,1)$ and the radius is 5.

35. The equation in standard form is:

$$y^2 + 6x = 0$$

$$y^2 = -6x$$

$$y^2 = 4(-\tfrac{3}{2}x)$$

Therefore, the vertex of the parabola is $(0,0)$, the focus is at the point $(0,-\tfrac{3}{2})$, the directix is $x = \tfrac{3}{2}$, and the graph opens to the negative x axis and the graph is:

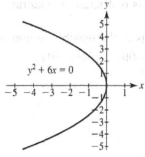

37. The equation in standard form is:

$$4y^2 - 4y + 12x + 37 = 0$$

$$y^2 - y + 3x + \tfrac{37}{4} = 0$$

$$y^2 - y + \tfrac{1}{4} = -3x - \tfrac{37}{4} + \tfrac{1}{4}$$

$$(y - \tfrac{1}{2})^2 = -3x - 9$$

$$(y - \tfrac{1}{2})^2 = 4(-\tfrac{3}{4}(x+3))$$

Therefore, the vertex of the parabola is $(-3,\tfrac{1}{2})$, the focus is at the point $(-\tfrac{15}{4},\tfrac{1}{2})$, the directix is $x = -\tfrac{9}{4}$, and the graph opens to the negative x axis and the graph is:

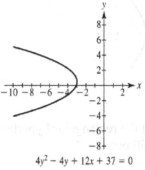

$$4y^2 - 4y + 12x + 37 = 0$$

39. The equation of the ellipse in general form will be:

$$\frac{(x-2)^2}{16} + \frac{(y+1)^2}{4} = 1$$

$$\frac{(x-2)^2}{4^2} + \frac{(y+1)^2}{2^2} = 1$$

The ellipse has its major axis on the x axis, the center will be $(2,-1)$, and the major axis endpoints will be $(-2,-1)$ and $(6,-1)$. The graph is:

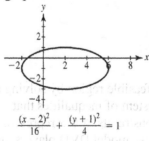

$$\frac{(x-2)^2}{16} + \frac{(y+1)^2}{4} = 1$$

41. The equation of the hyperbola in general form will be:

$$\frac{x^2}{9} - \frac{y^2}{25} = 1$$

$$\frac{x^2}{3^2} - \frac{y^2}{5^2} = 1$$

The hyperbola is horizontal with the center at $(0,0)$, the vertices of $(3,0)$ and $(-3,0)$, and asymptotes through the center with slopes of $\frac{5}{3}$ and $-\frac{5}{3}$. The graph is:

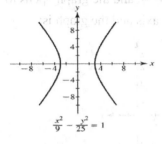

43. The equation of the rectangular hyperbola in general form will be:

$$xy = -12$$

or

$$y = -\frac{12}{x}$$

The graph of the equation will be:

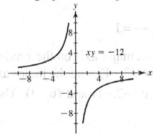

Chapter 8 Review

44. Determine the feasible region by solving and graphing the system of inequalities that represent the constraints. Let x represent the number of Deluxe model DVD players, and let y represent the number of Superior model DVD players.

$$\begin{cases} 3x + 2y \le 1800 \\ 40x + 60y \le 36{,}000 \\ x \ge 0, y \ge 0 \end{cases}$$

Rewriting yields

$$\begin{cases} y \le \dfrac{1800 - 3x}{2} \\ y \le \dfrac{36{,}000 - 40x}{60} \\ x \ge 0, y \ge 0 \end{cases}$$

Graphing the system yields

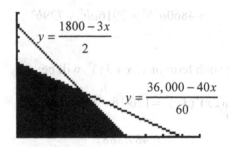

$[0, 1000]$ by $[0, 1000]$

The intersection point between the lines is $(360,360)$.

To determine the solution region, pick a point to test. Pick $(1,1)$. When substituted into the inequalities that form the system, the point $(1,1)$ creates true statements in all cases.

Since all the inequalities that form the system are true at the point $(1,1)$, the region that contains $(1,1)$ is the solution region.

Note that since all the inequalities contain an "equal to", all the boundary lines are solid, and the solution is restricted to Quadrant I.

The corner points of the feasible region are $(0,0),(0,600),(600,0),(360,360)$.

Testing the corner points of the feasible region to maximize the objective function, $f = 30x + 40y$, yields

At $(0,0)$:

$f = 30(0) + 40(0) = \$0$

At $(600,0)$:

$f = 30(600) + 40(0) = \$18,000$

At $(0,600)$:

$f = 30(0) + 40(600) = \$24,000$

At $(360,360)$:

$f = 30(360) + 40(360) = \$25,200$

The maximum value is $25,200 occurring at $(360,360)$. To produce a maximum profit of $25,200 the company should produce 360 Deluxe models and 360 Superior models.

45. a. Determine the feasible region by solving and graphing the system of inequalities that represent the constraints. Let x represent the number of days of production at the Pottstown plant, and let y represent the number of days of production at the Ethica plant.

$$\begin{cases} 20x + 40y \geq 1600 \\ 60x + 40y \geq 2400 \\ x \geq 0, y \geq 0 \end{cases}$$

Rewriting yields

$$\begin{cases} y \geq \dfrac{1600 - 20x}{40} \\ y \geq \dfrac{2400 - 60x}{40} \\ x \geq 0, y \geq 0 \end{cases}$$

Graphing the system yields

$$y = \frac{2400 - 60x}{40}$$

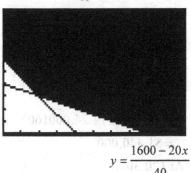

$$y = \frac{1600 - 20x}{40}$$

[0, 100] by [0, 100]

The intersection point between the lines is $(20,30)$.

To determine the solution region, pick a point to test. Pick $(20,40)$. When substituted into the inequalities that form the system, the point $(20,40)$ creates true statements in all cases.

Since all the inequalities that form the system are true at the point $(20,40)$, the region that contains $(20,40)$ is the solution region.

Note that since all the inequalities contain an "equal to", all the boundary lines are solid, and the solution is restricted to Quadrant I.

The corner points of the feasible region are $(80,0),(0,60),(20,30)$.

Testing the corner points of the feasible region to minimize the objective function, $g = 20{,}000x + 24{,}000y$, yields

At $(80,0)$:

$g = 20{,}000(80) + 24{,}000(0)$

   $= \$1{,}600{,}000$

At $(0,60)$:

$g = 20{,}000(0) + 24{,}000(60)$

   $= \$1{,}440{,}000$

At $(20,30)$:

$g = 20{,}000(20) + 24{,}000(30)$

   $= \$1{,}120{,}000$

The minimum cost is $\$1{,}120{,}000$ occurring at $(20,30)$. Therefore, operating the Pottstown plant for 20 days and the Ethica plant for 30 days produces the minimum manufacturing cost.

b. The minimum cost is $\$1{,}120{,}000$.

46. Determine the feasible region by solving and graphing the system of inequalities that represent the constraints. Let x represent the number of pounds of Feed A, and let y represent the number of pounds of Feed B.

$$\begin{cases} 2x + 8y \geq 80 \\ 4x + 2y \geq 132 \\ x \geq 0, y \geq 0 \end{cases}$$

Rewriting yields

$$\begin{cases} y \geq \dfrac{80 - 2x}{8} \\ y \geq 66 - 2x \\ x \geq 0, y \geq 0 \end{cases}$$

Graphing the system yields

$y = 66 - 2x$

$[0, 50]$ by $[0, 70]$

The intersection point between the lines is $(32,2)$.

To determine the solution region, pick a point to test. Pick $(32,4)$. When substituted into the inequalities that form the system, the point $(32,4)$ creates true statements in all cases.

Since all the inequalities that form the system are true at the point $(32,4)$, the region that contains $(32,4)$ is the solution region.

Note that since all the inequalities contain an "equal to", all the boundary lines are solid, and the solution is restricted to Quadrant I.

The corner points of the feasible region are $(0,66), (32,2), (40,0)$.

Testing the corner points of the feasible region to minimize the objective function, $g = 1.40x + 1.60y$, yields

At $(0,66)$:

$g = 1.40(0) + 1.60(66) = \$105.60$

At $(32,2)$:

$g = 1.40(32) + 1.60(2) = \$48$

At $(40,0)$:

$g = 1.40(40) + 1.60(0) = \$56$

The minimum value is $48 occurring at $(32,2)$. To minimize the $48 cost of the feed, the laboratory should use 32 pounds of Feed A and 2 pounds of Feed B.

47. Determine the feasible region by solving and graphing the system of inequalities that represent the constraints. Let x represent the number of leaf blowers, and let y represent the number of weed wackers.

$$\begin{cases} x+y \le 460 \\ x \le 260 \\ y \le 240 \\ x \ge 0, y \ge 0 \end{cases}$$

Rewriting yields

$$\begin{cases} y \le 460-x \\ x \le 260 \\ y \le 240 \\ x \ge 0, y \ge 0 \end{cases}$$

Graphing the system yields

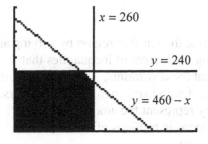

$x = 260$

$y = 240$

$y = 460 - x$

[0, 600] by [0, 500]

The intersection point between $y = 240$ and $y = 460-x$ is $(220,240)$, and between $y = 460-x$ and $x = 260$ is $(260,200)$.

To determine the solution region, pick a point to test. Pick $(1,1)$. When substituted into the inequalities that form the system, the point $(1,1)$ creates true statements in all cases.

Since all the inequalities that form the system are true at the point $(1,1)$, the region that contains $(1,1)$ is the solution region.

Note that since all the inequalities contain an "equal to", all the boundary lines are solid.

The corner points of the feasible region are $(0,0),(0,240),(260,0),(220,240),$ $(260,200)$.

Testing the corner points of the feasible region to maximize the objective function, $f = 5x + 10y$, yields

At $(0,0)$:
$f = 5(0)+10(0) = \$0$
At $(0,240)$:
$f = 5(0)+10(240) = \$2400$
At $(260,0)$:
$f = 5(260)+10(0) = \$1300$
At $(220,240)$:
$f = 5(220)+10(240) = \$3500$
At $(260,200)$:
$f = 5(260)+10(200) = \$3300$

The maximum value is $3500 occurring at $(220,240)$. To produce a maximum profit the company should produce 220 leaf blowers and 240 weed wackers.

48. Determine the feasible region by solving and graphing the system of inequalities that represent the constraints. Let x represent the two-bedroom apartments rented, and let y represent the number of three bedroom apartments rented.

$$\begin{cases} 2x+3y \le 180 \\ x \le 60 \\ y \le 40 \\ x \ge 0, y \ge 0 \end{cases}$$

Rewriting yields

$$\begin{cases} y \le \dfrac{180-2x}{3} \\ x \le 60 \\ y \le 40 \\ x \ge 0, y \ge 0 \end{cases}$$

Graphing the system yields

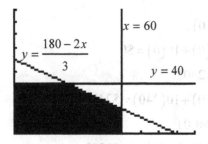

[0, 100] by [0, 100]

The intersection point between $y = 40$ and

$y = \dfrac{180-2x}{3}$ is $(30,40)$, and between

$y = \dfrac{180-2x}{3}$ and $x = 60$ is $(60,20)$.

To determine the solution region, pick a point to test. Pick $(1,1)$. When substituted into the inequalities that form the system, the point $(1,1)$ creates true statements in all cases.

Since all the inequalities that form the system are true at the point $(1,1)$, the region that contains $(1,1)$ is the solution region.

Note that since all the inequalities contain an "equal to", all the boundary lines are solid, and the solution is restricted to Quadrant I.

The corner points of the feasible region are $(0,0),(0,40),(60,0),(30,40),(60,20)$.

Testing the corner points of the feasible region to maximize the objective function, $f = 800x + 1150y$, yields

At $(0,0)$:

$f = 800(0)+1150(0) = \$0$

At $(0,40)$:

$f = 800(0)+1150(40) = \$46,000$

At $(60,0)$:

$f = 800(60)+1150(0) = \$48,000$

At $(30,40)$:

$f = 800(30)+1150(40) = \$70,000$

At $(60,20)$:

$f = 800(60)+1150(20) = \$71,000$

The maximum value is \$71,000 occurring at $(60,20)$. To produce a maximum profit the woman should rent 60 two-bedroom apartments and 20 3-bedroom apartments.

49. Determine the feasible region by solving and graphing the system of inequalities that represent the constraints. Let x represent the amount of auto loans in millions of dollars, and let y represent the amount of home equity loans in millions of dollars.

$$\begin{cases} x+y \le 30 \\ x \ge 2y \\ x \ge 0, y \ge 0 \end{cases}$$

Rewriting yields

$$\begin{cases} y \le 30-x \\ y \le \dfrac{x}{2} \\ x \ge 0, y \ge 0 \end{cases}$$

Graphing the system yields

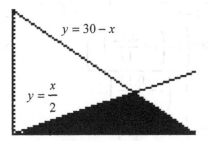

$[0, 30]$ by $[0, 30]$

The intersection point between the lines is $(20,10)$.

To determine the solution region, pick a point to test. Pick $(20,5)$. When substituted into the inequalities that form the system, the point $(20,5)$ creates true statements in all cases.

Since all the inequalities that form the system are true at the point $(20,5)$, the region that contains $(20,5)$ is the solution region.

Note that since all the inequalities contain an "equal to", all the boundary lines are solid.

The corner points of the feasible region are $(0,0),(30,0),(20,10)$.

Testing the corner points of the feasible region to maximize the objective function, $f = 0.08x + 0.07y$, yields

At $(0,0)$: $f = 0.08(0) + 0.07(0) = \$0$

At $(30,0)$: $f = 0.08(30) + 0.07(0) = \$2.4$

At $(20,10)$: $f = 0.08(20) + 0.07(10) = \$2.3$

The maximum value is $2.4 million occurring at $(30,0)$. To produce a maximum profit the finance company should make $30 million in auto loans and no home equity loans.

50. a. Job 1: $a_n = a_1 + (n-1)d$

$$a_n = 20,000 + (n-1)(1000)$$

$$a_n = 1000n + 19,000$$

$$a_5 = 1000(5) + 19,000 = 24,000$$

Job 2: $a_n = a_1 + (n-1)d$

$$a_n = 18,000 + (n-1)(1600)$$

$$a_n = 1600n + 16,400$$

$$a_5 = 1600(5) + 16,400 = 24,400$$

During the fifth year of employment, Job 2 produces a higher salary.

b. Job 1: $s_n = \dfrac{n(a_1 + a_n)}{2}$

$$s_5 = \dfrac{5(a_1 + a_5)}{2}$$

$$s_5 = \dfrac{5(20,000 + 24,000)}{2}$$

$$s_5 = 110,000$$

Job 2: $s_n = \dfrac{n(a_1 + a_n)}{2}$

$$s_5 = \dfrac{5(a_1 + a_5)}{2}$$

$$s_5 = \dfrac{5(18,000 + 24,400)}{2}$$

$$s_5 = 106,000$$

Over the first five years of employment, the total salary earned from Job 1 is higher than the total salary earned for Job 2.

51. Note that the given series is geometric with a common ratio of 0.6.

$$s_n = \dfrac{a_1(1 - r^n)}{1 - r}$$

$$s_{21} = \dfrac{400\left(1 - (0.6)^{21}\right)}{1 - (0.6)}$$

$$s_{21} = 999.978063$$

The level of the drug in the bloodstream after 21 doses over 21 days is approximately 999.98 mg.

52. a. Note that the question is modeled by a geometric series with a common ratio of 2 and an initial value of 1.

$a_1 = 1, r = 2$

$a_n = a_1 r^{n-1}$

$a_n = 1(2)^{n-1}$

$a_{64} = 1(2)^{64-1} = (2)^{63} \approx 9.22337 \times 10^{18}$

b. $a_1 = 1, r = 2$

$s_n = \dfrac{a_1(1 - r^n)}{1 - r}$

$s_{64} = \dfrac{1(1 - (2)^{64})}{1 - (2)}$

$s_{64} \approx 1.84467 \times 10^{19}$

53. $a_1 = 20,000, r = 1.06$

$a_n = a_1 r^{n-1}$

Note that after 5 years, $n = 6$.

$a_6 = 20,000(1.06)^{6-1}$

$a_6 = 20,000(1.06)^5$

$a_6 = 26,764.51$

After earning interest for five years, the value of the investment is $26,764.51.

54. $s_n = \dfrac{a_1(1 - r^n)}{1 - r}$

$s_{60} = \dfrac{300\left(1 - \left(1 + \dfrac{0.12}{12}\right)^{60}\right)}{1 - \left(1 + \dfrac{0.12}{12}\right)}$

$s_{60} = \dfrac{300\left(1 - (1.01)^{60}\right)}{1 - (1.01)}$

$s_{60} \approx 24,500.90$

The future value of the annuity is $24,500.90.

55. Recall that to earn a profit, revenue must exceed cost.

$R(x) > C(x)$

$177.50x > 3x^2 + 1228$

Graphing the functions yields

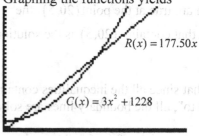

[0, 100] by [0, 15,000]

The two intersection points are approximately (51.18, 9082.08) and (8, 1420)

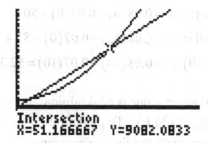

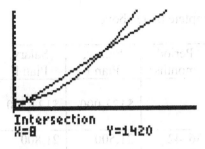

Intersection
X=8   Y=1420

Since x is the number of units in hundreds, when $800 < x < 5200$, $R(x) > C(x)$ and profit is achieved.

56. If the cables low point is at the origin, the right side of the cable that is connected to the tower would be considered the point $(400, 160)$. Use this point and the low point of the cable at $(0, 0)$ to find the equation:

$$x^2 = a(y - 0)$$
$$(400)^2 = a(160 - 0)$$
$$160,000 = 160a$$
$$a = 1000$$

Therefore, the equation of this cable will be $x^2 = 1000y$.

Now substitute $x = 300$ (100 feet from the tower) into the equation:

$$(300)^2 = 1000y$$
$$90,000 = 1000y$$
$$90 = y$$

Therefore, the height of the cable 100 feet from the tower is 90 feet.

57. The arch is the upper half of an ellipse with a width of 50 feet and a height of 20 feet. The major axis is the width (or the horizontal axis) and the minor axis is the height. Therefore the equation of the ellipse is given by:

$$\frac{x^2}{a^2} + \frac{y^2}{b^2} = 1$$
$$\frac{x^2}{25^2} + \frac{y^2}{20^2} = 1$$
$$\frac{x^2}{625} + \frac{y^2}{400} = 1$$

The distance of 10 feet from the center will be used because the truck is 10 feet wide and the center line in the road will be considered 0 and that means that $x = 10$. Use the above equation to solve for y:

$$\frac{(10)^2}{625} + \frac{y^2}{400} = 1$$
$$400(10)^2 + 625y^2 = 250,000$$
$$40,000 + 625y^2 = 250,000$$
$$625y^2 = 210,000$$
$$y^2 = 336$$
$$y \approx 18.33$$

Thus, the height of the arch is approximately 18.3 feet when you are 10 feet from the center so the truck can go under the archway without crossing into the other lane.

58. The equation of the hyperbola in standard form is given by:

$$625y^2 - 900x^2 = 562,500$$
$$\frac{y^2}{900} - \frac{x^2}{625} = 1$$
$$\frac{y^2}{30^2} - \frac{x^2}{25^2} = 1$$

The closest distance will be at twice the value of a or $(2)(30) = 60$ feet.

Chapter 8 Group Activities/Extended Application

Year	Period (months)	Salary, Plan I	Salary, Plan II
1	0–6	20,000	20,000
	6–12	20,000	20,300
2	12–18	20,500	20,600
	18–24	20,500	20,900
3	24–30	21,000	21,200
	30–36	21,000	21,500
Total for 3 years		$123,000	$124,500

1. The raises in Plan I total $3000.

2. The raises in Plan II total $4,500.

3. Plan II is clearly better for the employee. It yields an extra $1500 over the 3-year period.

4. See complete table above.

Year	Period (months)	Salary, Plan I	Salary, Plan II
Total for 3 years		$123,000	$124,500
Year 4	36–42	21,500	21,800
	42–48	21,500	22,100
Total for 4 years		$166,000	$168,400

Plan II continues to be better. It yields $8,400 in raises, whereas Plan I yields $6000 in raises.

5. For at least the first four years, the school board will not save money by awarding $300 raises every six months instead of $1000 annual raises. The school board did not take into consideration that the semi-annual raises would be awarded more frequently and therefore would compound faster than annual raises.

6. The difference in the raises over four years is $8400 - 6000 = \$2400$.

   200 teachers × $2400 per teacher = $480,000

   The school district will spend an additional $480,000 over the first four years.